Springer Complexity

Springer Complexity is an interdisciplinary program publishing the best research and academic-level teaching on both fundamental and applied aspects of complex systems - cutting across all traditional disciplines of the natural and life sciences, engineering, economics, medicine, neuroscience, social and computer science.

Complex Systems are systems that comprise many interacting parts with the ability to generate a new quality of macroscopic collective behavior the manifestations of which are the spontaneous formation of distinctive temporal, spatial or functional structures. Models of such systems can be successfully mapped onto quite diverse "real-life" situations like the climate, the coherent emission of light from lasers, chemical reaction-diffusion systems, biological cellular networks, the dynamics of stock markets and of the internet, earthquake statistics and prediction, freeway traffic, the human brain, or the formation of opinions in social systems, to name just some of the popular applications.

Although their scope and methodologies overlap somewhat, one can distinguish the following main concepts and tools: self-organization, nonlinear dynamics, synergetics, turbulence, dynamical systems, catastrophes, instabilities, stochastic processes, chaos, graphs and networks, cellular automata, adaptive systems, genetic algorithms and computational intelligence.

The two major book publication platforms of the Springer Complexity program are the monograph series "Understanding Complex Systems" focusing on the various applications of complexity, and the "Springer Series in Synergetics", which is devoted to the quantitative theoretical and methodological foundations. In addition to the books in these two core series, the program also incorporates individual titles ranging from textbooks to major reference works.

Editorial and Programme Advisory Board

Understanding Complex Systems

Founding Editor: J.A. Scott Kelso

Future scientific and technological developments in many fields will necessarily depend upon coming to grips with complex systems. Such systems are complex in both their composition - typically many different kinds of components interacting simultaneously and nonlinearly with each other and their environments on multiple levels - and in the rich diversity of behavior of which they are capable.

The Springer Series in Understanding Complex Systems series (UCS) promotes new strategies and paradigms for understanding and realizing applications of complex systems research in a wide variety of fields and endeavors. UCS is explicitly transdisciplinary. It has three main goals: First, to elaborate the concepts, methods and tools of complex systems at all levels of description and in all scientific fields, especially newly emerging areas within the life, social, behavioral, economic, neuroand cognitive sciences (and derivatives thereof); second, to encourage novel applications of these ideas in various fields of engineering and computation such as robotics, nano-technology and informatics; third, to provide a single forum within which commonalities and differences in the workings of complex systems may be discerned, hence leading to deeper insight and understanding.

UCS will publish monographs, lecture notes and selected edited contributions aimed at communicating new findings to a large multidisciplinary audience.

Sifeng Liu and Yi Lin

Grey Systems

Theory and Applications

Springer

Authors

Sifeng Liu
College of Economics and Management Science
Nanjing University of Aeronautics and Astronautics
29 Imperial Street, Nanjing 210016, P.R. China
E-mail: sfliu@nuaa.edu.cn

Yi Lin
Department of Mathematics
Slippery Rock University
Slippery Rock, PA 16057, USA
E-mail: jeffrey.forrest@sru.edu

ISBN 978-3-642-16157-5 e-ISBN 978-3-642-16158-2

DOI 10.1007/978-3-642-16158-2

Understanding Complex Systems ISSN 1860-0832

Library of Congress Control Number: 2010937345

Typeset & Cover Design: Scientific Publishing Services Pvt. Ltd., Chennai, India.

Printed on acid-free paper

9 8 7 6 5 4 3 2 1

springer.com

Abstract

This book systematically presents the fundamental theory, methods, and techniques of practical application of grey systems theory. It stands for the crystallization of the authors' many years of theoretical exploration, real-life application, and teaching. It absorbs most of the recent theoretical and applied advances of the theory achieved by scholars from across the world so that it vividly presents the reader with the overall picture of this new theory and its frontal research activities.

This book contains 10 chapters, including Introduction to Grey Systems Theory, Basic Building Blocks of Grey Systems Theory, Grey Relationships, Grey Systems Modeling, Discrete Grey Prediction Models, Combined Grey Models, Grey Models for Decision Making, Grey Game Models, Grey Control, and others. It includes a computer software package developed for grey systems modeling. The contents, such as the axiomatic system of buffer operators and a series of weakening and strengthening operators, the axioms for measuring the greyness of grey numbers, the general grey incidences (grey absolute incidence, grey relative incidence, grey comprehensive incidence, grey analogy incidence, grey nearness incidence), discrete grey models, fixed weight grey cluster evaluation and grey evaluation methods based on triangular whitenization weight functions, multi-attribute intelligent grey target decision models, the applicable range of the G(1,1) model, grey econometric models (G-E), grey Cobb-Douglass model (G-C-D), grey input-output model (G-I-O), grey Markov models (G-M), grey game models (G-G), etc., are introduced by the authors the first time.

This book will be appropriate as a reference and/or textbook for graduate students or high level undergraduate students, majoring in areas of science, technology, agriculture, medicine, astronomy, earth science, economics, and management. It can also be utilized as a reference book by researchers and technicians in research institutions, business entities, and government agencies.

Acknowledgement

The relevant works done and contained in the book are supported by National Natural Science Foundation of China (grant No. 90924022, 70473037, 70971064, 70701017, 70901040, and 70901041), the Key Project of Philosophic and Social Sciences of the China (No. 08AJY024), the Key Project of Soft Science Foundation of China (grant No. 2008GXS5D115), the Foundation for Doctoral Programs (grant No. 20020287001, 200802870020, 20093218120032) and the Foundation for Humanities and Social Sciences of the Chinese National Ministry of Education (grant No.08JA630039), the Key Project of Natural Science Foundation of Jiangsu Province (grant No. BK2003211), the Key Project of Soft Science Foundation of Jiangsu Province (grant No. BR2008081), the Foundation for Humanities and Social Sciences of Jiangsu Province (grant No.07EYA017). At the same time, the authors would like to acknowledge the partial support of the Science Foundation for the Excellent and Creative group in Science and Technology of Nanjing University of Aeronautics and Astronautics and Jiangsu Province (No.Y0553-091).

Over the years, our works have been highly commented and praised by many first class scholars of our modern time, such as Hermann Haken, founder of synergetics, Julong Deng, founder of grey systems theory, Robert Vallee, president of the World Organization of Cybernetics and Systems, Jifa Gu, former president of the International Federation for Systems Research, Qun Lin, Da Chen, academicians of Chinese Academy of Sciences, Zhongtuo Wang, member of the Chinese Academy of Engineers, Guozhi Xu, a deceased member of the Institute of Systems Science of the Chinese Academy of Science, and others. Xuesen Qian, a world renowned scientist, specifically delivered to us his congratulations on our achievements. Many colleagues and administrators of different levels have provided us various needed support.

In the process of writing this book, the authors have consulted widely, referred to published literature by many scholars. In particular, they are deeply in debt to Professor Hans Kuijper, who initially proposed the prospective of this book, and Dr. Thomas Ditzinger, who provides the great supports to the authors with the final publication of this book. Using this opportunity, the authors would like to express their appreciation to them all.

Authors' Preface

Answering the calls of the readers of our previous publications, this book systematically presents the main advances in grey systems theory and applications. Considering the readers' feedbacks and suggestions, this volume introduces the most recent research results based on what is presented in our earlier book, "Grey Information: Theory and Practical Applications," Springer, 2006. In particular, the following materials, which represent the authors' recent achievements, are highlighted in the book: the characteristics of unascertained systems, the simplicity principle of science, the grey algebraic system developed on the concepts of cores and degrees of greyness, a series of new types of practical buffer operators, grey incidence models developed on the basis of similarity and closeness, the grey evaluation models based on the central-point triangular whitenization weight functions and comparisons of two classes of these evaluation models, discrete grey models, multi-variable discrete grey models, optimal discrete grey models, combined grey-rough models, multi-attribute intelligent grey target decision models, robust stability of grey systems, etc. The attached software designed for grey systems modeling is developed by Zeng Bo using Visual C#, the widely employed tool of the C/S software. This user friendly software possesses the characteristics that users can conveniently input and/or upload data and clearly distinguish module functions. Also the software has the ability to present users with operational details and periodic and partial results. Additionally, users can adjust the levels of computational accuracy based on their practical needs.

During the writing of this book, we insisted on achieving theoretical simplicity and clarity, while maintaining the ease for the reader to follow the threads. With a good number of practical applications, we intended to illustrate the methodology of grey systems theory and modeling techniques so that we could successfully emphasize on the practical applicability of the grey systems thinking. At the same time, we paid close attention on the construction of mathematical foundations, the establishment of axiom systems, and the rigor and preciseness in the mathematical deductions. By absorbing the most recent developments of various research groups from around the world, we tried to use the fewest possible number of pages to present the most complete picture and its frontier activities in this new area of scientific endeavor.

The overall planning and organization of topics contained in this book were jointly done by Sifeng Liu and Yi Lin. Particularly, Sifeng Liu authored Chapters 1 – 3 and Section 7.3; Yi Lin detailed Chapter 4 and Appendices A, B, and C; Xie Naiming composed Chapter 5; Jian Lirong composed Chapter 6; Dang Yaoguo wrote Chapter 7; Fang Zhigeng was responsible with Chapter 8; Su Chunhua worked out Chapter 9; and Chapter 10 and the attached computer software were done by Zeng Bo. The software has been perfected by Wei Meng and Yingjie Yang. Both Sifeng Liu and Yi Lin were responsible for unifying the terms used throughout the book and finalized all the details.

If you have any comment, please communicate with us, because only by working together as a team, we can grow and mature onto a higher level of learning. Shoyld you desire, we could be reached at sfliu@nuaa.edu.cn (for Sifeng Liu) and jeffrey.forrest@sru.edu or Jeffrey.forrest@iigss.net (for Yi Lin).

January 2010

Sifeng Liu
Yi Lin

Dr. Hermann Haken's Preface

With human knowledge maturing and scientific exploration deepening and largely expanding in the course of time, mankind finally realizes the fundamental fact that due to both internal and external disturbances and limitations of human and technical sensing organs, all information received or collected contains some kind of uncertainty. Accompanying the progress of science and technology and the aforementioned realization, our understanding about various kinds of uncertainties has gradually been deepened. Attesting to this end, in the second half of the 20th century, the continual appearance of several influential and different types of theories and methods on unascertained systems and information has become a major aspect of the modern world of learning. Each of these new theories was initiated and followed-up by some of the best minds of our modern time.

In their recent book, entitled "Grey Information: Theory and Practical Applications," published in its traditionally excellent way by Springer, Professors Sifeng Liu and Yi Lin presented in a systematic fashion the theory of grey systems, which was first proposed by J. L. Deng in early 1980s and enthusiastically supported by hundreds of scientists and practitioners in the following years. Based on the hard work of these scholars in the past (nearly) thirty years, scholars from many countries currently are studying and working on the theory and various applications of this fruitful scientific endeavor. With this book published by such a prestigious leading publisher of the world, it can be expected that more scientific workers from different parts of the world will soon join hands and together make grey systems and information a powerful theory capable of bringing forward practically beneficial impacts to the advancement of the human society.

This book focuses on the study of such unascertained systems that are known with small samples or "poor information." Different of all other relevant theories on uncertainties, this work introduces a system of many methods on how to deal with grey information. Starting off with a brief historical introduction, this book carries the reader through all the basics of the theory. And, each important method studied is accompanied with a real-life project the authors were involved in during their professional careers.

Many of the methods and techniques the reader will learn in this book were originally introduced by the authors. They show how from our knowledge based on partially and poorly known information can be obtained to accurate descriptions and effective controls of the systems of interest. Because this book shows how the theory of grey systems and information was established and how each method could be practically applied, this book can easily be used as a reference by scholars who are interested in either theoretical exploration or practical applications or both. I recommend this book highly to anyone who has either a desire or a need to learn.

July 26, 2007

Professor Dr. Dr. h.c. mult. Hermann Haken
Founder of Synergetics

Note: Professor Hermann Haken prepared this note for the earlier book by the same authors, published in 2006. With his permission, it is printed here as a preface for this current book.

Dr. Robert Vallee's Preface

I am much interested and impressed by Dr. Sifeng Liu and Dr. Yi Lin's recently published monograph on grey information, dealing with the theory and practical applications.

This book encompasses many aspects of mathematics under the aegis of uncertain information. I am greatly in favour of this attitude, concerning the uncertainty of information, which has been mine since a long time ago. Also, this book focuses on practice and aims at explorations of new knowledge. It is a comprehensive, all-in-one exposition, detailing not only with the theoretical foundation but also real-life applications. Because of this characteristic of quality and usefulness, Liu and Lin's book possesses the value of the widest possible range of reference by the workers and practitioners from all corners of natural and social sciences and technology.

In this book, Liu and Lin present the theory of grey information and systems starting on such background information as the relevant history, an attempt to establish an unified information theory, the basics of grey elements, and reaching all the most advanced topics of the theory. Complemented by many first-hand and practical project-successes, the authors developed an organic theory and methodology of grey information and grey systems, dealing with errors. In fact, there is much more to tell about error than about truth. Error (inexactitude) can be met everywhere and truth (exactitude) nowhere. But inexactitude contains a part of the truth. Greyness is the field we live in. Extremes, as whiteness and blackness, are inaccessible, but very useful, ideal concepts.

With the publication of such a book that contains not only a theory, aspects of magnificent real-life implications and explorations of new research, but also the history, the theorization of various difficult concepts, and directions for future works, there is no doubt that Drs. Liu and Lin have made a remarkable contribution to the development and applications of systems science.

July 25, 2007

Professor Robert Vallée
Université Paris-Nord
President of the World Organisation of Cybernetics and Systems

Note: This note is a book review written by Professor Robert Vallée for the earlier book by the same authors, published in 2006. It is originally published in the Emerald's *Kybernetes: The International Journal of Cybernetics, Systems and Management Science*, vol. 37, no. 1, in 2008. With his and Emerald's permission, it is printed here as a preface for this current book.

Contents

Chapter 1
Introduction to Grey Systems Theory

1.1 Appearance and Growth of Grey Systems Research

1.1.1 The Scientific Background

On the basis of dividing the spectrum of scientific and technological endeavors into finer sections, the overall development of modern science has shown the tendency of synthesis at a higher level. This higher level synthesis has caused the appearance of the various studies of systems science with their specific methodological and epistemological significance. Systems science reveals the much deeper and more intrinsic connections and interlockings of objects and events and has greatly enriched the overall progress of science and technology. Many of the historically difficult problems in different scientific fields have been resolved successfully along with the appearance of systems science and its specific branches. And because of the emergence of various new areas in systems science, our understanding of nature and the laws that govern objective evolutions has been gradually deepened. At the end of the 1940s, there appeared systems theory, information theory, cybernetics. Toward the end of 1960s and the start of 1970s, there appeared the theory of dissipative structures, synergics, catastrophe, and *bifurcations*. During the middle and toward the end of the 1970s, there appeared one by one such new transfield and interfiled theories of systems science as the ultracircular theory, dynamic systems, pansystems, etc.

When investigating systems, due to both the existence of internal and external disturbances and the limitation of our understanding, the available information tends to contain various kinds of uncertainty and noises. Along with the development of science and technology and the progress of the mankind, our understanding of uncertainties of systems has been gradually deepened and the research of uncertain systems has reached at a new height. During the second half of the 20th century, in the areas of systems science and systems engineering, the seemingly non-stoppable emergence of various theories and methodologies of unascertained systems has been a great scene. For instance, L. A. Zadeh established fuzzy mathematics in the 1960s, Julong Deng developed grey systems theory and

Z. Pawlak advanced rough set theory in the 1980s, Guongyun Wang created uncertainty mathematics in the 1990s, etc. All these works represent some of the most important efforts in the research of uncertain systems of this time period. From different angles, these works provide the theories and methodologies for describing and dealing with uncertain information.

The grey systems theory, established by Julong Deng in 1982, is a new methodology that focuses on the study of problems involving small samples and poor information. It deals with uncertain systems with partially known information through generating, excavating, and extracting useful information from what is available. So, systems' operational behaviors and their laws of evolution can be correctly described and effectively monitored. In the natural world, uncertain systems with small samples and poor information exist commonly. That fact determines the wide range of applicability of grey systems theory.

1.1.2 The Development History and Current State

In 1982, *Systems & Control Letters*, an international journal by North-Holland, published the first paper in grey systems theory, "The Control Problems of Grey Systems," by Julong Deng. In the same year, the *Journal of Huazhong University of Science and Technology* published the first paper, also by Julong Deng, on grey systems theory in Chinese language. The publication of these papers signaled the official appearance of the cross disciplinary grey systems theory. As soon as these works appeared, they immediately caught the attention of many scholars and scientific practitioners from across the world. Numerous well-known scientists strongly supported the validity and livelihood of such research. Many young scholars actively participated in the investigation of grey systems theory. With great enthusiasm these young men and women carried the theoretical aspects of the theory to new heights and employed their exciting results to various fields of application. In particular, successful applications in great many fields have won the attention of the international world of learning. Currently, a great number of scholars from China, United States, England, Germany, Japan, Australia, Canada, Austria, Russia, Turkey, the Netherlands, Iran, and others, have been involved in the research and application of grey systems theory. In 1989, the British journal, *The Journal of Grey System*, was launched. Currently, this publication is indexed by INSPEC (formerly Science Abstracts) of England, Mathematical Review of the United States, Science Citation Index, and other important indexing agencies from around the world. In 1997, a Chinese publication, named *Journal of Grey System*, is launched in Taiwan. It is later in 2004 that this publication becomes all English. Additionally, a new journal, named *Grey Systems: Theory and application*, edited by the faculty of Institute for Grey Systems Studies at NUAA, will be launched by Emerald in 2011. There are currently over one thousand different professional journals in the world that have accepted and published papers in grey systems theory. As of this writing, a journal of the Association for Computing Machinery (USA), *Communications in Fuzzy Mathematics* (Taiwan), *Kybernetes: The International Journal of Cybernetics, Systems and Management Science*, have respectively published special issues on grey systems theory.

Many finest universities around the world offer courses in grey systems theory. For instance, Nanjing University of Aeronautics and Astronautics (NAUU) not only offers such courses to PhD level and master level students, but also provides a service course on grey systems to all undergraduate students in different majors. Huazhong University of Science and Technology, NUAA, Wuhan University of Technology, Fuzhou University, and several universities in Taiwan recruit and produce PhD students focusing on the research in grey systems. It is estimated that well-over several thousands of graduate students from around the world employ the thinking logic and methodology of grey systems in their research and the writing of their dissertations.

Many publishers from around the world, such as Science Press (mainland China), Press of National Defense Industry (mainland China), Huazhong University of Science and Technology Press (mainland China), Jiangsu Press of Science and Technology (mainland China), Shangdong People's Press (mainland China), Literature Press of Science and Technology (mainland China), Taiwan Quanhua Science and Technology Books, Taiwan Guaoli Books Limited, Science and Engineering Press of Japan, the IIGSS Academic Press and Taylor and Francis Group (USA), Springer-Verlag (Germany), etc., have published over 100 different kinds of monographs in grey systems. A whole array of brand new hybrid branches of study, such as grey hydrology, grey geology, grey theory and methods of breeding, grey medicine, grey systems analysis of regional economics, etc., have appeared along with the opportunity presented by grey systems theory. Agencies at national, provincial, and local governments all actively sponsored research works in grey systems. Each year a very good number of theoretical and applied works on grey systems are financially supported by various foundations. It is estimated that throughout China more than 200 research outcomes of grey systems were recognized officially by national, provincial, or ministerial agencies. In 2002, several Chinese scholars received recognition from the World Organization of Cybernetics and Systems.

Based on our incomplete statistics, it is found that such internationally recognized indexing sources as SCI, EI, ISTP, SA, MR, MA, and others, have collected research works in grey systems theory published by Chinese scholars alone more than 10,000 times. According to Chinese Science Citation Data base (CSCD), Science Time of China, November 26, 1997, the works by Julong Deng of Huazhong University of Science and Technology were cited most times consecutively for many years among all publications by Chinese scholars. In 1993, the Blue Book of Chinese Science and Technology (no. 8), edited and published by Chinese Ministry of Science and Technology, recognized grey systems theory as a new methodology of soft science established by Chinese scholars. In 2008, the Reports of Scientific Developments and Researches (2007 – 2008), edited and published by Chinese Association of Science, spent a good amount of space on introducing grey systems theory as one of the major innovative achievements in management science and engineering disciplines.

In 2006, the conference of grey systems theory and applications was held successfully in Beijing with financial sponsorship provided by Chinese Center for Advanced Science and Technology, headed by Li Zhengdao (a Nobel Laureate),

Zhou Guangzao and Lu Yongxian (academicians of Chinese Academy of Sciences). In 2008, the 16th National Conference of Grey Systems once again was financially supported by Chinese Center for Advanced Science and Technology.

Many important international conferences, such as the Conference on Uncertain Systems Modeling, Systems Prediction and Control, Workshops of the International Institute for General Systems Studies, Congress of World Organization of Cybernetics and Systems, IEEE conferences on System, Man, and Control, International Conferences of Computers and Industrial Engineering, etc., have listed grey systems theory as a special topic of interest. For instance in March 2002, the joint 12th Congress of the World Organization of Systems and Cybernetics (WOSC) and the 4th Workshop of the IIGSS, held in Pittsburgh of the United States arranged six special sessions for grey systems theory. In August 2003, the 32nd International Conference on Computers and Industrial Engineering, held in Ireland, opened four different sessions for grey systems theory. The IEEE International Conference on Systems, Man, and Control, held in the Holland in October 2003, the IEEE Conference on Networks, Remote Sensing and Control, held in Arizona, USA, in March 2005, the 13th Congress of the WOSC, held in Slovak in July 2005, the IEEE Conference on Systems, Man and Control, held in Hawaii, USA., in October 2005, the IEEE Conference on Systems, Man, Control, held in Taiwan, China, in October 2006, the IEEE Conference on Systems, Man and Control, held in Montreal, Canada, in October 2007, the 14th Congress of the WOSC, held in Poland in September 2008, the IEEE Conference on Systems, Man and Control, held in Singapore in October 2008, the IEEE Conference on Systems, Man and Control, held in San Antonio, USA, in October 2009, etc. all arranged special topic sessions for grey systems research. Grey systems theory has caught the attention of many important international conferences and become a center of discussion at many international events, which no doubt will play an important and active role for the world of systems researchers to get better acquainted with grey systems theory. During November 18 – 20, 2007, the inaugural IEEE International Conference on Grey Systems and Intelligent Services (IEEE GSIS) was successfully held in Nanjing, China. Nearly 300 scholars from around the world participated in this event. The IEEE headquarter approved the organization of this event, and National Natural Science Foundation (China), Nanjing University of Aeronautics and Astronautics (NUAA), and Grey Systems Association (China) jointly sponsored the actual details of the conference. The college of economics and management of the NUAA hosted the event. Professor Liu Sifeng, vice president of Chinese Association of Optimization and Economic Mathematics (CAEM), chair of the special grey systems committee of the CAEM, dean of the college of economics and management of NUAA, and president of the Institute for Grey Systems Studies, presided this event. Scholars from China, USA, England, Japan, South Africa, Russia, Turkey, Malaysia, Iran, Taiwan, Hong Kong, and other geographic regions submitted over 1,019 papers for possible presentation at this event. Eventually, 332 of the submissions were accepted.

As recommended by the participants, the program and organizing committees recognized Professor Julong Deng for his initial works of establishing grey systems theory, and the NUAA's college of economics and management for its outstanding works of organizing the important event. Five works presented at this conference were given a first prize, two second prizes and two third prizes. Dr. James Tien, the honorary chair of the conference, member of the National Academy of Engineers (USA), former executive chair of IEEE, presented the certificate to Professor Julong Deng; Dr. Robin Qiu, founding president of International Institute of Service Sciences, presented the certificate to the NUAA's college of economics and management; and Professor Julong Deng, founder of grey systems theory, presented certificates to all the best-paper award winners.

As decided by the conference organizing committee, the IEEE GSIS will be organized biannually. Scholars from the USA, England, Taiwan, Turkey, South Africa, and Japan, expressed their interest of hosting such events in the future. During November 10 – 12, 2009, the second IEEE GSIS was once again held in Nanjing. In early 2008, IEEE Technical Committee on Grey systems was organized. On the basis of such an international platform as IEEE, grey systems theory will surely be widely recognized and studied by many scholars from around the world in the years to come.

1.1.3 Characteristics of Unascertained Systems

The fundamental characteristic of uncertain systems is the incompleteness and inadequacy in their information. Due to the dynamics of system evolutions, the biological limitations of the human sensing organs, and the constraints of relevant economic conditions and technological availabilities, uncertain systems exist commonly.

1.1.3.1 Incomplete Information

Incompleteness in information is one of the fundamental characteristics of uncertain systems. The situation involving incomplete system information can have the following four cases:

(1) The information about the elements (parameters) is incomplete;
(2) The information about the structure of the system is incomplete;
(3) The information about the boundary of the system is incomplete; and
(4) The information on the system's behaviors is incomplete.

The situation of incomplete information is often seen in our social, economic, and scientific research activities. For instance, in agricultural productions, even if we know all the exact information regarding the areas of plantation, seeds, fertilizers, irrigations, due to the uncertainties in areas like labor quality, natural environments, weather conditions, the commodity markets, etc., it is still extremely difficult to precisely predict the production output and the consequent economic values. For biological prevention systems, even if we clearly know the relationship between insects and their natural enemies, it is still very difficult for us to achieve the expected prevention effects due to our ignorance with the knowledge on the relationships between the insects and the baits, their natural enemies and the baits, and a specific kind of natural enemy with another kind of natural enemy. As for the adjustment and reform of the price system, it is often difficult for the policy makers to take actions because of the lack of the information about how much psychological pressure the consumers could bear and how price change on a certain commodity would affect the prices of other commodities. On the security markets, even the brightest market analysts can not be assured of winning constantly due to their inability to correctly predict economic policy and interest rate changes, management changes at various companies, the direction of the political winds, investors' behavioral changes in the international markets, and price changes in one block of commodities on other blocks. For the general social-economic system, because there are not clear relationships between the "inside" and the "outside" and between the system itself and its environment, and because the boundary between the inside of the system and the outside is difficult to define, it is hard to analyze the effect of input on the output.

Incompleteness in available information is absolute, while completeness in information is relative. Man employs his limited cognitive ability to observe the infinite universe so that it is impossible for him to obtain the so-called complete information. The concept of large samples in statistics in fact represents the degree of tolerance man has to give to incompleteness. In theory, when a sample contains at least 30 objects, it is considered "large." However, for some situations, even when the sample contains thousands or several tens of thousands of objects, the true statistical laws still cannot be successfully uncovered.

1.1.3.2 Inaccuracies in Data

Another fundamental characteristic of uncertain systems is the inaccuracy naturally existing in the available data. The meanings of uncertain and inaccurate are roughly the same. They both stand for errors or deviations from the actual data values. From the essence of how uncertainties are caused, they can be categorized into three types: the conceptual, level, and prediction types.

(1) The Conceptual Type
Inaccuracies of the conceptual type item from the expression about a certain event, object, concept, or wish. For instance, all such frequently used concepts as "large," "small," "many," "few," "high," "low," "fat," "thin," "good," "bad," "young," "beautiful," etc., are inaccurate due to the lack of clear definition. It is very difficult to use exact quantities to express these concepts. As a second example, suppose that a job seeker with an MBA degree wishes to get an offer of an annual salary of no less than \$150,000. A manufacturing firm plans to control its rate of deficient products to be less than 0.1%. These are all cases of inaccurate wishes.

(2) The Level Type
This kind of inaccuracy of data is caused by a change in the level of research or observation. The available data, when seen on the level of the system of concern, that is the macroscopic level, or the level of the whole, or the cognitive conceptual level, might be accurate. However, when they are seen on a lower level, that is a microscopic level or a partial localized level of the system, they generally become inaccurate. For example, the height of a person can be measured accurately to the

unit of centimeters or millimeters. However, if the measurement has to be accurate to one ten-thousandth level, the earlier accurate reading will become extremely inaccurate.

(3) The Prediction Type (The Estimation Type)
Because it is difficult to complete understand the laws of evolution, prediction of the future tends to be inaccurate. For instance, it is estimated that two years from now, the GDP of a certain specified area will surpass $10 billion; it is estimated that a certain bank will attract as much savings from individual residents in an amount between 70,000 thousand to 90,000 thousand for the year in 2015; it is predicted that in the coming years the temperature during the month of October will not go beyond 30° C; etc. All these examples provide the uncertain numbers of the prediction type. In statistics, it is often the case that samples are collected to estimate the whole. So, many statistical data are inaccurate. As a matter of fact, no matter what method is used, it is very difficult for anyone to obtain the absolutely accurate (estimated) value. When we draw out plans for the future and make decisions about what course of action to take, we in general have to reply on not completely accurate predictions and estimates.

1.1.3.3 The Scientific Principle of Simplicity

In the development history of science, achieving simplicity has been a common goal of almost all scientists. In the time frame of as early as in the sixth century BC, natural philosophers had a common wish in their efforts of understanding the material nature: Build the knowledge of the material world on the basis of a few common, simple elements. The ancient Pythagoras of the Greece introduced the theory of four elements (earth, water, fire, and gas) at around 500 BC. They believed that all material matters in the universe were composed of these four simple elements. During around the same time, ancient Chinese also had the theory of five elements of water, fire, wood, gold, and earth. These are the most primitive and elementary thoughts about simplicity.

The scientific principle of simplicity originates from the simplicity of thinking employed in the process of understanding nature. As the natural science matures over time, simplicity becomes the foundation for man to fathom the world and becomes a guiding principle of scientific research.

Newtonian laws of motion unify the macroscopic phenomena of objective movements in their form of extreme simplicity. In his *Mathematical Principles of Natural Philosophy*, Newton pointed out that nature does not do useless works; because nature is fond of simplicity, it does not like to employ extra reasons to flaunt itself. During the ear of relativity, Albert Einstein introduced two criteria for testing a theory: external confirmation and internal completeness, that is, the logical simplicity. He believed that from the angle of employing scientific theories to reflect the harmony and orderliness of nature, a true scientific theory must comply with the principle of simplicity. In the 1870s, Ampere, Weber, Maxwell, and others one after another established theories to explain the phenomenon of electromagnetism based on their different assumptions: Because Maxwell's theory is the one that best complies to the principle of simplicity, it becomes well accepted. Another example is the well-known Kepler's third law of planetary motion: $T^2=D^3$, in form it looks very simple.

According to the slaving principle of synergetics, one can transform the original high-dimensional equation into a low-dimensional evolution equation of order-parameters by eliminating the fast-relaxing variables in the high-dimensional nonlinear equation that describes the evolution process of a system. Because the order-parameters dominate the dynamic characteristics of the system near the boundary points, through solving the evolution equation of order-parameters, one can obtain the system's time structure, space structure or time-space structure so that he can materialize efficient control over the system's behavior.

The simplicity of scientific models is actualized by employing simple expressions and by ignoring unimportant factors of the system of concern. In the area of economics, the methods of using Gini coefficient to describe differences among consumers' incomes and of employing Cobb-Dauglas production function to measure the contribution of advancing technology in the economic growth are all introduced on the basis of simplifying realistic systems. The model F. Modigliani uses to describe the average propensity to consume

$$\frac{C_t}{y_t}=a+b\frac{y_0}{y_t}, a>0, b>0$$

the curve Alban W. Phillips employs to describe the relationship between the rate of inflation $\frac{\Delta p}{p}$ and the unemployment rate x

$$\frac{\Delta p}{p}=a+b\frac{1}{x}$$

and the well-known capital asset pricing model

$$E[r_i]=r_f+\beta_i(E[r_m]-r_f)$$

can all be essentially reduced to the simplest linear regression model with some straightforward transformations.

1.1.3.4 Precise Models Suffer from Inaccuracies

When the available information is incomplete and the collected data inaccurate, any pursuit of chasing after precise models in general becomes meaningless. This fact has been well described by Lao Tzu more than two thousand years ago. The mutually antagonistic principle proposed by L. A. Zadeh, the founder of fuzzy mathematics, also clearly addresses this end: When the complexity of a system increases, our ability to precisely and meaningfully describe the characteristics of the system decreases accordingly until such a threshold that as soon as it is surpassed, the preciseness and meaningfulness become two mutually excluding characteristics. This mutually antagonistic principle reveals that single minded pursuit after preciseness could reduce the operationality and meaningfulness of the cognitive outcome. So, precise models are not necessarily an effective means to deal with complex matters.

In 1994, Jiangping Qiu and Xisheng Hua respectively established a theoretically delicate statistical regression model and relatively coarse grey model based on the deformation data and leakage data of a certain large scale hydraulic dam. Their work shows that their grey model provided better fit than the statistical regression model. When comparing the errors between the predictions of the two models with the actual observations, it is found that the prediction accuracy of the grey model is generally better than that of the regression model, see Table 1.1 for details.

Table 1.1 Comparison between the prediction errors of a statistical model and a grey model

Order No	Type	Average error	
		Statistical model	Grey model
1	Horizontal displacement	0.862	0.809
2	Horizontal displacement	0.446	0.232
3	Vertical displacement	1.024	1.029
4	Vertical displacement	0.465	0.449
5	Water level of pressure measurement hole	6.297	3.842
6	Water level of pressure measurement hole	0.204	0.023

In 2001, Dr. Haiqing Guo, academician Zhongru Wu et al. respectively established a statistical regression model and a grey time series combined model using the observational data of displacement in the vertical direction of a certain large clay-rock filled dam of inclined walls. They compared the data fitting and predictions of the two models and the actual observations and found that the data fitting of the grey combined model is much more superior than that of the statistical regression model.

Xiaobing Li, Haiyan Sun et al. employed fuzzy prediction functions to dynamically trace and precisely control the fuel oil feeding temperature for anode baking. The control effect was clearly better than that obtained by utilizing the traditional PID control method.

Academician Caixing Sun and his research group made use of grey incidence analysis, grey clustering, and various new types of grey prediction models to diagnose and predict insulation related accidents of electric transformers. Their huge amount of results indicates that these relatively coarse methods and models are more operational and provide more efficient results.

1.1.4 Comparison of Several Studies of Uncertain Systems

Probability and statistics, fuzzy mathematics, and grey systems theory are three mostly seen research methods employed for the investigation of uncertain systems. Their research objects all contain certain kinds of uncertainty, which represents their commonality. It is exactly the differences among the uncertainties in the research objects that these three theories of uncertainty are different from each other with their respective characteristics.

Fuzzy mathematics emphasizes on the investigation of problems with cognitive uncertainty, where the research objects possess the characteristic of clear intension and unclear extension. For instance, "young man" is a fuzzy concept, because each person knows the intension of "young man." However, if you are going to determine the exact range within which everybody is young and outside which each person is not young, then you will find yourself in a great difficulty. That is because the concept of young man does not have a clear extension. For this kind of problem of cognitive uncertainty with clear intension and unclear extension, the situation is dealt with in fuzzy mathematics by making use of experience and the so-called membership function.

Probability and statistics study the phenomena of stochastic uncertainty with emphasis placed on revealing the historical statistical laws. They investigate the chance for each possible outcome of the stochastic uncertain phenomenon to occur. Their starting point is the availability of large samples that are required to satisfy a certain typical form of distribution.

The focus of grey systems theory is on the uncertainty problems of small samples and poor information that are difficult for probability and fuzzy mathematics to handle. It explores and uncovers the realistic laws of evolution and motion of events and materials through information coverage and through the works of sequence operators. One of its characteristics is construct models with small amounts of data. What is clearly different of fuzzy mathematics is that grey systems theory emphasizes on the investigation of such objects that process clear extension and unclear intension. For example, by the year of 2050, China will control its total population within the range of 1.5 to 1.5 billion people. This range from 1.5 billion to 1.6 billion is a grey concept. Its extension is definite and clear. However, if one inquires further regarding exactly which specific number within the said range it will be, then he will not be able to obtain any meaningful and definite answer.

Based on what is discussed above, we summarize the differences among these three most studies subject matters in Table 1.2.

Table 1.2 Comparison between the three methods of uncertainty research

Object	Grey systems	Prob. Statistics	Fuzzy math
Research objects	Poor information	stochastics	Cognitive uncertainty
Basic sets	Grey hazy sets	Cantor sets	Fuzzy sets
Methods	Information coverage	Mapping	Mapping
Procedures	Sequence operator	Frequency distribution	Cut set
Data requirement	Any distribution	Typical distribution	Known membership
Emphasis	Intension	intension	extension
Objective	Laws of reality	Historical laws	Cognitive expression
Characteristics	Small sample	Large sample	Experience

1.1.5 Emerging Studies on Uncertain Systems

Fuzzy mathematics, grey systems theory, and rough set theory are currently the most actively studied systems theories of uncertainty. From the ISI and EI Compendex data bases it is found that the number of research papers with one of the keywords "fuzzy set," "grey system," and "rough set" increases rapidly, see Table 1.3 for more details.

Table 1.3 Search outcomes of ISI and EI Compendex data bases

Keyword	Fuzzy set	Grey system	Rough set
# appears in ISI data base of past 5 years	5,947	1,517	2,637
# appear in EI Compendex data base (1990 – 2008)	52,988	4,027	9,280

A search of the Chinese Data Base of Scholarly Periodicals (CNKI) shows that from 1990 to 2008, the number of scholarly publications with one of the keywords "fuzzy mathematics," "grey systems," and "rough set," also shows an up-trending development. See Tables 1.4 to 1.6 for more details.

Table 1.4 The search outcomes of "fuzzy mathematics"

Time	1990	1991	1992	1993	1994	1995	1996	1997	1998	1999
# of papers	345	373	401	346	543	575	574	551	514	530
Time	2000	2001	2002	2003	2004	2005	2006	2007	2008	Total
# of papers	605	583	598	714	720	799	908	1006	933	11618

Table 1.5 The search outcomes of "grey systems"

Time	1990	1991	1992	1993	1994	1995	1996	1997	1998	1999
# of papers	149	181	195	203	517	477	481	483	448	456
Time	2000	2001	2002	2003	2004	2005	2006	2007	2008	Total
# of papers	418	435	512	556	550	576	652	730	762	8781

Table 1.6 The search outcomes of "rough set"

Time	1990	1991	1992	1993	1994	1995	1996	1997	1998	1999
# of papers	0	0	0	0	0	0	1	1	9	19
Time	2000	2001	2002	2003	2004	2005	2006	2007	2008	Total
# of papers	50	102	142	267	412	553	710	779	919	3964

The research of uncertain (fuzzy, grey, and rough) systems can be categorized into the following three aspects:

(1) The mathematical foundation of the uncertain systems theories;
(2) The modeling of uncertain systems and computational schemes, including various uncertain system modeling, modeling combined with other relevant methods, and related computational methods; and
(3) The wide-range applications of uncertain systems theories in natural and social sciences.

Currently, the theoretical studies of uncertain (fuzzy, grey, rough) systems have been widely applied in all areas of natural science, social science, and engineering, including aviation, spaceflight, civil aviation, information, metallurgy, machinery, petroleum, chemical industry, electrical power, electronics, light industries, energy resources, transportation, medicine, health, agriculture, forestry, geography, hydrology, seismology, meteorology, environment protection, architecture, behavioral science, management science, law, education, military science, etc. These practical applications have brought forward definite and noticeable social and economic benefits.

Both the theoretical and applied research of uncertain (fuzzy, grey, rough) systems has been extremely active. However, all these works have shown the emphasis on applications without much enough effort placed on further developing the theory and establishing innovative methods. In particular, not enough attention is given to investigate the differences and commonalities between the various available uncertain systems theories so that not enough works on melting together the traditional theories and methods with the newly appearing uncertain systems theories and methods were published. This fact more or less affected the development of uncertain (fuzzy, grey, rough) systems theories.

As a matter of fact, each traditional or newly emerging uncertain systems theory and method complements each other and cannot be clearly separated from each other. When facing various kinds of uncertainty problems different uncertain systems theory, different uncertain system theory and method have its strengths, which complement and supplement each other instead of repelling each other. Many complex, dynamic uncertainty problems are really way beyond the scope and capacity of any single uncertain systems theory and method. It requires the researcher to combine various kinds of classical theories with the newly developed uncertain systems theories and methods. It is a must for this kind of interaction, exchange, and combination of the relevant theories and methods to occur for the further, healthy development of science.

1.1.6 Position of Grey Systems Theory in Cross-Disciplinary Researches

Corresponding to differences in how we look at the objective world and matters, the scientific system is divided up differently. In the 17th century, Francis Bacon divided human knowledge into three major blocks: history, poetry and arts, and philosophy, based on his belief that such division should correspond to human capacity of memory, imagination, judgment. Later, Saint Simon and Georg Wilhelm Friedrich Hegel respectively proposed the ideas of dividing disciplines of knowledge according to metaphysics and materialism. During the latter half of the 19th century, Friedrich Engels proposed to define disciplines according to different forms of motion of materials and their natural order. This proposal helped to lay down the structural system of science and the theoretical foundation for the division of knowledge into various compartments.

In China, the totality of knowledge tends to be divided into either two major blocks with one block of humanities and the other science, or into three areas of natural science, mathematics, and social science. As for the foundations of natural sciences, they have been customarily divided into the following six blocks: mathematics, physics, chemistry, astronomy, earth science, and biology. Mr. Xuesheng Qian suggested to the entire system of science and technology into natural science, social science, systems science, science of thoughts, science of human body, and mathematical science, each of which possesses three layers: the foundations, technology, and engineering.

In this book, we divide the totality of knowledge on the basis of the classification of the problems addressed. Firstly, we classify problems according to their complexity and uncertainty. And then, we point out the corresponding disciplines of specific epistemological significance based on the problems' characteristics so that the position of grey system theory in the spectrum of cross-disciplinary studies is clarified.

Let us use a rectangle Ω to represent the totality of all matters in the world and circles A, B, C, and D to represent respectively the sets of simple matters, complex matters, deterministic matters, and indeterminate matters. So, we obtain the Venn diagram for the classification of scientific problems in Figure 1.1, where when the methods of resolving the corresponding problems are labeled, we obtain the Venn diagram for cross-disciplinary studies in Figure 1.2.

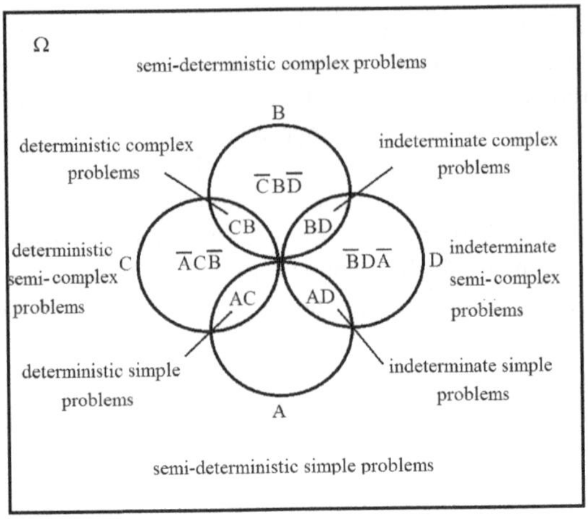

Fig. 1.1 The Venn diagram for classification of scientific problems

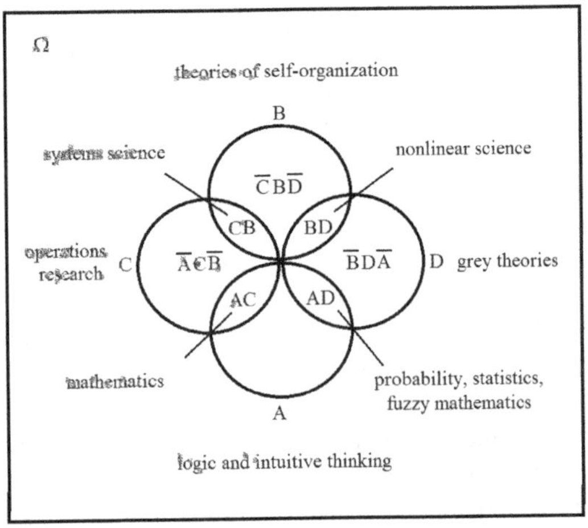

Fig. 1.2 The Venn diagram for classification of cross-disciplines

From a comparison of Figures 1.1 and 1.2, it can be seen that as the scientific methodology of resolving indeterminate semi-complex problems, grey systems theory is established as a jump from probability and statistics that are developed to resolve simple indeterminate problems. As for successful resolutions of complex indeterminate problems, one has to wait for new breakthroughs in the study of nonlinear science.

1.2 Basics of Grey Systems

1.2.1 Elementary Concepts of Grey Systems

Many social, economic, agricultural, industrial, ecological, biological, etc., systems are named by considering the features of classes of the research objects, while grey systems are labeled using the color of the systems of concern.

In the theory of control, people often make use of colors to describe the degree of clearness of the available information. For instance, Ashby refers the objects with unknown internal information to as black boxes. This terminology has been widely accepted in the scientific community. As an another example, as a society moves toward democracy, the citizens gradually demand more information regarding the formation of policies and more in depth meaning of the policies. That is, the citizens want to have an increased degree of transparency. We use "black" to indicate unknown information, "white" the completely known information, and "grey" the partially known and partially unknown information. Accordingly, the systems with completely known information will be regarded as white, those systems with completely unknown information black, and the systems with partially known information and partially unknown information will be seen as grey.

Table 1.7 Extensions of the concept of "grey"

conept / situation	Black	Grey	White
From information	unknown	incomplete	Completely known
From appearance	dark	blurred	clear
From processes	new	Changing	old
From properties	chaotic	multivariate	order
From methods	negation	change for better	confirmation
From attitude	letting go	tolerant	rigorous
From the outcomes	no solution	multi-solutions	unique solution

At this junction, we need to pay attention to the difference between "systems" and "boxes." Usually, "boxes" are used when one does not pay much attention on or does not attempt to utilize the information regarding the interior while focusing on the external characteristics. In this case, the researcher generally investigates the properties and characteristics of the object through analyzing the input-output relation. Other the other hand, "systems" are employed to indicate the study of the object's structure and functions through analyzing the existing organic connections between the object, relevant factors, and its environment and the related laws of change.

The research objects of grey systems theory consist of such uncertain systems that they are known only partially with small samples and poor information. The theory focuses on the generation and excavation of the partially known information to materialize the accurate description and understanding of the material world.

Based on the discussion above, the situation of incomplete system information can exist in one of the following four cases:

(i) The information about the elements (parameters) is incomplete;
(ii) The information about the structure of the system is incomplete;
(iii) The information about the boundary of the system is incomplete; and
(iv) The information on the system's behaviors is incomplete.

Incompleteness in information is the fundamental meaning of being "grey." From different angles and in varied situations, the meaning of "grey" can be expanded or stretched. For this end, see the details in Table 1.7.

1.2.2 Fundamental Principles of Grey Systems

In the process of establishing the grey systems theory, Julong Deng discovered and extracted the following fundamental principles of grey systems. It is readily for the reader to see that these principles contain intrinsic philosophical intensions.

Axiom 1.1 (Principle of informational differences). "Difference" implies the existence of information. Each piece of information must carry some kind of "difference".

Axiom 1.2 (Principle of Non-Uniqueness). The solution to any problem with incomplete and indeterminate information is not unique.

Axiom 1.3 (Principle of Minimal Information). One characteristic of grey systems theory is that it makes the most and best use of the available "minimal amount of information."

Axiom 1.4 (Principle of Recognition Base). Information is the foundation on which people recognize and understand (nature).

Axiom 1.5 (Principle of New Information Priority). The function of new pieces of information is greater than that of old pieces of information.

Axiom 1.6 (Principle of Absolute Greyness). "Incompleteness" of information is absolute.

These principles of grey systems are listed here for the completeness of this book. If the reader likes to see more deliberations on these principles, please consult with (Liu and Lin, 2006, pp. 5 – 7).

1.2.3 Main Components of Grey Systems Theory

Through nearly thirty years of development, grey systems theory has been built up as a newly emerging scientific discipline with its very own theoretical structure consisting of systems analysis, evaluation, modeling, prediction, decision-making, control, and techniques of optimization. Its main contents contain

a) The theoretical system developed on the basis of grey algebraic system, grey equations, grey matrices, etc.;
b) The methodological system established on the basis of sequence operators and generations of grey sequences;
c) The analysis and evaluation system constructed on the basis of grey incidence spaces and grey cluster evaluations;
d) The prediction model system centered around GM(1.1);
e) The decision-making model system represented by multi-attribute intelligent grey target decision models;
f) The system of combined grey models innovatively developed for producing new and practically useful results; and
g) The optimization model system, consisting mainly of grey programming, grey input-output analysis, grey game theory, and grey control.

Grey algebraic system, grey matrices, grey equations, etc., constitute the foundation of grey systems theory. In terms of the theoretical beauty and completeness of the theory, there are still a lot of problems left open in this area.

In this book, generations of grey sequences are merged into the concept of sequence operators, which mainly include buffer operators (weakening buffer operators, strengthening operators), mean generation operators, ratio generation operators, stepwise ratio generators, accumulating generators, inverse accumulating generators, etc.

Grey incidence analysis includes such materials as grey incidence axioms, degree of grey incidence, generalized degree of grey incidence (absolute degree, relative degree, synthetic degree), the degrees of grey incidence based on either similar visual angles or approximate visual angles, grey incidence order, superiority analysis, and others.

Grey cluster evaluation includes such contents as grey variable weight clustering, grey fixed weight clustering, cluster evaluations based on (center-point or end-point) triangular whitenization weight functions, and other related materials.

Through grey generations or the effect of sequence operators to weaken the randomness, grey prediction models are designed to excavate the hidden laws; and through the interchange between difference equations and differential equations, a practical jump of using discrete data sequences to establish continuous dynamic differential equations is materialized. Here, GM(1,1) is the central model that has been most widely employed; and discrete grey models are a class of new models we initially developed.

In terms of grey predictions, they produce quantitative forecasts on the basis of the GM model. Based on their functions and characteristics, grey predictions can be grouped into sequence predictions, interval predictions, disaster predictions, seasonal disaster predictions, stock-market-like predictions, system predictions, etc.

The grey combined models include grey econometric models (G-E), grey Cobb-Douglass models (G-C-D), grey Markov models (G-M), grey-rough mixed models, etc.

Grey decision-making includes multi-attribute intelligent grey target models, grey incidence decision-making, grey cluster decision-making, grey situation decisions, grey stratified decisions, etc.

The main contents of grey control include the control problems of essential grey systems, the controls composed of grey systems methods, such as grey incidence control, GM(1,1) prediction control, etc.

Considering all the feedbacks from the readers of our earlier monograph, *Grey Information: Theory and Practical Applications* (Springer, 2006), we have paid special attention to organize some of the most recent new results obtained by colleagues from around the world in this volume. Also, for the convenience of practical applications, this book is accompanied with a computer software on grey systems modeling, which is designed by Zeng Bo of our research group.

Chapter 2
Basic Building Blocks

2.1 Grey Numbers, Degree of Greyness, and Whitenization

2.1.1 Grey Numbers

Such a number instead of its range whose exact value is unknown is referred to as a *grey number*. In applications, a grey number in fact stands for an indeterminate number that take its possible value within an interval or a general set of numbers. This grey number is generally written using the symbol "⊗." There are several types of grey numbers.

1. Grey numbers with only a lower bound: This kind of grey number $\otimes$ is written as $\otimes \in [\underline{a}, \infty]$ or $\otimes(\underline{a})$, where $\underline{a}$ stands for the definite, known lower bound of the grey number $\otimes$. The interval $[\underline{a}, \infty]$ is referred to as the field of $\otimes$.

2. Grey numbers with only an upper bound: This kind of grey number $\otimes$ is written as $\otimes \in (-\infty, \overline{a}]$ or $\otimes(\overline{a})$, where $\overline{a}$ stands for the definite, known upper bound of $\otimes$.

3. *Interval grey numbers*: This kind of grey number $\otimes$ has both a lower $\underline{a}$ and an upper bound $\overline{a}$, written $\otimes \in [\underline{a}, \overline{a}]$.

4. *Continuous* and *discrete grey numbers*: A grey number that only takes a finite number or a countable number of potential values is known as discrete. If a grey number can potentially take any value within an interval, then it is known as continuous.

5. *Black* and *white numbers*: When $\otimes \in (-\infty, +\infty)$, that is, when $\otimes$ has neither any upper nor lower bound, then $\otimes$ is known as a black number. When $\otimes \in [\underline{a}, \overline{a}]$ and $\underline{a} = \overline{a}$, $\otimes$ is known as a white number.

6. *Essential* and *non-essential grey numbers*: The former stands for such a grey number that temporarily cannot be represented by a white number; and the

latter such grey numbers each of which can be represented by a white number obtained either through experience or certain method. The definite white number is referred to as the whitenization (value) of the grey number, denoted $\tilde{\otimes}$. Also, we use $\otimes(a)$ to represent the grey number(s) with a as its whitenization.

Of course, based on how the involved uncertainty is caused, grey numbers can also be categorized as conceptual type, layer type, and prediction type. In what follows, let us look at the operations of interval grey numbers.

Given grey numbers $\otimes_1 \in [a,b], a<b$, and $\otimes_2 \in [c,d], c<d$, let us use $*$ represent an operation between $\otimes_1$ and $\otimes_2$. If $\otimes_3 = \otimes_1 * \otimes_2$, then $\otimes_3$ should also be an interval number satisfying $\otimes_3 \in [e,f], e<f$, and for any $\tilde{\otimes}_1$ and $\tilde{\otimes}_2$, $\tilde{\otimes}_1 * \tilde{\otimes}_2 \in [e,f]$. In particular, for the operation of addition, we have

$$\otimes_1 + \otimes_2 \in [a+c, b+d] \tag{2.1}$$

The additive inverse is given by

$$-\otimes \in [-b, -a] \tag{2.2}$$

Therefore, the subtraction is given by

$$\otimes_1 - \otimes_2 = \otimes_1 + (-\otimes_2) \in [a-d, b-c] \tag{2.3}$$

The multiplication of $\otimes_1$ and $\otimes_2$ is defined as follows:

$$\otimes_1 \cdot \otimes_2 \in [\min\{ac, ad, bc, bd\}, \max\{ac, ad, bc, bd\}] \tag{2.4}$$

The reciprocal of $\otimes_1 \in [a,b], a<b, a\neq 0, b\neq 0, ab>0$ is defined as

$$\otimes_1^{-1} \in \left[\frac{1}{b}, \frac{1}{a}\right] \tag{2.5}$$

Accordingly, for $\otimes_1 \in [a,b], a<b$; $\otimes_2 \in [c,d], c<d$, 且 $c\neq 0, d\neq 0, cd>0$, the division of $\otimes_1$ by $\otimes_2$ is given as follows:

$$\otimes_1/\otimes_2 = \otimes_1 \times \otimes_2^{-1} \in \left[\min\left\{\frac{a}{c}, \frac{a}{d}, \frac{b}{c}, \frac{b}{d}\right\}, \max\left\{\frac{a}{c}, \frac{a}{d}, \frac{b}{c}, \frac{b}{d}\right\}\right] \tag{2.6}$$

For scalar multiplication, let $\otimes \in [a,b]$, $a<b$, and k a positive real number, then we have

$$k \cdot \otimes \in [ka, kb] \tag{2.7}$$

For exponents, let $\otimes \in [a,b], a<b, k$ k a positive real number, then we have

$$\otimes^k \in [a^k, b^k] \tag{2.8}$$

which is known as the *k*th power of the grey number $\otimes$.

Historically, the study on operations and the algebraic system of grey numbers has been active in the research of grey systems. However, as of this writing, satisfactory results have not been achieved. In the 1980s, we once proposed the concept of average whitenization for grey numbers in order to develop an operation system for grey numbers. However, due to the difficulty of satisfactorily dealing with disturbed grey elements, no definite result was established. In what follows, let us look at the concept of cores of grey numbers. On this basis and the degree of greyness, we will establish an axiomatic system for the operations of grey numbers and the corresponding grey algebraic system. Within this approach, operations of grey numbers are reduced to those of real numbers so that some of the difficulties met in the construction of grey number operations and the corresponding grey algebraic system are eased to some degree.

Definition 2.1 (Sifeng Liu). Let $\otimes \in [\underline{a}, \overline{a}], \underline{a} < \overline{a}$. Assume that the information of distribution of the grey number $\otimes$ is unknown.

1. If $\otimes$ is continuous, then $\hat{\otimes} = \frac{1}{2}(\underline{a} + \overline{a})$ is the core of the grey number $\otimes$;
2. If $\otimes$ is discrete, $a_i \in [\underline{a}, \overline{a}](i = 1, 2, \cdots n)$ are all the possible values $\otimes$ might take on, then $\hat{\otimes} = \frac{1}{n}\sum_{i=1}^{n} a_i$ is the *core* of the grey number.

   (Note: If some $a_k(\otimes)$ is a grey element such that $a_k(\otimes) \in [\underline{a}_k, \overline{a}_k], \underline{a}_k < \overline{a}_k$, then take $a_k = \hat{a}_k$).

Definition 2.2. Let $\otimes \in [\underline{a}, \overline{a}], \underline{a} < \overline{a}$ be a random grey number with known distribution. Then, $\hat{\otimes} = E(\otimes)$ is the core of the grey number $\otimes$.

The core $\hat{\otimes}$ of the grey number $\otimes$, as a representative of the grey number $\otimes$, plays an irreplaceable role in the process of transforming grey number operations into those of real numbers. As a matter of fact, the core $\hat{\otimes}$ of the grey number $\otimes$, as a real number, can be operated upon by a whole series of operations, such as addition, subtraction, multiplication, and division. Additionally, The outcomes of these operations on the cores of grey numbers can be naturally seen as the results of the cores of outcomes of the corresponding grey number operations.

Definition 2.3. Let $\hat{\otimes}$ and g° be respectively the core and the *degree of greyness* of a grey number $\otimes$. Then $\hat{\otimes}_{(g^{\circ})}$ is seen as a simple form of the grey number $\otimes$.

According to the definition of degrees of greyness introduced in Subsection 2.1.3 in this book, the simplified form $\hat{\otimes}_{(g^\circ)}$ contains all the information of the grey number $\otimes \in [\underline{a}, \overline{a}], \underline{a} < \overline{a}$.

Proposition 2.1. For interval grey numbers, there is an one-to-one correspondence between the simplified forms $\hat{\otimes}_{(g^\circ)}$ and grey numbers $\otimes \in [\underline{a}, \overline{a}], \underline{a} < \overline{a}$.

In fact, for any chosen grey number $\otimes \in [\underline{a}, \overline{a}], \underline{a} < \overline{a}$, one can compute $\hat{\otimes}_{(g^\circ)}$ through both $\hat{\otimes}$ and g°. On the other hand, when $\hat{\otimes}_{(g^\circ)}$ is given, one can determine the position of $\otimes$ from $\hat{\otimes}$. So, from the degree of greyness g°, as defined in Subsection 2.1.3, one can compute the measure of the grey number $\otimes$ and consequently the upper and lower bounds $\overline{a}$ and $\underline{a}$, which provides the detailed information for $\otimes \in [\underline{a}, \overline{a}], \underline{a} < \overline{a}$.

Definition 2.4. Assume that Ω is the *discourse field* of a grey number $\otimes$. If $\mu(\Omega) = 1$, then $\otimes$ is called a *standard grey number*. The simplified form of a standard grey number is known as the standard form of the grey number.

Proposition 2.2. If $\otimes$ is a standard grey number, then $g^\circ(\otimes) = \mu(\otimes)$.

In terms of standard grey numbers, the concept of degrees of greyness is identical to that of measures of grey numbers. If we further limit the discourse field Ω to the interval [0,1], then $\mu(\otimes$ stands for the length of the small interval over [0,1]. In this case, recovering the standard form of a grey number to its general form becomes quite simply.

Axiom 2.1 (Non-decreasing Greyness, Sifeng Liu). When two grey numbers are added, subtracted, multiplied, or divided, the degree of greyness of the result is not smaller the larger degree of greyness of the original grey numbers.

For convenience, we usually take the larger degree of greyness of the grey numbers that participate in the operation as the degree of greyness of the operational outcome. From Axiom 2.1, one can see the following:

Corollary 2.1. When a white number and a grey number are added, subtracted, multiplied, or divided, the degree of greyness of the outcome equals that of the grey number.

When several grey numbers are involved in arithmetic operations, we first perform the operations using the corresponding cores, and then take the largest degree of greyness of the grey numbers to produce the simplified form the operational outcome. Based on the simplified form $\hat{\otimes}_{(g^\circ)}$ of grey numbers, we can obtain the following properties of the arithmetic operations of grey numbers:

$$\hat{\otimes}_{1(g_1^\circ)} = \hat{\otimes}_{2(g_2^\circ)} \Leftrightarrow \hat{\otimes}_1 = \hat{\otimes}_2 \text{ and } g_1^\circ = g_2^\circ \tag{2.9}$$

$$\hat{\otimes}_{1(g_1^\circ)} + \hat{\otimes}_{2(g_2^\circ)} = (\hat{\otimes}_1 + \hat{\otimes}_2)_{(g_1^\circ \vee g_2^\circ)} \tag{2.10}$$

$$-\hat{\otimes}_{(g^\circ)} = (-\hat{\otimes})_{(g^\circ)} \tag{2.11}$$

$$\hat{\otimes}_{1(g_1^\circ)} - \hat{\otimes}_{2(g_2^\circ)} = (\hat{\otimes}_1 - \hat{\otimes}_2)_{(g_1^\circ \vee g_2^\circ)} \tag{2.12}$$

$$\hat{\otimes}_{1(g_1^\circ)} \times \hat{\otimes}_{2(g_2^\circ)} = (\hat{\otimes}_1 \times \hat{\otimes}_2)_{(g_1^\circ \vee g_2^\circ)} \tag{2.13}$$

$$\text{If } \hat{\otimes} \neq 0 \text{, then } 1/\hat{\otimes}_{(g^\circ)} = (1/\hat{\otimes})_{(g^\circ)} \tag{2.14}$$

$$\text{If } \hat{\otimes}_2 \neq 0 \text{, then } \hat{\otimes}_{1(g_1^\circ)} / \hat{\otimes}_{2(g_2^\circ)} = (\hat{\otimes}_1 / \hat{\otimes}_2)_{(g_1^\circ \vee g_2^\circ)} \tag{2.15}$$

$$\text{Let } k \text{ be a real number, then } k \cdot \hat{\otimes}_{(g^\circ)} = (k \cdot \hat{\otimes})_{(g^\circ)} \tag{2.16}$$

$$\text{Let } k \text{ be a real number, then } (\hat{\otimes}_{(g^\circ)})^k = (\hat{\otimes})^k{}_{(g^\circ)} \tag{2.17}$$

Because when $g_1^\circ = g_2^\circ = g^\circ$, $g_1^\circ \vee g_2^\circ = g^\circ$, the addition, subtraction, multiplication, and division operations of grey numbers with equal degree of greyness will be special cases of equs. (2.11), (2.13), (14), and (2.16). The details are omitted. Also, as expected, the arithmetic operations of grey numbers can be generalized to ny finite number of grey numbers.

For a set $F(\otimes)$ of grey numbers, if for any $\otimes_i, \otimes_j \in F(\otimes)$, each of $\otimes_i + \otimes_j, \otimes_i - \otimes_j$, $\otimes_i \cdot \otimes_j$, and $\otimes_i / \otimes_j$, when all of these operations are defined, still belongs to $F(\otimes)$, then the set $F(\otimes)$ is known as a field of grey numbers. It is not hard to check the following result:

Theorem 2.1. The set of all interval grey numbers forms a field.

Definition 2.5. Assume that $R(\otimes)$ is a set of interval grey numbers. If for any $\otimes_i$, $\otimes_j$, $\otimes_k \in R(\otimes)$, the following hold true:

1° $\otimes_i + \otimes_j = \otimes_j + \otimes_i$;

2° $(\otimes_i + \otimes_j) + \otimes_k = \otimes_i + (\otimes_j + \otimes_k)$;

3° There is a zero element $0 \in R(\otimes)$ such that $\otimes_i + 0 = \otimes_i$;

4° For any $\otimes \in R(\otimes)$, there is a $-\otimes \in R(\otimes)$ such that $\otimes + (-\otimes) = 0$;

5° $\otimes_i \cdot (\otimes_j \cdot \otimes_k) = (\otimes_i \cdot \otimes_j) \cdot \otimes_k$;

6° There is unit element $1 \in R(\otimes)$ such that $1 \cdot \otimes_i = \otimes_i \cdot 1 = \otimes_i$;

7° $(\otimes_i + \otimes_j) \cdot \otimes_k = \otimes_i \cdot \otimes_k + \otimes_j \cdot \otimes_k$; and

8° $\otimes_i \cdot (\otimes_j + \otimes_k) = \otimes_i \cdot \otimes_j + \otimes_i \cdot \otimes_k$,

then $R(\otimes)$ is referred to as a *grey linear space*.

Theorem 2.2. The totality of all interval grey numbers constitutes a grey linear space.

Any algebraic equation that contains grey parameter(s) (or called grey element(s)) is called a *grey algebraic equation*. Each differential equation involving grey parameter(s) will be accordingly called a *grey differential equation*. Any matrix containing grey entry(-ies) will be referred to as a *grey matrix*, denoted $A(\otimes)$, where $\otimes(i, j) = [a_{ij}, b_{ij}]$ stands for the entry at the (*i*,*j*) cell, whose simplified form is written as $\otimes(i, j) = \hat{\otimes}_{(g_{ij}^{\circ})}(i, j)$.

2.1.2 Whitenization of Grey Numbers and Degree of Greyness

One type of grey number vibrates around a certain fixed value. The *whitenization* of this kind of grey number is relatively easy. One can simply use that fixed value as its whitenization. A grey number that vibrates around a can be written as $\otimes(a) = a + \delta_a$ or $\otimes(a) \in (-, a, +)$, where δ_a stands for the vibration. In this case, the whitenized value is $\tilde{\otimes}(a) = a$. For the general interval grey number $\otimes \in [a, b]$, let us take its whitenization value $\tilde{\otimes}$ as follows:

$$\tilde{\otimes} = \alpha a + (1 - \alpha) b, \alpha \in [0,1] \tag{2.18}$$

which is known as the equal-weight whitenization. If the weight $\alpha = \frac{1}{2}$, the resultant whitenization is known as *equal-weight mean whitenization*. When the distribution of an interval grey number is unknown, the equal-weight mean whitenization is often employed.

For two given interval grey numbers $\otimes_1 \in [a, b]$, $\otimes_2 \in [a, b]$, satisfying that $\tilde{\otimes}_1 = \alpha a + (1 - \alpha) b$, $\alpha \in [0,1]$, and $\tilde{\otimes}_2 = \beta a + (1 - \beta) b$, $\beta \in [0,1]$, if $\alpha = \beta$, we say that both $\otimes_1$ and $\otimes_2$ take the same value. If $\alpha \neq \beta$, we say that

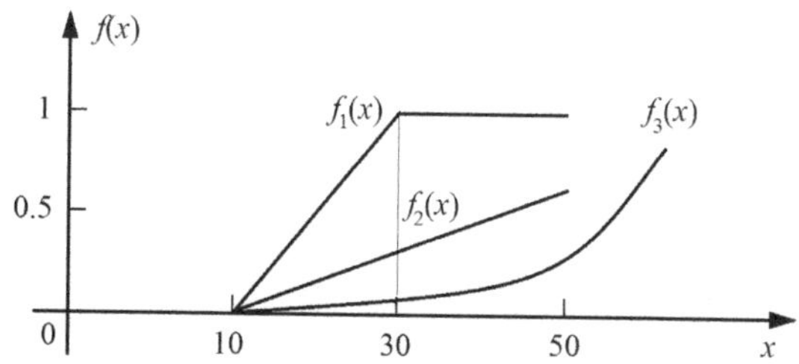

Fig. 2.1 Different types of whitenization weight functions

the grey numbers $\otimes_1$ and $\otimes_2$ do not take the same value. When two grey numbers $\otimes_1$ and $\otimes_2$ take their values in the same interval [a,b], then only when they take the same value, one can possibly have $\otimes_1 = \otimes_2$.

When the distribution of an interval grey number is known, it is natural to take such a whitenization that is generally not equl-weight. For instance, A certain person's age is possible between 30 to 45 years old. So, $\otimes \in [30,45]$ is a grey number. What is also known is that the person finished his 12 years of pre-college education and entered college in the 1980s. Hence, the chance for the person to be 38 years old or in the interval from 36 to 40 years ages in age in 2005 is quite good. For this grey number, it will not be reasonable for us to employ equal-weight whitenization. To describe what is known, we use a *whitenization weight function* to describe the preference the grey number has over the potential values it might take.

For any conceptual type grey number that represents wishes, its whitenization weight function generally is monotonically increasing. In figure 2.1, the whitenization weight function $f(x)$ stands for, say, the grey number of loan amount (in ten thousand dollars) and its degree of preference. A straight line stands for the "normal desire," that is, the degree of preference is directly proportional to the amount of the loan with different slopes representing different intensities of desire. In particular, $f_1(x)$ represents a relatively mild intensity of desire, where a loan amount of \$100,000 is not enough, a loan in the amount of \$200,000 will be more satisfying, and a loan of \$300,000 will be quite adequate. $f_2(x)$ stands for a desire with more intensity, where a loan in the amount of \$350,000 is only about 40% satisfactory. The curve of $f_3(x)$ means that even for a loan in the amount of \$400,000, the degree of satisfaction is only about 20%. To be satisfied, the loan amount has to be somewhere around \$800,000.

Generally speaking, the whitenization weight function of a grey number is designed according to what is known to the researcher. So, it does not have a fixed form. The start and end of the curve should have its significance. For instance, in a trade negotiation, there is a process of changing from a grey state to a white state. The eventual agreed upon deal will somewhere between the ask and bit. So, the relevant whitenization weight function should start at the level of ask (or bit) and end at the level of bit (or ask).

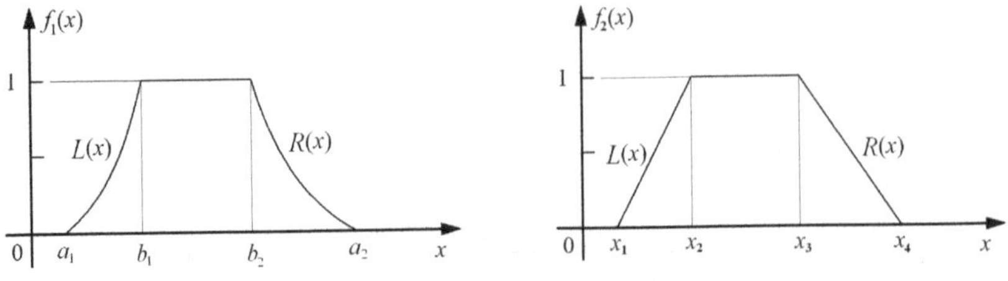

(a) (b)

Fig. 2.2 Typical whitenization weight functions

The typical whitenization weight function is a continuous function with fixed starting and ending points so that the left-hand side increases and right-hand side decreases, Figure 2-2(a), where

$$f_1(x)=\begin{cases} L(x), & x\in[a_1,b_1) \\ 1, & x\in[b_1,b_2] \\ R(x), & x\in(b_2,a_2] \end{cases}$$

In practical applications for the convenience of computer programming and computation, the left- and right-hand functions $L(x)$ and $R(x)$ generally simplified into straightlines, Figure 2.2(b), so that we have

$$f_2(x)=\begin{cases} L(x)=\dfrac{x-x_1}{x_2-x_1}, & x\in[x_1,x_2) \\ 1, & x\in[x_2,x_3] \\ R(x)=\dfrac{x_4-x}{x_4-x_3}, & x\in(x_3,x_4] \end{cases}$$

Theorem 2.3. Let X be the real number space, $Y\subseteq[0,1]$, and the mapping $f:X\to Y$, $x\mapsto f(x)$ satisfies the conditions of the typical whitenization weight function. Then f possesses the following properties:

1° $f(\phi)=\phi$;

2° $f(X)=Y$;

3° $\forall A,B\subseteq X$, if $A\subseteq B$, then $f(A)\subseteq f(B)$;

4° if, $A\neq\phi$, then $f(A)\neq\phi$;

5° $f(A\cup B)=f(A)\cup f(B)$; and

6° $f(A\cap B)\subseteq f(A)\cap f(B)$.

For the whitenization weight function shown in Figure 2.2(a), the following

$$g^{\circ} = \frac{2|b_1 - b_2|}{b_1 + b_2} + \max\left\{\frac{|a_1 - b_1|}{b_1}, \frac{|a_2 - b_2|}{b_2}\right\} \tag{2.19}$$

is referred (Julong Deng) to as the *degree of greyness* of $\otimes$. The expression of g° is a sum of two parts. The first part represents the greyness of the grey number as affected by the size of the peak area under the curve of the whitenization weight function, while the second part shows the effect of the size of the area under the curves of $L(x)$ and $R(x)$. Generally, the greater the peak area and the area under $L(x)$ and $R(x)$, the greater value g° takes. When $\max\left\{\frac{|a_1 - b_1|}{b_1}, \frac{|a_2 - b_2|}{b_2}\right\} = 0$, $g^{\circ} = \frac{2|b_1 - b_2|}{b_1 + b_2}$. In this case, the whitenization weight function is a horizontal line. When $\frac{2|b_1 - b_2|}{b_1 + b_2} = 0$, grey number $\otimes$ is a grey number with its basic value $b = b_1 = b_2$. When g° = 0, $\otimes$ is a white number.

2.1.3 Degree of Greyness Defined by Using Axioms

In 1996, by using the *length* $l(\otimes)$ of the grey number interval and its mean whitenization $\hat{\otimes}$, Sifeng Liu (1996) established an axiomatic definition for the degree of greyness:

$$g^{\circ}(\otimes) = \frac{l(\otimes)}{\hat{\otimes}} \tag{2.20}$$

which is valid on the basis of the postulates of non-negativity, zero greyness, infinite greyness, and scalar multiplication. However, the concept of greyness as defined in either equs. (2.19) or (2.20) suffers from the following problems:

1) When the length $l(\otimes)$ of grey interval approaches infinity, the degree of greyness as defined in either (2.19) or (2.20) is likely to approach infinity.
2) Grey numbers centered at zero will not have greyness. In this case, in equ. (2.19), one has $b_1 = b_2 = 0$; and in equ. (2.20), one faces $\hat{\otimes} = 0$. That is, in this case, neither (2.19) nor (2.20) is meaningful.

Grey numbers are a form of expressing the behavioral characteristics of the grey system of concern (Julong Deng, 1990). The greyness of grey numbers reflects the degree the researcher understands the uncertainty involved (Sifeng Liu, 1999; Shiliang Chen, 2001). Therefore, the magnitude of the greyness of a grey number should be closely related the background on which the grey number is created or the field of discourse within which the relevant number becomes grey. If this background or the field of discourse and its characteristics of the grey system of our concern are not detailed, we in fact will have no means to discuss the degree of greyness of the relevant grey number. With this understanding in place, let Ω be the field of discourse on which grey number $\otimes$ is created and $\mu(\otimes)$ the measure of the number field from which $\otimes$ takes its value. Then, the degree of greyness $g^{\circ}(\otimes)$ of the grey number $\otimes$ satisfies the following axioms:

Axiom 2.2. $0 \le g^{\circ}(\otimes) \le 1$. That is, the degree of greyness of any grey number has to be within the range from 0 to 1.

Axiom 2.3. For any $\otimes \in [\underline{a}, \overline{a}], \underline{a} \le \overline{a}$, when $\underline{a} = \overline{a}$, $g^{\circ}(\otimes) = 0$. That is, each white number does not contain any ambiguity so that its degree of greyness is 0.

Axiom 2.4. $g^{\circ}(\Omega) = 1$. That is because the background from which the grey number Ω is created is generally known, it does not contain any useful information leading to the greatest level of uncertainty.

Axiom 2.5. $g^{\circ}(\otimes)$ is directly proportional to $\mu(\otimes)$ and inversely proportional $\mu(\Omega)$.

Based on these four axioms, Sifeng Liu introduced the following definition of the degree of greyness for grey number $\otimes$:

$$g^{\circ}(\otimes) = \mu(\otimes) / \mu(\Omega) \tag{2.21}$$

where Ω is the field of discourse of the grey number $\otimes$ and $\mu(\otimes)$ the measure of $\otimes$'s field.

Theorem 2.4. The degree of greyness of grey numbers satisfy the following properties:

1) If $\otimes_1 \subset \otimes_2$, then $g^{\circ}(\otimes_1) \le g^{\circ}(\otimes_2)$.
2) $g^{\circ}(\otimes_1 \cup \otimes_2) \ge g^{\circ}(\otimes_k)$, $k = 1, 2$, where $\otimes_1 \cup \otimes_2 = \{\xi | \xi \in [a,b] or \xi \in [c,d]\}$ is the union of the grey numbers $\otimes_1 \in [a,b], a < b$ and $\otimes_2 \in [c,d], c < d$.
3) $g^{\circ}(\otimes_1 \cap \otimes_2) \le g^{\circ}(\otimes_k)$, $k = 1, 2$, where $\otimes_1 \cap \otimes_2 = \{\xi | \xi \in [a,b] and \xi \in [c,d]\}$ is interaction of the grey numbers $\otimes_1 \in [a,b], a < b$ and $\otimes_2 \in [c,d], c < d$.

4) If $\otimes_1 \subset \otimes_2$, then $g^\circ(\otimes_1 \cup \otimes_2) = g^\circ(\otimes_2)$, $g^\circ(\otimes_1 \cap \otimes_2) = g^\circ(\otimes_1)$

5) If $\mu(\Omega) = 1$ and the measures of $\otimes_1$ and $\otimes_2$ are independent of μ, then

   1° $g^\circ(\otimes_1 \cap \otimes_2) = g^\circ(\otimes_1) \cdot g^\circ(\otimes_2)$; and

   2° $g^\circ(\otimes_1 \cup \otimes_2) = g^\circ(\otimes_1) + g^\circ(\otimes_2) - g^\circ(\otimes_1) \cdot g^\circ(\otimes_2)$.

Proof. All the details for conclusions 1) – 4) are omitted. For 5 1°), from $\mu(\Omega) = 1$ and the assumption that measures of $\otimes_1$ and $\otimes_2$ are independent of μ, we have

$$g^\circ(\otimes_1 \cap \otimes_2) = \mu(\otimes_1 \cap \otimes_2) = \mu(\otimes_1) \cdot \mu(\otimes_2) = g^\circ(\otimes_1) \cdot g^\circ(\otimes_2).$$

Similarly for 2°, we have

$$\begin{aligned} g^\circ(\otimes_1 \cup \otimes_2) &= \mu(\otimes_1 \cup \otimes_2) = \mu(\otimes_1) + \mu(\otimes_2) - \mu(\otimes_1) \cdot \mu(\otimes_2) \\ &= g^\circ(\otimes_1) + g^\circ(\otimes_2) - g^\circ(\otimes_1) \cdot g^\circ(\otimes_2). \text{ QED.} \end{aligned}$$

How grey numbers are combined affects the degree of greyness and the reliability of the resultant grey number. Generally, when grey numbers are "unioned" together, the resultant degree of greyness and the reliability of the new information increase; when grey numbers are intersected together, the resultant degree of greyness drops and the reliability of the combined information decreases. When solving practical problems and having to process a large amount of grey numbers, it is advisable to combine the numbers at several different levels so that useful information can be extracted at individual levels. And in the process of combining grey numbers, "union" and "intersection" operations should be done within both individual levels and across different levels in order to guarantee that the extracted information satisfy the pre-determined requirements in terms of reliability and degree of greyness.

2.2 Sequence Operators

One of the main tasks of grey systems theory is to uncover the mathematical relationships between different system variables and the laws of change of certain system variables themselves based on the available data of characteristic behaviors of social, economic, ecological, etc., systems. Grey systems theory looks at each stochastic variable as a grey quantity that varies within a fixed region and within a certain timeframe, and each stochastic process as a grey process.

When investigating the behavioral characteristics of a system, what is available is often a sequence of definite white numbers. There is no substantial difference between whether we treat the sequence as a trajectory or actualization of a stochastic process or whitenized values of a grey process. However, how to uncover the laws of evolution of the systems' behavioral characteristics, different methods are developed using different thinking logics. For instance, the theory of stochastic investigates the statistical laws on the basis of the probabilities borrowed from prior knowledge. This methodology generally requires large amounts of data. However, sometimes even with large amounts of data available, there is no guarantee that any of the desired laws can be successfully uncovered. That is because the number of basic forms of distribution considered in this methodology is very limited. It is often extremely difficult to deal with non-typical distribution processes. On the other hand, grey systems theory uncovers laws of change by excavating and organizing the available raw data, representing an approach of finding data out of data. This approach is referred to as grey sequence generation. The grey systems theory believes that although the expression of an objective system might be complicated, its data chaotic, the system possesses its overall functions and properties. So, there must be some internal laws governing the existence of the system and its operation. The key is how the investigator chooses his appropriate method to excavate the laws and make use of the laws. For any given grey sequence, its implicit pattern can always be revealed through weakening the explicit randomness.

2.2.1 *Systems under Shocking Disturbances and Sequence Operators*

The prediction of the behaviors of problems under the influence of *shock disturbances* has always been a difficult problem. For such predictions, any theory on how to choose models would lose its validity. It is because the problem to be address here is not about which model is better than others; instead, when a system is severely impacted by shocks, the available behavioral data of the past no long represent the current state of the system. In this case, the available data of the system's behavior can no longer truthfully reflect the law of change of the system.

Assume that $X^{(0)}=\left(x^{(0)}(1),x^{(0)}(2),\cdots,x^{(0)}(n)\right)$ stands for a sequence of a system's true behaviors. Then the observed behaviors of the system $X = (x(1), x(2), \ldots, x(n)) = (x^{(0)}(1) + \varepsilon_1, x^{(0)}(2) + \varepsilon_2, \ldots, x^{(0)}(n) + \varepsilon_n) = X^{(0)} + \varepsilon$, where $\varepsilon = (\varepsilon_1, \varepsilon_2, \ldots, \varepsilon_n)$ is a term for the shocking disturbance, represent a disturbed sequence. To correctly uncover and recognize the true behavior sequence $X^{(0)}$ of the system from the shock-disturbed sequence X, one first has to go over the hurdle ε. If we directly established our model and made our predictions using the severely affected data X without first cleaning up the disturbance, then our predictions would most likely fail, because what the model described was not the true state $\mathrm{X}^{(0)}$ of change of the underlying system.

The wide spread existence of severely shocked systems often causes quantitative predictions disagree with the outcomes of intuitive qualitative analyses. Hence, seeking an organic equilibrium between quantitative predictions and qualitative analyses by eliminating shock wave disturbances in order to recover the true state of the systems' behavioral data so that the accuracy of the consequent predictions can be greatly improved is one of the most important tasks placed in the front each and every scientific practitioner. To this end,

The discussion in this section is centered around the overall goal of reaching $X^{(0)}$ from X.

2.2.2 Axioms That Define Buffer Operators

Definition 2.6. Assume that $X = (x(1), x(2), \ldots, x(n))$ is a system's behavior data sequence.

(1) If $\forall k = 2, 3, \ldots, n$, $x(k) - x(k-1) > 0$, then X is referred to as a *monotonic increasing sequence*;

(2) If the inequality sign in (1) is inversed, then X is referred to as a *monotonic decreasing sequence*;

(3) If there are k, $k' \in \{2, 3, \ldots, n\}$ such that $x(k) - x(k-1) > 0, x(k') - x(k'-1) < 0$, then X is referred to as a random vibrating or fluctuating sequence. If $M = \max\{x(k) | k = 1,2,\cdots,n\}$ and $m = \min\{x(k) | k = 1,2,\cdots,n\}$, then $M - m$ is referred to as the *amplitude* of the sequence X.

Assume that X is a data sequence of a system's behavior, D an operator to work on X, and after being applied by the operator D, X becomes the following sequence:

$$XD = (x(1)d, x(2)d, \cdots, x(n)d)$$

where D is referred to as a *sequence operator* and XD the *first order sequence* (of the operator D). If D_1, D_2, and D_3 are all sequence operators, then D_1D_2 is referred to as a second order sequence operator, and

$$XD_1D_2 = (x(1)d_1d_2, x(2)d_1d_2, \cdots, x(n)d_1d_2)$$

a second order sequence of D_1D_2. Similarly, $D_1D_2D_3$ is referred to as a third order sequence operator and

$$XD_1D_2D_3 = (x(1)d_1d_2d_3, x(2)d_1d_2d_3, \cdots, x(n)d_1d_2d_3)$$

a third order sequence of $D_1D_2D_3$; etc.

Axiom 2.6 (Fixed Point). Assume that X is a data sequence of a system's behavior and D a sequence operator. Then D satisfies $x(n)d = x(n)$.

This fixed point axiom means that under the effect of a sequence operator, the data point $x(n)$, the last entry of the sequence, of the system's behavior data sequence stays unchanged. Based on the conclusions of relevant qualitative analysis, we can also make several of the last entries of the data unchanged by the operator D, say, $x(j)d \neq x(j)$ and $x(i)d = x(i)$, for $j = 1, 2, \ldots, k-1$; $i = k, k+1, \ldots, n$.

Axiom 2.7 (Sufficient usage of information). Each entry value $x(k)$, $k= 1, 2, \ldots, n$, in the data sequence X of the system's behavior should sufficiently participate in the entire process of application of the operator.

This axiom requires that any sequence operator should be defined by using the known information of the given sequence. It cannot be produced without referencing the available raw data.

Axiom 2.8 (Analyticity and normality). Each $x(k)d$, $k = 1, 2, \ldots, n$, is expressed by a uniform, elementary analytic representation of $x(1), x(2), \ldots, x(n)$.

This last axiom requires the procedure of applying sequence operators be clear, normalized, and uniform so that it can be conveniently carried out on computers.

Any sequence operator satisfying these three axioms is referred to as a *buffer operator*; the first order, second order, third order, …, sequences obtained by applying a buffer operator is referred to as the first order, second order, third order, …, *buffered sequences*. For a raw data sequence X and a buffer operator D, when X is respectively an increasing, decreasing, or fluctuating sequence, if the buffered sequence XD increases, decreases, or fluctuates slower or with smaller amplitude, respectively, than the original sequence X, then D is referred to as a *weakening operator*. If the buffered operator XD increases, decreases, or fluctuates faster or with larger amplitude, respectively, than the original sequence X, then D is referred to as a *strengthening operator*.

2.2.3 Properties of Buffer Operators

Theorem 2.5. Assume that X is a monotonic increasing sequence. Then, if D is a weakening operator, then $x(k)d \geq x(k)$, $k = 1, 2, \ldots, n$; if D is a strengthening operator, then $x(k)d \leq x(k)$, $k = 1, 2, \ldots, n$.

Theorem 2.6. Assume that X is a monotonic decreasing sequence. Then, if D is a weakening operator, then $x(k)d \leq x(k)$, $k = 1, 2, \ldots, n$; if D is a strengthening operator, then $x(k)d \geq x(k)$, $k = 1, 2, \ldots, n$.

Theorem 2.7. Assume that X is a fluctuating sequence and XD a buffered sequence. Then, if D is a weakening operator, then $\max_{1\leq k\leq n}\{x(k)\} \geq \max_{1\leq k\leq n}\{x(k)d\}$ and $\min_{1\leq k\leq n}\{x(k)\} \leq \min_{1\leq k\leq n}\{x(k)d\}$; if D is a strengthening operator, then $\max_{1\leq k\leq n}\{x(k)\} \leq \max_{1\leq k\leq n}\{x(k)d\}$ and $\min_{1\leq k\leq n}\{x(k)\} \geq \min_{1\leq k\leq n}\{x(k)d\}$.

For the detailed proofs and relevant discussions of these theorems, please consult with (Liu and Lin, 2006, pp. 64 – 67). What Theorem 2.5 implies that each monotonic increasing sequence expands under the effect of weakening operator and shrinks under strengthening operator. What Theorem 2.6 indicates that each monotonic decreasing sequence shrinks under the effect of weakening operator and expands under strengthening operator.

2.2.4 Construction of Practically Useful Buffer Operators

Given a raw data sequence $X=(x(1),x(2),\cdots,x(n))$, define $XD=(x(1)d,x(2)d,\cdots,x(n)d)$, where

$$x(k)d=\frac{1}{n-k+1}[x(k)+x(k+1)+\cdots+x(n)],\ k=1,2,\cdots,n \tag{2.22}$$

it can be shown that no matter whether X is monotonic increasing, decreasing, or vibrating, D is always a weakening operator. This operator is referred to as an *average (or mean) weakening buffer operator* (AWBO).

Next, let us define another sequence operator D as follows:

$$x(k)d=\frac{x(1)+x(2)+\cdots+x(k-1)+kx(k)}{2k-1};k=1,2,\cdots,n-1 \tag{2.23}$$

It can be readily shown that when the raw data sequence X is either monotonic increasing or decreasing, D is a strengthening buffer operator.

For the given raw data sequence X, let D_i be a sequence operator defined by

$$x(k)d_i=\frac{x(k-1)+x(k)}{2};k=2,3,\cdots,n;i=1,2 \tag{2.24}$$

where $x(1)d_1=\alpha x(1),\alpha\in[0,1]$, $x(1)d_2=(1+\alpha)x(1),\alpha\in[0,1]$, and $x(n)d_i=x(n),i=1,2$, then D_1 is a strengthening operator for monotonic increasing sequences, and D_2 a weakening buffer operator for monotonic decreasing sequences.

Define a sequence operator D on the raw data sequence X by

$$x(k)d=\frac{kx(k)+(k+1)x(k+1)+\cdots+nx(n)}{(n+k)(n-k+1)/2};k=1,2,\cdots,n \tag{2.25}$$

Then, we have the following result:

Theorem 2.8. (Naiming Xie). No matter whether the raw data sequence X is monotonic increasing, decreasing, or vibrating, the butter operator D, as defined in equ. (2.25), is always weakening.

Proof. We will only look at the case when X is monotonic increasing. As for the other two cases, the proofs are similar and omitted. From the assumption of X, it follows that

$$x(k)d - x(k) = \frac{kx(k) + (k+1)x(k+1) + \cdots nx(n)}{(n+k)(n-k+1)/2} - x(k)$$

$$= \frac{(k+1)[x(k+1) - x(k)] + (k+2)[x(k+2) - x(k)] + \cdots + n[x(n) - x(k)]}{(n+k)(n-k+1)/2} \geq 0$$

Therefore, $x(k)d \geq x(k)$. That is, *D* is a weakening butter operator. QED.

This sequence operator *D* is referred to as a *weighted average (or mean) weakening buffer operator* (WAWBO).

Theorem 2.9. Assume that $X = (x(1), x(2), \cdots, x(n))$ is nonnegative, that is, $x(i) \geq 0$, i = 1, 2, …, n. Define a sequence operator *D* as follows:

$$x(k)d = [x(k) \cdot x(k+1) \cdots x(n)]^{\frac{1}{n-k+1}} = \left[\prod_{i=k}^{n} x(i)\right]^{\frac{1}{n-k+1}}, \quad (k = 1,2,\cdots,n) \tag{2.26}$$

then *D* is always a weakening buffer operator, no matter whether *X* is monotonic increasing, decreasing, or fluctuating.

Proof. It is ready to show that *D* satisfies the three axioms of a buffer operator.

Case 1. When X is monotonic increasing, from

$$x(k)d_2 = [x(k) \cdot x(k+1) \cdots x(n)]^{\frac{1}{n-k+1}} \geq [x(k) \cdot x(k) \cdots x(k)]^{\frac{1}{n-k+1}} = x(k)$$

for $k = 1,2,\cdots,n$, it follows that *D* is a weakening buffer operator.

Case 2. Similarly, it can be shown that when *X* is either monotonic decreasing or vibrating, D is also a weakening buffer operator. QED.

This operator D, as defined in equ. (2.26) is referred to as a *geometric average (or mean) weakening buffer operator* (GAWBO).

Due to reasons, in some applications the importance of different entries in the data sequence varies with time. For these situations, assume that accompanying the available raw data sequence, there is also a vector $\omega = (\omega_1, \omega_2, \cdots, \omega_n)$ of associated weights.

Theorem 2.10. Let $X = (x(1), x(2), \cdots, x(n))$ be a nonnegative data sequence with its accompanying weight vector $\omega = (\omega_1, \omega_2, \cdots, \omega_n)$. If *D* is a sequence operator defined by

$$x(k)d = \frac{\omega_k x(k) + \omega_{k+1} x(k+1) + \cdots + \omega_n x(n)}{\omega_k + \omega_{k+1} + \cdots + \omega_n} = \frac{1}{\sum_{i=k}^{n} \omega_i} \sum_{i=k}^{n} \omega_i x(i), \tag{2.27}$$

$k=1,2,\cdots,n$, then D is always a weakening buffer operator no matter whether X is monotonic increasing, decreasing, or vibrating.

Proof. It is ready to show that D satisfies the three axioms for buffer operators.

Case 1. When X is monotonic increasing, for each k one has

$$x(k)d=\frac{1}{\omega_k+\omega_{k+1}+\cdots+\omega_n}\left[\omega_k x(k)+\omega_{k+1}x(k+1)+\cdots+\omega_n x(n)\right]$$
$$\geq\frac{1}{\omega_k+\omega_{k+1}+\cdots+\omega_n}\left[\omega_k x(k)+\omega_{k+1}x(k)+\cdots+\omega_n x(k)\right]=x(k)$$

So, D is a weakening buffer operator.

Case 2. Similarly, it can be shown that D is a weakening buffer operator when X is monotonic decreasing or fluctuating. QED.

When $\omega=(1,1,\cdots,1)$, that is, $\omega_i=1$, we have

$$\frac{1}{\sum_{i=k}^{n}\omega_i}\sum_{i=k}^{n}\omega_i x(i)=\frac{1}{n-k+1}\sum_{i=k}^{n}x(i)$$

which implies that average weakening buffer operators (AWBO) are special cases of weighted average weakening buffer operators (WAWBO).

If $\omega=(1,2,\cdots,n)$, that is, $\omega_i=i$, we have

$$\frac{1}{\sum_{i=k}^{n}\omega_i}\sum_{i=k}^{n}\omega_i x(i)=\frac{1}{(n+k)(n-k+1)}\sum_{i=k}^{n}ix(i)$$

which implies that the weighted average weakening buffer operator given in Theorem 2.8 is a special case of that given Theorem 2.10.

Theorem 2.11. Assume that $X=(x(1),x(2),\cdots,x(n))$ is a nonnegative data sequence of a system's behavior with entry weight vector $\omega=(\omega_1,\omega_2,\cdots,\omega_n)$, $\omega_k\geq 0$, $\omega_n\neq 0$ $(k=1,2,\cdots,n)$. Define a sequence operator D as following:

$$x(k)d=\left[x(k)^{\omega_k}\cdot x(k+1)^{\omega_{k+1}}\cdots x(n)^{\omega_n}\right]^{\frac{1}{\omega_k+\omega_{k+1}+\cdots+\omega_n}}=\left[\prod_{i=k}^{n}x(i)^{\omega_i}\right]^{\frac{1}{\sum_{i=k}^{n}\omega_i}} \tag{2.28}$$

$k=1,2,\cdots,n$. Then D is a weakening buffer operator no matter if X is monotonic increasing, decreasing, or vibrating.

Proof. It is ready to show that D satisfies the axioms of buffer operators.

Case (1): If X is a monotonic increasing sequence, then for any k, we have

$$x(k)d = \left[x(k)^{\omega_k} \cdot x(k+1)^{\omega_{k+1}} \cdots x(n)^{\omega_n}\right]^{\frac{1}{\omega_k+\omega_{k+1}+\cdots+\omega_n}} = \left[\prod_{i=k}^{n} x(i)^{\omega_i}\right]^{\frac{1}{\sum_{i=k}^{n}\omega_i}}$$

$$\geq \left[x(k)^{\omega_k} \cdot x(k)^{\omega_{k+1}} \cdots x(k)^{\omega_n}\right]^{\frac{1}{\omega_k+\omega_{k+1}+\cdots+\omega_n}} = x(k).$$

So, D is a weakening buffer operator.

Case (2): Similarly, it can be shown that when X is a monotonic decreasing or vibrating sequence, D is a weakening buffer operator. QED.

This sequence operator D is referred to as a *weighted geometric average weakening buffer operator* (WGAWBO).

If $\omega = (1,1,\cdots,1)$, that is, $\omega_i = 1$, then we have

$$\left[x(k)^{\omega_k} \cdot x(k+1)^{\omega_{k+1}} \cdots x(n)^{\omega_n}\right]^{\frac{1}{\omega_k+\omega_{k+1}+\cdots+\omega_n}} = \left[\prod_{i=k}^{n} x(i)^{\omega_i}\right]^{\frac{1}{\sum_{i=k}^{n}\omega_i}}$$

$$= \left[x(k) \cdot x(k+1) \cdots x(n)\right]^{\frac{1}{n-k+1}}$$

That implies that each geometric average weakening buffer operator (GAWBO) is a special case of weighted geometric average weakening buffer operator (WGAWBO).

Theorem 2.12 (Naiming Xie). For a given raw data sequence X, define a sequence operator D as follows:

$$x(k)d = \frac{x(k)}{x(n)} \cdot x(k); k = 1,2,\cdots,n \tag{2.29}$$

then D is a strengthening operator, no matter whether X is monotonic increasing, decreasing, or vibrating.

Proof. It suffices to see how to show the case of monotonic increasing. For the other cases, the argument is similar. If X is monotonic increasing, then the following holds true:

$$x(k)d - x(k) = \frac{x(k)}{x(n)} \cdot x(k) - x(k) = \frac{x(k)}{x(n)} \cdot (x(k) - x(n)) \leq 0$$

Therefore, $x(k)d \leq x(k)$ and so D is a strengthening operator. QED.

Theorem 2.13 (Naiming Xie). For a given increasing or decreasing sequence X of raw data, the operator D defined as follows:

$$x(k)d = \frac{[x(k)+x(k+1)+\cdots x(n)]/(n-k+1)}{x(n)} \cdot x(k); k=1,2,\cdots n \tag{2.30}$$

is a strengthening buffer operator.

Proof. (omitted).

Theorem 2.14. For a data sequence X of a system's behavior, the operator defined below

$$x(k)d = \frac{(n-k+1)(x(k))^2}{x(k)+x(k+1)+\cdots+x(n)} = \frac{(n-k+1)(x(k))^2}{\sum_{i=k}^{n} x(i)} \tag{2.31}$$

where $\sum_{i=k}^{n} x(i) \neq 0$ and $k=1,2,\cdots,n$, is a strengthening buffer operator, no matter X is monotonic increasing, decreasing, or vibrating.

Proof. It is ready to show that D satisfies the three axioms for buffer operators.

Case 1, when X is monotonic increasing, one has

$$\begin{aligned} x(k)d &= \frac{(n-k+1)(x(k))^2}{x(k)+x(k+1)+\cdots+x(n)} \\ &\leq \frac{(n-k+1)(x(k))^2}{x(k)+x(k)+\cdots+x(k)} \\ &= x(k), (k=1,2,\cdots,n) \end{aligned}$$

So, D is a strengthening buffer operator.

Case 2, similarly, it can be shown that D is also a strengthening buffer operator for both the situations that X is monotonic decreasing or vibrating. QED.

This sequence operator, as defined in equ. (2.31), is referred to as an *average (or mean) strengthening buffer operator* (ASBO).

Theorem 2.15. Assume that X is a nonnegative data sequence of a system's behavior. Then the operator D defined below

$$x(k)d = \frac{(x(k))^2}{[x(k)\cdot x(k+1)\cdots x(n)]^{\frac{1}{n-k+1}}} = \frac{(x(k))^2}{\left[\prod_{i=k}^{n} x(i)\right]^{\frac{1}{n-k+1}}}, \quad (k=1,2,\cdots,n) \tag{2.32}$$

is a strengthening buffer operator, no matter whether X is monotonic increasing, decreasing, or vibrating.

Proof. The reader is urged to show that D satisfies the axioms of buffer operators.

Case 1, when X is monotonic increasing, one has

$$
\begin{aligned}
x(k)d &= \frac{(x(k))^2}{\left[x(k)\cdot x(k+1)\cdots x(n)\right]^{\frac{1}{n-k+1}}} \\
&\le \frac{(x(k))^2}{\left[x(k)\cdot x(k)\cdots x(k)\right]^{\frac{1}{n-k+1}}} \\
&= x(k),\quad (k=1,2,\cdots,n)
\end{aligned}
$$

So, D is a strengthening buffer operator.

Case 2, for the situations that X is monotonic decreasing or vibrating, similar proofs can be constructed. QED.

This operator D, as defined in equ. (2.32), is referred to as a *geometric average strengthening buffer operator* (GASBO).

Theorem 2.16. Let X be a sequence of raw data and ω the accompanying weight vector of the entries of X. Then, the operator D defined by

$$
x(k)d = \frac{(\omega_k+\omega_{k+1}+\cdots+\omega_n)(x(k))^2}{\omega_k x(k)+\omega_{k+1}x(k+1)+\cdots+\omega_n x(n)} = \frac{\sum_{i=k}^{n}\omega_i (x(k))^2}{\sum_{i=k}^{n}\omega_i x(i)} \tag{2.33}
$$

$k=1,2,\cdots,n$, is a strengthening buffer operator, no matter whether X is monotonic increasing, decreasing, or vibrating.

Proof. The argument for D to satisfy the axioms of buffer operators is straightforward and is omitted.

Case 1, when X is monotonic increase, one has

$$
\begin{aligned}
x(k)d &= \frac{(\omega_k+\omega_{k+1}+\cdots+\omega_n)(x(k))^2}{\omega_k x(k)+\omega_{k+1}x(k+1)+\cdots+\omega_n x(n)} \\
&\le \frac{(\omega_k+\omega_{k+1}+\cdots+\omega_n)(x(k))^2}{\omega_k x(k)+\omega_{k+1}x(k)+\cdots+\omega_n x(k)} \\
&= x(k),\quad (k=1,2,\cdots,n)
\end{aligned}
$$

So, D is a strengthening buffer operator.

Case 2, similar arguments hold true for the cases that X is monotonic decreasing or vibrating. The details are omitted. QED.

This operator D, as defined in equ. (2.33), is referred to as a *weighted average strengthening buffer operator* (WASBO).

If $\omega=(1,1,\cdots,1)$, that is, $\omega_i=1$, we have

$$\frac{\sum_{i=k}^{n}\omega_i(x(k))^2}{\sum_{i=k}^{n}\omega_i x(i)}=\frac{(n-k+1)(x(k))^2}{\sum_{i=k}^{n}x(i)}$$

which implies that each average strengthening buffer operator is a special weighted average strengthening buffer operator.

Theorem 2.17. Let X be a nonnegative data sequence of a system's behavior with its weight vector $\omega=(\omega_1,\omega_2,\cdots,\omega_n)$, $\omega_k\geq 0$, $\omega_n\neq 0$ $(k=1,2,\cdots,n)$ for its entries. Then the operator D defined by

$$x(k)d=\frac{(x(k))^2}{\left[x(k)^{\omega_k}\cdot x(k+1)^{\omega_{k+1}}\cdots x(n)^{\omega_n}\right]^{\frac{1}{\omega_k+\omega_{k+1}+\cdots+\omega_n}}}=\frac{(x(k))^2}{\left[\prod_{i=k}^{n}x(i)^{\omega_i}\right]^{\frac{1}{\sum_{i=k}^{n}\omega_i}}} \tag{2.34}$$

$k=1,2,\cdots,n$, is a strengthening buffer operator, no matter whether X is monotonic increasing, decreasing, or vibrating.

Proof. All the details of showing D satisfy the axioms of buffer operators are omitted.

Case 1, if X is monotonic increasing, for any k, one has

$$x(k)d=\frac{(x(k))^2}{\left[x(k)^{\omega_k}\cdot x(k+1)^{\omega_{k+1}}\cdots x(n)^{\omega_n}\right]^{\frac{1}{\omega_k+\omega_{k+1}+\cdots+\omega_n}}}$$

$$\leq\frac{(x(k))^2}{\left[x(k)^{\omega_k}\cdot x(k)^{\omega_{k+1}}\cdots x(k)^{\omega_n}\right]^{\frac{1}{\omega_k+\omega_{k+1}+\cdots+\omega_n}}}=x(k)$$

So, D is a strengthening buffer operator.

Case 2, for the situations that X is monotonic decreasing or vibrating, similar arguments can be constructed. QED.

This operator D, as defined in equ. (2.34), is referred to as a weighted geometric average strengthening buffer operator (WGASBO).

If $\omega=(1,1,\cdots,1)$, that is, $\omega_i=1$, then we have

$$\frac{(x(k))^2}{\left[x(k)^{\omega_k}\cdot x(k+1)^{\omega_{k+1}}\cdots x(n)^{\omega_n}\right]^{\frac{1}{\omega_k+\omega_{k+1}+\cdots+\omega_n}}}=\frac{(x(k))^2}{\left[x(k)\cdot x(k+1)\cdots x(n)\right]^{\frac{1}{n-k+1}}}$$

That implies that each GASBO is a special WGASBO.

To limit the length of this presentation, we will no longer mention other forms of practical buffer operators. The concept of butter operators have been employed not only in grey systems modeling, but also in other different kinds of model buildings. Generally, before building a mathematical model, based on the qualitative analysis and its conclusions, one applies buffer operators on the original data sequence in order to soften or eliminate the effect of shock-disturbances on the behaviors of the system of our concern. By doing so, expected results are often obtained.

Example 2.1. The GDP values for the years of 1983 - 1986 contributed by private business entities located in Ge County, Henan Province, China, are given as follows:

$$X = (10155, 12588, 23480, 35388).$$

These data values represent a rapid economic growth in the regional private sector. In particular, during the time period from 1983 to 1986, the average annual growth is 51.6%; and for the years 1984 – 1986, the average annual growth amounts to 67.7%. All the members of the county economic planning committee, including politicians, scholars, and representatives from various sectors of the society, unanimously agreed that it would be impossible for the regional private businesses to develop at such a high rate of growth in the coming years. If these data were directly employed to build models, the resultant predictions would not be acceptable. After careful analysis and discussions, it was recognized that such high levels of growth were mainly due to the low economic base, which was mainly caused by a lack of adequate policies that encouraged private economic development. To weaken the increasing rate of growth existing in the available sequence, we need to artificially include the currently advantageous policies into the data values of the past years. To achieve this end, let us apply the weakening operator in equ. (2.22) twice on the given dats sequence to produce

$$XD^2 = (27260, 29547, 32411, 35388)$$

We then established a model based on this sequence XD^2 of data and made our predictions: For the time period from 1986 to 2000, this county's private economic sector would grow at an annual rate of 9.4%. This result was obtained in 1987 and had been very well confirmed by the actual data collected for the second half of the eighth five-year economic development period and the entire ninth five-year economic development period.

Example 2.2. For the time period from 1996 to 1999, the annual gross revenues produced by the agricultural, forestry, animal husbandry, and fishery sectors in Nanjing City area were given below (in 0.1 billion yuan):

$$X = (91.9895, 94.2439, 96.9644, 98.9199).$$

The growth rate shown in X is very slow for an average of about only 2.4% annually. Such a slow rate of growth was not appropriate comparing to the fast advances of the overall economic development of the area. If the slow growth continued in these economic sectors, it would surely cause imbalances in the development of the overall economic structure of the region so that the sustained regional economic growth would be adversely affected. Starting in 2000, Nanjing City gradually adjusted the economic structure of the countryside so that the situation of slow economic growth was improved. In order to accurately control in a timely fashion the economic development tendency, there is a need to produce scientifically reasonable economic forecasts. To achieve this goal, we have to deal with the available data where slow growth was recorded so that the resultant predictions will possess the practical value for the purpose of foretelling the future so that potential government interference can be planned ahead of time. By applying the strengthening operator in Theorem 2.12 twice on the available sequence of data, we obtain second order buffered data sequence

$$XD^2 = (79.5513, 85.5446, 93.1686, 98.9199).$$

Establishing a GM(1,1) model (for details, see (Liu and Lin, 2006) or Section 4.1 in this book) based on this buffered sequence provides

$$\frac{dX^{(1)}}{dt} - 0.0720X^{(1)} = 77.1389$$

with the time response function as follows:

$$\hat{X}^{(1)}(k+1) = 1150.7003e^{0.0720k} - 1071.1503$$

Based on this equation, the computational simulation results, the effectiveness of the data fit, and the prediction efficacy are given in Tables 2.1 and 2.2.

Table 2.1 The effectiveness of the simulation results

Year	Strengthened data $x^{(0)}(k)$	Simulated data $\hat{x}^{(0)}(k)$	error $\varepsilon(k) = \hat{x}^{(0)}(k) - x^{(0)}(k)$	Relative error $\Delta_k = \frac{\lvert\varepsilon(k)\rvert}{x^{(0)}(k)}$
1997	85.5446	85.9245	0.3799	0.4441%
1998	93.1686	92.3407	-0.8279	0.8886%
1999	98.9199	99.2359	0.316	0.3195%

Table 2.2 The efficacy of the predictions

year	Actual data $x^{(0)}(k)$	predictions $\hat{x}^{(0)}(k)$	Error $\varepsilon(k) = \hat{x}^{(0)}(k) - x^{(0)}(k)$	Relative error $\Delta_k = \frac{\lvert\varepsilon(k)\rvert}{x^{(0)}(k)}$
2000	106.3412	106.6460	0.3048	0.2866%
2001	113.29	114.6094	1.3194	1.1646%
2005		152.8703		

From Tables 2.1 and 2.2, it can be seen that by employing the buffered data using a strengthening operator to establish our model, the simulated results and the corresponding predictions are quite good. In particular, for the years of 2000 and 2001, the predicted values when compared to the actual data of these years have reached a rate of accuracy of over 98%.

2.3 Generation of Grey Sequences

2.3.1 Average Generator

When collecting data, due to various obstacles that are difficult to overcome, the available data sequences might contain missing entries. On the other hand, although some data sequences are complete without any missing entry, due to a sudden change in the system's behavior at some time moment the corresponding entries in the sequences are out of ordinary, creating great difficulties for the researcher. In this case, if the abnormal entries are removed, there will be blank entries created. Hence, how to effectively fill blanks in data sequences naturally becomes one of the first questions one has to address in his processing of available data. Generation by using averages is one of the often employed methods to create new data, fill a vacant entry in the available data sequence, and construct new sequences.

Assume that $X = (x(1), x(2), \ldots, x(k), x(k+1), \ldots, x(n))$ is a sequence of raw data. Then, entry $x(k)$ is referred to the preceding value and $x(k+1)$ the succeeding value. If $x(n)$ stands for a piece of new information, then for any $k \le n-1$, $x(k)$ will be seen as a piece of old information. If the sequence X has a blank entry at location k, denoted $\varnothing(k)$, then the entries $x(k-1)$ and $x(k+1)$ will be referred to as $\varnothing(k)$'s boundary values with $x(k-1)$ being the preceding boundary and $x(k+1)$ the succeeding boundary. If a value $x(k)$ at the location of $\varnothing(k)$ is generated on the basis of $x(k-1)$ and $x(k+1)$, then the established value $x(k)$ is referred to as an internal point of the interval $[x(k-1), x(k+1)]$.

Definition 2.7. Assume that $x(k-1)$ and $x(k)$ are two neighboring entries in a data sequence X. If $x(k-1)$ stands for a piece of old information and $x(k)$ a piece of new information, and $x^*(k) = \alpha x(k) + (1-\alpha)x(k-1)$, for $\alpha \in [0,1]$, then this new value $x^*(k)$ is referred to as generated by the new and old information under the generation coefficient (weight) α. When $\alpha > 0.5$, the generation of $x^*(k)$ is seen with more weight placed on the new information than the old information. When $\alpha < 0.5$, the generation of $x^*(k)$ is seen with more weight placed on the old information than the new information. If $\alpha = 0.5$, then the value $x^*(k)$ is seen as generated without preference.

In terms of stable time series, the exponential smoothing method, employed in the smooth prediction, focuses on the generation of predictions with more preference given to old information than new information. It is because the smoothing value

$$s_k^{(1)} = \alpha x_k + (1-\alpha)s_{k-1}^{(1)}$$

stands for a weighted sum of the old and new information with the weight α taking value from the range of 0.1 – 0.3.

Assume that sequence X has a blank entry $\varnothing(k)$ at location k. If this blank entry $\varnothing(k)$ is filled by using $x^*(k) = 0.5x(k-1) + 0.5x(k+1)$, where $x(k-1)$ and $x(k+1)$ are the adjacent neighbors of the location k, then the resultant sequence will be referred to as generated by using the non-adjacent neighbor mean. If $x(k+1)$ stands for a piece of new information, then non-adjacent neighbor mean generation is an equal weight generation of new and old information. This kind of generation is employed when it is difficult to determine the degree of influence of the new and old information on the missing value $x(k)$.

For a given sequence $X = (x(1), x(2), \ldots, x(n))$, the value $x^*(k) = 0.5x(k) + 0.5x(k-1)$ is referred to as generated by adjacent neighbor mean. The sequence consisting of adjacent neighbor means is referred to as generated by adjacent neighbor means. In grey systems modeling, the generation of adjacent neighbor mean information is often employed. It provides a method of constructing new sequences based on available time series data.

For the sequence X of length n, as given above, if Z stands for the sequence generated by adjacent neighbor means, then the length of $Z = (z(2), z(3), \ldots, z(n))$ is $n-1$, where $z(1)$ cannot be generated based on what is given in X.

2.3.2 Smoothness of Sequences

Differentiable everywhere is a characteristic of smooth continuous functions. However, each data sequence is made up of discrete individual points. So, in the traditional sense, for sequences of data, there is no possibility for us to speak of their differentiability. That is, we cannot apply the concept of derivatives to investigate the smoothness of sequences. Due to this reason, let us look at the characteristics of smooth continuous functions from another angle. If a sequence possesses those properties similar to those of a smooth continuous function, then we treat the sequence as smooth.

Assume that $X(t)$ is a monotonic increasing continuous curve, as shown in Figure 2.3. Let us insert $n+1$ dividing points into the time interval [a, b] of

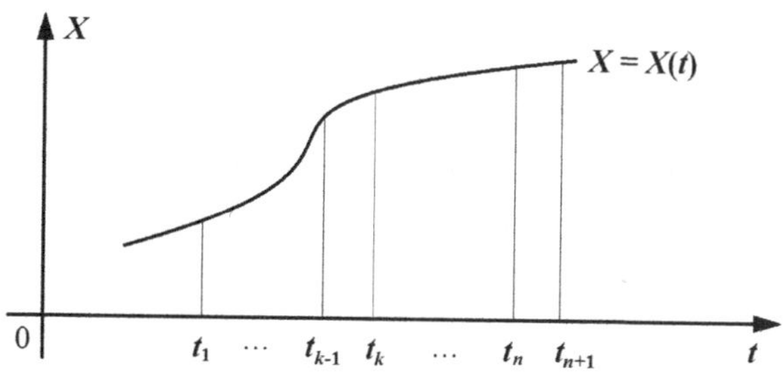

Fig. 2.3 Characteristics of a smooth curve

consideration: $t_1 < t_2 < \ldots < t_k < t_{k+1} < \ldots < t_{n+1}$. Denote $\Delta t_k = t_{k+1} - t_k$, $k = 1, 2, \ldots, n$. Accordingly, we obtain n segments of $X(t)$. Let us pick an arbitrary point $x(k)$ within the kth segment $[x(t_k), x(t_{k+1})]$, $k = 1, 2, \ldots, n$. Now, we obtain two sequences: $X = (x(1), x(2), \ldots, x(n))$ and $X_0 = (x(t_1), x(t_2), \ldots, x(t_n))$, where sequence X_0 is made up of all the lower boundary points of all the segments of the curve $X(t)$. If $X(t)$ is a smooth and continuous function, then when the previous subdivision of the time interval is fine enough, we should have the following properties:

1. Any two sequences of internal points will be sufficiently close to each other; and
2. Any sequence of internal points will be sufficiently close to the sequence of lower boundaries.

With this analysis in place, we can define smooth sequences as follows.

Let $\Delta t = \max\{\Delta t_k: 1 \le k \le n\}$, d a distance function in the n-dimensional space $\boldsymbol{R}^n$, and X^* a representative sequence of a prefixed differentiable function. If when Δt approaches 0 no matter how the interval [a, b] is partitioned and how the internal points from the subintervals are chosen, the following hold true:

i. For any sequences of internal points X_i, X_j, $d(X^*, X_i) = d(X^*, X_j)$; and
ii. $d(X^*, X) = d(X^*, X_0)$;

then $X(t)$ is referred to as a smooth continuous function.

Now, let us look at how we can define smooth sequences. Assume that $X = (x(1), x(2), \ldots, x(k), x(k+1), \ldots, x(n), x(n+1))$ is a given sequence and $Z = (z(1), z(2), \ldots, z(n))$ the sequence of adjacent neighbor means of X. That is, $z(k) = 0.5x(k) + 0.5x(k+1), k = 1, 2, \cdots, n$. Let X^* be a representative differentiable function and d a distance function of $\boldsymbol{R}^n$. And let us continue to use the symbol X to represent the sequence X with the entry $x(n+1)$ deleted.

Definition 2.8. If sequence X satisfies:

1) When k is sufficiently large, $x(k) < \sum_{i=1}^{k-1} x(i)$;

2) $\max_{1\le k\le n}\left|x^*(k) - x(k)\right| \ge \max_{1\le k\le n}\left|x^*(k) - z(k)\right|$,

then sequence X is referred to as *a smooth sequence*.

Definition 2.9. Assume that X is a smooth sequence, Z its adjacent neighbor mean sequence, X^* a differentiable function, and d a distance function of $\boldsymbol{R}^n$. If there is $\varepsilon \in [0,1]$ such that $\left|d(X^*, X) - d(X^*, Z)\right| \le \varepsilon$, then X is referred to as a sequence of smoothness greater than $1/\varepsilon$, the number $S(d) = \left|d(X^*, X) - d(X^*, Z)\right|^{-1}$ the degree of X's smoothness. When $\left|d(X^*, X) - d(X^*, Z)\right| = 0$, that is, when $S(d) = \infty$, the sequence X is said to be *infinitely smooth*.

It can be readily seen that the greater a pattern appears in X, the closer X is to a smooth continuous function and the greater its degree of smoothness. In particular, if the data in X is uniformly distributed, then X is an infinitely smooth sequence.

Definition 2.10. The ratio

$$\rho(k)=\frac{x(k)}{\sum_{i=1}^{k-1}x(i)};k=2,3,\cdots,n \tag{2.35}$$

is referred to as the *smoothness ratio* of the sequence X.

The concept of smoothness ratio reflects the smoothness of sequences from a different angle. In particular, it uses the ratio $\rho(k)$ of the kth data value $x(k)$ over the sum $\sum_{i=1}^{k-1}x(i)$ of the previous values to check whether or not the changes in the data points of X are stable. It can be seen that the more stable the changes of the data points in sequence X are, the smaller the smoothness ratio $\rho(k)$.

Definition 2.11. If a sequence X satisfies

1) $\frac{\rho(k+1)}{\rho(k)}<1;k=2,3,\cdots,n-1$;
2) $\rho(k)\in[0,\varepsilon];k=3,4,\cdots,n$; and
3) $\varepsilon<0.5$,

then X is referred to as a *quasi-smooth sequence*.

Definition 2.12. Assume that X is a sequence of missing entries. If a new sequence generated based on X is quasi-smooth, then this new sequence is referred to as quasi-smooth generated.

2.3.3 Stepwise Ratio Generator

If the first entry $x(1)$ and last entry $x(n)$ of a sequence are blank, that is, $x(1)=\varnothing(1)$ and $x(n)=\varnothing(n)$, then we will not be able to fill these missing entries by using the method of adjacent neighbor mean generation. In this case, the method of *stepwise ratio generation* is often employed. In particular, given a sequence $X=(x(1), x(2), \ldots, x(n))$,

$$\sigma(k)=\frac{x(k)}{x(k-1)};k=2,3,\cdots,n \tag{2.36}$$

are referred to as the *stepwise ratios* of X. Now, the missing entry $x(1)=\varnothing(1)$ can be generated by using the stepwise ratio of its right-hand side neighbors, and $x(n)=\varnothing(n)$ its left-hand side neighbors. The sequence obtained by filling all its missing entries using stepwise ratios is referred to as stepwise ratio generated.

Proposition 2.3. The stepwise ratio $\sigma(k+1)$ and the smoothness ratio, as defined in Subsection 2.3.2, satisfy the following

$$\sigma(k+1)=\frac{\rho(k+1)}{\rho(k)}(1+\rho(k)); k=2,3,\cdots,n \tag{3.37}$$

Proposition 2.4. If X = ($x(1)$, $x(2)$, …, $x(n)$) is an increasing sequence satisfying that for any $k=2,3,\cdots,n$, $\sigma(k)<2$ and $\frac{\rho(k+1)}{\rho(k)}<1$, then for any $\varepsilon \in [0,1]$ and $k=2,3,\cdots,n$, when $\rho(k)\in[0,\varepsilon]$, $\sigma(k+1)\in[1,1+\varepsilon]$.

2.3.4 *Accumulating and Inverse Accumulating Generators*

Accumulating generation is a method employed to whitenize a grey process. It plays an extremely important role in grey systems theory. Through accumulation, one can potentially uncover a development tendency existing in the process of accumulating grey quantities so that the characteristics and laws of integration hidden in the chaotic original data can be sufficiently revealed. For instance, when looking at the financial outflows of a family, if we do our computations on a daily basis, we might not see any obvious pattern. However, if our calculations are done on a monthly basis, some pattern of spending, which is somehow related to the monthly income of the family, will possibly emerge.

The inverse accumulating generation is often employed to acquire additional insights from the available small amount of information. It plays the role of recovery from the acts of accumulating operators and is the inverse operation of the accumulating process. In particular, for an original sequence $X^{(0)}$ = ($x^{(0)}(1)$, $x^{(0)}(2)$,…, $x^{(0)}(n)$), when the *accumulating* (*generation*) *operator D* is applied on $X^{(0)}$, we obtain $X^{(0)}D$ = ($x^{(0)}(1)d$, $x^{(0)}(2)d$,…, $x^{(0)}(n)d$), where

$$x^{(0)}(k)d=\sum_{i=1}^{k}x^{(0)}(i); k=1,2,\cdots,n \tag{2.38}$$

If the accumulating operator D is applied r times on $X^{(0)}$, we obtain

$$X^{(0)}D^{r}=X^{(r)}=\left(x^{(r)}(1),x^{(r)}(2),\cdots,x^{(r)}(n)\right)$$

where

$$x^{(r)}(k)=\sum_{i=1}^{k}x^{(r-1)}(i); k=1,2,\cdots n \tag{2.39}$$

Corresponding to the accumulating operator, the *inverse accumulating operator D* is defined as follows: For sequence $X^{(0)}$ = ($x^{(0)}(1)$, $x^{(0)}(2)$,…, $x^{(0)}(n)$), $X^{(0)}D$ is given by

$$x^{(0)}(k)d = x^{(0)}(k) - x^{(0)}(k-1); k = 2,\cdots,n \tag{2.40}$$

If the inverse accumulating operator D is applied r times on $X^{(0)}$, we write conventionally

$$X^{(0)}D^r = \alpha^{(r)}X^{(0)} = \left(\alpha^{(r)}x^{(0)}(1), \alpha^{(r)}x^{(0)}(2), \cdots, \alpha^{(r)}x^{(0)}(n)\right)$$

where $\alpha^{(r)}x^{(0)}(k) = \alpha^{(r-1)}x^{(0)}(k) - \alpha^{(r-1)}x^{(0)}(k-1); k = 1,2,\cdots,n$.

Proposition 2.5. Assume that $X^{(0)} = (x^{(0)}(1), x^{(0)}(2),\ldots, x^{(0)}(n))$ is a nonnegative sequence satisfying $x^{(0)}(k) \in [a,b]; k = 1,2,\cdots,n$. If $X^{(r)} = \left(x^{(r)}(1), x^{(r)}(2), \cdots, x^{(r)}(n)\right)$ is the r-th accumulated sequence of $X^{(0)}$, then when r is sufficiently large, for any $\varepsilon > 0$, there is natural number N such that for any k, $N < k \le n$,

$$\frac{x^{(r)}(k)}{\sum_{i=1}^{k-1} x^{(r)}(i)} < \varepsilon .$$

What this proposition says is that for each nonnegative sequence, applying the accumulating operator multiple times can make the sequence sufficiently smooth with the smoothness ratio $\rho(k) \to 0 (k \to \infty)$.

Proposition 2.6. Assume $X^{(0)}$ is the same as in Proposition 2.5. Let $X^{(1)}$ = $\left(x^{(1)}(1), x^{(1)}(2), \cdots, x^{(1)}(n)\right)$ be the sequence generated from applying the accumulating operator once and $Z^{(1)} = \left(z^{(1)}(2), \cdots, z^{(1)}(n)\right)$ the sequence generated by adjacent neighbor means of $X^{(1)}$. Then, for any $\varepsilon > 0$ there is a positive natural number $N = N(\varepsilon)$ such that for any k, $N < k \le n$,

$$\frac{x^{(0)}(k)}{z^{(1)}(k)} < \rho(k) = \frac{x^{(0)}(k)}{\sum_{i=1}^{k-1} x^{(0)}(i)} < \varepsilon \cdot$$

Proof. It suffices to show $z^{(1)}(k) \ge \sum_{i=1}^{k-1} x^{(0)}(i)$. From definition, $z^{(1)}(k) = 0.5x^{(1)}(k)$ + $0.5x^{(1)}(k-1)$, $\sum_{i=1}^{k-1} x^{(0)}(i) = x^{(1)}(k-1) = 0.5x^{(1)}(k-1) + 0.5x^{(1)}(k-1)$, and $x^{(1)}(k)$

$= \sum_{i=1}^{k-1} x^{(0)}(i) = \sum_{i=1}^{k-1} x^{(0)}(i) + x^{(0)}(k) = x^{(1)}(k-1) + x^{(0)}(k)$. So, from $x^{(0)}(k) \geq 0$, it follows that $z^{(1)}(k) \geq \sum_{i=1}^{k-1} x^{(0)}(i)$. QED.

2.4 Exponentiality of Accumulating Generations

After applying the accumulating operator a few times, the general nonnegative quasi-smooth sequence will show with decreased randomness the pattern of exponential growth. It is found that the smoother the original sequence is, the more obvious an exponential growth pattern in the first order accumulation generated sequence will appear.

Assume that $X^{(0)} = (x^{(0)}(1), x^{(0)}(2),\ldots, x^{(0)}(n))$ is a sequence of raw data and its inverse accumulation generated sequence is $\alpha^{(1)}X^{(0)} = (\alpha^{(1)}x^{(0)}(1), \alpha^{(1)} x^{(0)}(2),\ldots, \alpha^{(1)}x^{(0)}(n))$. If there is k such that $\alpha^{(1)}x^{(0)}(k) = x^{(0)}(k) - x^{(0)}(k-1) > 0$, then the sequence $X^{(0)}$ experiences an increase at step k. Otherwise, the sequence $X^{(0)}$ experiences a decrease at step k. If there are k_1, k_2 such that $\alpha^{(1)}x^{(0)}(k_1) > 0$ and $\alpha^{(1)}x^{(0)}(k_2) < 0$, then $X^{(0)}$ is referred to as a *random sequence*. If $X^{(0)}$ is either monotonic increasing or decreasing, while $\alpha^{(1)}X^{(0)}$ is random, then $X^{(0)}$ is referred to as a *first-order* weak random sequence. If for each $i = 0, 1, 2, \ldots, r-1$, $\alpha^{(i)}X^{(0)}$ is monotonic (either increasing or decreasing), while $\alpha^{(r)}X^{(0)}$ is random, then $X^{(0)}$ is referred to as an *r-th-order* weak random sequence. If for any natural number r, $\alpha^{(r)}X^{(0)}$ is monotonic, then $X^{(0)}$ is referred to as a non-random sequence.

Theorem 2.18. Assume that $X^{(0)}$ is a positive sequence and $X^{(r)}$ the r-th accumulation generated sequence of $X^{(0)}$, then $X^{(r)}$ must be an r-th order weak random sequence.

The exponential function $X(t) = ce^{at} + b, c, a \neq 0$, is referred to as *homogeneous*, if $b = 0$, non-homogeneous, if $b \neq 0$. And, if a sequence $X = (x(1), x(2), \ldots, x(n))$ satisfies $x(k) = ce^{ak}$, c, $a \neq 0$, for $k = 1, 2, \ldots, n$, then X is referred to as a *homogeneous exponential sequence*. If $x(k) = ce^{ak} + b$, $c, a, b \neq 0$, for $k = 1, 2, \ldots, n$, then X is referred to as a *non-homogeneous sequence*.

Theorem 2.19. A sequence X is a homogeneous exponential, if and only if for $k = 1, 2, \ldots, n$, $\sigma(k)$ is a constant.

Proof. For details, see page 82 of (Liu and Lin, 2006). QED.

For the given sequence $X = (x(1), x(2), \ldots, x(n))$, if $\forall k$, $\sigma(k) \in (0, 1)$, then X is referred to as satisfying the *law of negative grey exponent*; if $\forall k$, $\sigma(k) \in (1, b)$, for some $b > 1$, then X is referred to as satisfying the *law of positive grey exponent*; if $\forall k$, $\sigma(k) \in [a, b]$, $b - a = \delta$, then X is referred to as satisfying the *law of grey exponent* with the absolute degree of greyness δ; and if $\delta < 0.5$, then X is referred to as satisfying the *law of quasi-exponent.*

Theorem 2.20. Assume that $X^{(0)}$ is a nonnegative quasi-smooth sequence. Then, the sequence $X^{(1)}$, generated by applying accumulating generation once on $X^{(0)}$, satisfies the law of quasi-exponent.

Proof. See page 82 of (Liu and Lin, 2006) for details. QED.

This theorem is the theoretical foundation for grey systems modeling. As a matter of fact, because economic, ecological, agricultural systems and others can all be seen systems of energy, while the accumulation and release of energy generally satisfy an exponential law, that explains why the idea of exponential modeling of grey systems theory has found an extremely wide range of applications.

Theorem 2.21. Assume that $X^{(0)}$ is a nonnegative sequence. If $X^{(r)}$ satisfies a law of exponent and the stepwise ratio of $X^{(r)}$ is given by $\sigma^{(r)}(k) = \sigma$, then

1) $\sigma^{(r+1)}(k)=\dfrac{1-\sigma^{k}}{1-\sigma^{k-1}}$;
2) When $\sigma \in (0, 1)$, $\lim_{k\to\infty}\sigma^{(r+1)}(k)=1$; and for each k, $\sigma^{(r+1)}(k)\in(1,1+\sigma]$;
3) When $\sigma > 1$, $\lim_{k\to\infty}\sigma^{(r+1)}(k)=\sigma$; and for each k, $\sigma^{(r+1)}(k)\in(\sigma,1+\sigma]$.

Proof. See page 83 of (Liu and Lin, 2006) for details. QED.

This theorem says that if the rth accumulating generation of $X^{(0)}$ satisfies an obvious law of exponent, additional application of the accumulating generator will destroy the pattern. In practical applications, if the rth accumulating generation of $X^{(0)}$ satisfies the law of quasi-exponent, we generally stop applying the accumulating generator any further. To this end, Theorem 2.20 implies that only one application of the accumulating generator is needed for nonnegative quasi-smooth sequences before establishing an exponential model.

Chapter 3
Grey Incidence and Evaluations

The general abstract system, such as a social system, economic system, agricultural system, ecological system, education system, etc., involves many different kinds of factors. It is the result of the mutual interactions of these factors that determines the development tendency and the behavior of the system. It is often the case that among all the factors, the investigators want to know which ones are primary and which ones are secondary; which ones have dominant effect, while others exert less influence, on the development of the system; which factors motivate the positive development of the system so that these factors should be strengthened; which ones constitute obstacles for the desirable development of the system so that these factors should be weakened … For instance, for the overall performance of an economic system, there are generally many influencing factors. In order to realize the desire of producing additional output with less input, the necessary systems analysis needs to be done carefully.

Regression analysis, difference equations, main component analysis, etc., are the most commonly employed methods for conducting systems analysis. However, all these methods suffer from the following weaknesses:

(1) Large samples are needed in order to produce reliable conclusions.
(2) The available data need to satisfy some typical type of probability distribution; linear relationships between factors and system behaviors are assumed, while between the factors it is required to have no interactions. These requirements in generally are difficult to satisfy.
(3) The amount of computation is large and generally done by using computers.
(4) It is possible that quantitative conclusions do not agree with qualitative analysis outcomes so that the laws governing the development of the system of concern are distorted or misunderstood.

What is especially important is that when the available data contain relatively large amount of greyness and experience large rises and major falls without following any conventional probability distribution, it will be extremely difficult to apply the traditional methods of statistics to analyze these data.

Grey incidence analysis provides a new method to analyze systems where statistical methods do not seem appropriate. It is applicable no matter whether the sample size is large or small and no matter if the data satisfy a certain conventional distribution or not. What is more advantageous is that the amount of computation involved is small and can be carried out conveniently without the problem of disagreement between quantitative and qualitative conclusions.

The basic idea of grey incidence analysis is to use the degree of similarity of the geometric curves of the available data sequences to determine whether their connections are close or not. The more similar the curves are, the closer incidence exists between the sequences; and vice versa.

3.1 Grey Incidence and Degree of Grey Incidences

3.1.1 Grey Incidence Factors and Set of Grey Incidence Operators

When analyzing a system, after choosing the quantity to reflect the characteristics of the system of concern, one needs to determine the factors that influence the behavior of the system. If a quantitative analysis is considered, one needs to process the chosen characteristic quantity and the effective factors using sequence operators so that the available data are converted to their relevant non-dimensional values of roughly equal magnitudes.

Assume that X_i is a system factor, its observation value at the ordinal position k is $x_i(k)$, $k=1,2,\cdots,n$, then $X_i=(x_i(1),x_i(2),\cdots,x_i(n))$ is referred to as the *behavioral sequence* of the factor X_i. If k stands for the time order, then $x_i(k)$ is referred to as the observational value of the factor X_i at time moment k, and $X_i=(x_i(1),x_i(2),\cdots,x_i(n))$ the behavioral time sequence (or series) of X_i. If k stands for an index ordinal number and $x_i(k)$ the observational value of the kth index of the factor X_i, then $X_i=(x_i(1),x_i(2),\cdots,x_i(n))$ is referred to as the *behavioral index sequence* of the factor X_i. If k stands for the ordinal number of the observed object and $x_i(k)$ the observed value of the kth object of the factor X_i, then $X_i=(x_i(1),x_i(2),\cdots,x_i(n))$ is referred to as the *horizontal sequence* of factor X_i's behavior.

For example, X_i represents an economic factor, k time, and $x_i(k)$ the observed value of the factor X_i at the time moment k, then $X_i=(x_i(1),x_i(2),\cdots,x_i(n))$ is a time series of economic behaviors. If k is the ordinal number of an index, then $X_i=(x_i(1),x_i(2),\cdots,x_i(n))$ is the index sequence of an economic behavior. If k represents the ordinal number of different economic regions or departments, then $X_i=(x_i(1),x_i(2),\cdots,x_i(n))$ is a horizontal sequence of an economic behavior. No matter what kinds of sequence data are available, they can be employed to do incidence analysis.

Let $X_i = (x_i(1), x_i(2), \cdots, x_i(n))$ be the behavioral sequence of factor X_i, D_1 a sequence operator such that $X_i D_1 = (x_i(1)d_1, x_i(2)d_1, \cdots, x_i(n)d_1)$, where

$$x_i(k)d_1 = x_i(k) / x_i(1),\ x_i(1) \neq 0,\ k = 1,2,\cdots,n \tag{3.1}$$

then D_1 is referred to as an *initialing operator* with $X_i D_1$ being its image, called *initial image* of X_i. If sequence operator D_2 satisfies $X_i D_2 = (x_i(1)d_2, x_i(2)d_2, \cdots, x_i(n)d_2)$ and

$$x_i(k)d_2 = \frac{x_i(k)}{\overline{X_i}},\ \overline{X_i} = \frac{1}{n}\sum_{k=1}^{n} x_i(k),\ k = 1,2,\cdots,n \tag{3.2}$$

then D_2 is referred to as an *averaging operator* with $X_i D_2$ being its image, called the *average image* of X_i. If sequence operator D_3 satisfies $X_i D_3 = (x_i(1)d_3, x_i(2)d_3, \cdots, x_i(n)d_3)$ and

$$x_i(k)d_3 = \frac{x_i(k) - \min_k x_i(k)}{\max_k x_i(k) - \min_k x_i(k)};\ k = 1,2,\cdots,n \tag{3.3}$$

then D_3 is referred to as an *interval operator* with $X_i D_3$ being its image, called the *interval image* of X_i. If the behavioral sequence of the factor X_i satisfies $x_i(k) \in [0,1]$, $i = 1,2,\cdots,n$, and sequence operator D_4 satisfies $X_i D_4 = (x_i(1)d_4, x_i(2)d_4, \cdots, x_i(n)d_4)$ and

$$x_i(k)d_4 = 1 - x_i(k),\ k = 1,2,\cdots,n \tag{3.4}$$

then D_4 is referred to as a *reversing operator* with $X_i D_4$ being its image, called the *reverse image* of X_i. If sequence operator D_5 satisfies $X_i D_5 = (x_i(1)d_5, x_i(2)d_5, \cdots, x_i(n)d_5)$ and

$$x_i(k)d_5 = 1 / x_i(k),\ x_i(k) \neq 0,\ k = 1,2,\cdots,n \tag{3.5}$$

then D_5 is referred to as a *reciprocating operator* with $X_i D_5$ being its image, called the *reciprocal image* of X_i. The set $D = \{D_i \mid i = 1,2,3,4,5\}$ is known as the *set of grey incidence operators*. If X stands for the set of all system factors and D the set of grey incidence operators, then (X,D) is referred to as the *space of grey incidence factors* of the system.

3.1.2 Metric Spaces

If each factor in the set of system factors is seen as a point in the Euclidean space, where the observed values of the factor are treated as the coordinates of the point, then we can investigate the relationships between the factors and the influence of the factors on the system's characteristics. And, we will be able to define degrees of the grey incidence by making use of the distance function of the n-dimensional space. In general, assume that X, Y, and Z are points of an abstract space. If a real-valued function $d(X,Y)$ satisfies $d(X,Y)\geq 0$, $d(X,Y)=0 \Leftrightarrow X=Y$, $d(X,Y)=d(Y,X)$, and $d(X,Z)\leq d(X,Y)+d(Y,Z)$, then $d(X,Y)$ is referred to as a *distance (function)* of the abstract space. The distance $d(X, O)$, where O is the origin of the space, is referred to as the *norm* of X, denoted $\|X\|$. In particular, it is ready to see that for any points $X=(x(1),x(2),\cdots,x(n))$ and $Y=(y(1),y(2),\cdots,y(n))$ from $\boldsymbol{R}^n$, the following are distance functions:

$$d_1(X,Y)=|x(1)-y(1)|+|x(2)-y(2)|+\cdots+|x(n)-y(n)|$$

$$d_2(X,Y)=[|x(1)-y(1)|^2+|x(2)-y(2)|^2+\cdots+|x(n)-y(n)|^2]^{\frac{1}{2}}$$

$$d_3(X,Y)=\frac{d_1(X,Y)}{1+d_1(X,Y)}$$

$$d_p(X,Y)=[|x(1)-y(1)|^p+|x(2)-y(2)|^p+\cdots+|x(n)-y(n)|^p]^{\frac{1}{p}}$$

$$d_\infty(X,Y)=\max_k\{|x(k)-y(k)|, k=1,2,\cdots,n\}$$

Accordingly, we have the following norms: $\|X\|_1=\sum_{k=1}^{n}|x(k)|$; $\|X\|_2=[\sum_{k=1}^{n}|x(k)|^2]^{1/2}$; $\|X\|_p=[\sum_{k=1}^{n}|x(k)|^p]^{1/p}$; and $\|X\|_\infty=\max_k\{|x(k)|\}$.

If $X(t)$ and $Y(t)$ are two continuous functions from $C[a,b]$, the set of all continuous functions defined on the interval $[a,b]$, then the following are distance functions of $C[a,b]$.

$$d_1(X(t),Y(t)) = \int_a^b |X(t)-Y(t)|\,dt$$

$$d_2(X(t),Y(t)) = [\int_a^b |X(t)-Y(t)|^2\,dt]^{1/2}$$

$$d_2(X(t),Y(t)) = [\int_a^b |X(t)-Y(t)|^2\,dt]^{1/2}$$

$$d_3(X(t),Y(t)) = \frac{d_1(X(t),Y(t))}{1+d_1(X(t),Y(t))}$$

$$d_4(X(t),Y(t)) = \int_a^b \frac{|X(t)-Y(t)|}{1+|X(t)-Y(t)|}dt$$

$$d_p(X(t),Y(t)) = [\int_a^b |X(t)-Y(t)|^p\,dt]^{1/p}$$

$$d_\infty(X(t),Y(t)) = \max_t\{|X(t)-Y(t)|, t\in[a,b]\}$$

If $d(X(t),O)$ stands for the distance from the continuous function $X(t)$ and the zero function, then corresponding to the previous distance functions we have the following commonly used norms of functions:

1. $\|X(t)\|_1 = \int_a^b |X(t)|\,dt$;

2. $\|X(t)\|_2 = [\int_a^b |X(t)|^2\,dt]^{1/2}$;

3. $\|X(t)\|_p = [\int_a^b |X(t)|^p\,dt]^{1/p}$;

4. $\|X(t)\|_\infty = \max_t\{|X(t)||, t\in[a,b]\}$

3.1.3 Degrees of Grey Incidences

Given a sequence $X=(x(1),x(2),\cdots,x(n))$, we can image the corresponding *zigzagged line* of the plane X = $\{x(k)+(t-k)(x(k+1)-x(k)) \mid k=1,2,\cdots,n-1; t\in[k,k+1]\}$. Without causing confusion, the same symbol is used for both the sequence and its zigzagged line. And for convenience, we will not distinguish the two in our discussions. Let X_0 be the sequence of a system's characteristic behavior, which is increasing, and X_i the behavior sequence of a relevant factor. If X_i is also an increasing sequence, then we say that both X_i and X_0 have a *positive* or *direct incidence relationship*. If X_i is a decreasing sequence, then we say that both X_i and X_0 have a *negative* or *inverse incidence relationship*. Because sequences of inverse incidence relationships can be transformed into those of direct incidence relationships by using either reversing or reciprocating operators, we will focus our attention on the study of positive incidence relationships.

For the given sequence $X = (x(1), x(2), \cdots, x(n))$, $\alpha = \frac{x(s) - x(k)}{s - k}$, $s > k$, $k = 1, 2, \cdots, n-1$, will be referred to as the *slope* of X on the interval $[k, s]$; and $\alpha = \frac{1}{n-1}(x(n) - x(1))$ the *mean slope* of X.

Theorem 3.1. Assume that X_i and X_j are nonnegative increasing sequences such that $X_j = X_i + c$, where c is a nonzero constant. Let D_1 be an initialing operator and $Y_i = X_i D_1$ and $Y_j = X_j D_1$. If α_i and α_j are respectively the mean slopes of X_i and X_j, and β_i and β_j the mean slopes of Y_i and Y_j. Then, the following must be true: $\alpha_i = \alpha_j$, and when $c < 0$, $\beta_i < \beta_j$; when $c > 0$, $\beta_i > \beta_j$.

Proof. For details, see (Liu and Lin, 2006, p. 94). QED.

What is meant here is that when the absolute amounts of increase of two increasing sequences are the same, the sequence with a smaller initial value will increase faster than the other. To maintain a same relative rate of increase, the absolute amount of increase of the sequence with the greater initial value must be greater than that of the sequence with a smaller initial value.

Definition 3.1. Let $X_0 = (x_0(1), x_0(2), \cdots, x_0(n))$ be a data sequence of a system's characteristic and $X_i = (x_i(1), x_i(2), \cdots, x_i(n))$, $i = 1, 2, ..., m$, relevant factor sequences. For given real numbers $\gamma(x_0(k), x_i(k))$, $i = 1, 2, ..., m$, $k = 1, 2, ..., n$, if

$$\gamma(X_0, X_i) = \frac{1}{n}\sum_{k=1}^{n} \gamma(x_0(k), x_i(k))$$

satisfies: 1) *Normality*: $0 < \gamma(X_0, X_i) \le 1$, $\gamma(X_0, X_i) = 1 \Leftrightarrow X_0 = X_i$; 2) *Wholeness*: $\forall X_i$, $X_j \in X = \{X_s \mid s = 0, 1, 2, \cdots, m; m \ge 2\}$,

$\gamma(X_i, X_j) \neq \gamma(X_j, X_i)$, $i \neq j$; 3) *Pair Symmetry*: $\forall X_i, X_j \in X$, $\gamma(X_i, X_j) = \gamma(X_j, X_i) \Leftrightarrow X = \{X_i, X_j\}$; and 4) *Closeness*: the smaller $|x_0(k) - x_i(k)|$ is, the greater $\gamma(x_0(k), x_i(k))$, then $\gamma(X_0, X_i)$ is referred to as a *degree of grey incidence* between X_i and X_0, $\gamma(x_0(k), x_i(k))$ the *incidence coefficient* of X_i and X_0 at point *k*.

Theorem 3.2. Given a system's behavioral sequences $X_0 = (x_0(1), x_0(2), \cdots, x_0(n))$ and $X_i = (x_i(1), x_i(2), \cdots, x_i(n))$, $i = 1,2,...,m$, for $\xi \in (0,1)$, define

$$\gamma(x_0(k), x_i(k)) = \frac{\min_i \min_k |x_0(k) - x_i(k)| + \xi \max_i \max_k |x_0(k) - x_i(k)|}{|x_0(k) - x_i(k)| + \xi \max_i \max_k |x_0(k) - x_i(k)|} \tag{3.6}$$

and

$$\gamma(X_0, X_i) = \frac{1}{n} \sum_{k=1}^{n} \gamma(x_0(k), x_i(k)) \tag{3.7}$$

then $\gamma(X_0, X_i)$ is a *degree of grey incidence* between X_0 and X_i, where ξ is known as the *distinguishing coefficient.*

Proof. For details see (Liu and Lin, 2006, p. 97). QED.

The degree $\gamma(X_0, X_i)$ of grey incidence is commonly written as γ_{0i} with the incidence coefficient $\gamma(x_0(k), x_i(k))$ as $\gamma_{0i}(k)$.

3.2 General Grey Incidences

3.2.1 Absolute Degree of Grey Incidence

Proposition 3.1. Let $X_i = (x_i(1), x_i(2), \cdots, x_i(n))$ be the data sequence of a system's behavior, $X_i - x_i(1)$ denote the zigzagged line $(x_i(1) - x_i(1), x_i(2) - x_i(1), \cdots, x_i(n) - x_i(1))$, and

$$s_i = \int_1^n (X_i - x_i(1))dt \tag{3.8}$$

then when X_i is increasing, $s_i \geq 0$; when X_i is decreasing, $s_i \leq 0$, and when X_i is vibrating, the sign of s_i various.

The intuition of the results in Proposition 3.1 is given in Figure 3.1, where (a) shows that case when the sequence is increasing; (b) the situation when X_i is decreasing, and (c) the scenario when X_i vibrate.

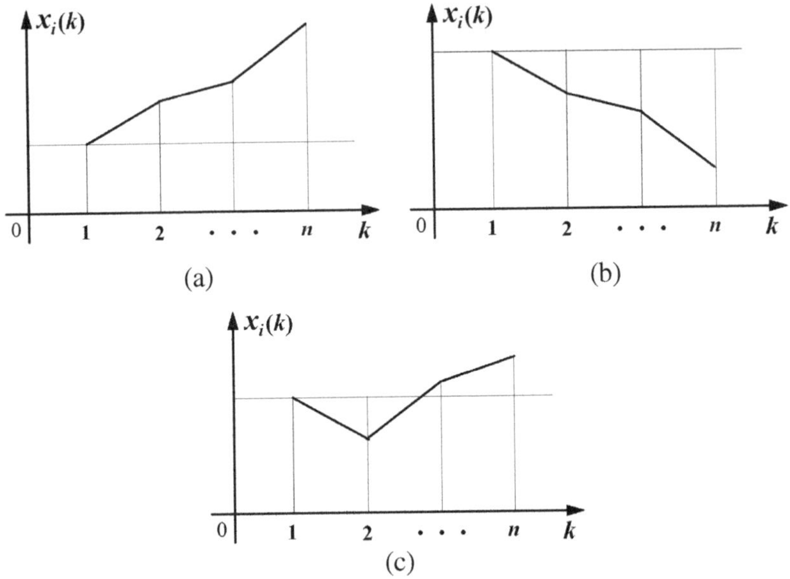

Fig. 3.1 The geometry of Proposition 3.1

Let $X_i = (x_i(1), x_i(2), \cdots, x_i(n))$ be the data sequence of a system's behavior and D a sequence operator satisfying that $X_i D = (x_i(1)d, x_i(2)d, \cdots, x_i(n)d)$ and $x_i(k)d = x_i(k) - x_i(1)$, $k = 1,2,\cdots,n$, then D is referred to as a *zero-starting point operator* with $X_i D$ being the image of X_i. What is often written is $X_i D$ = $X_i^0 = (x_i^0(1), x_i^0(2), \cdots, x_i^0(n))$.

Proposition 3.2. Assume that the images of the zero-starting point of two behavioral sequences X_i and X_j are respectively $X_i^0 = (x_i^0(1), x_i^0(2), \cdots, x_i^0(n))$ and $X_j^0 = (x_j^0(1), x_j^0(2), \cdots, x_j^0(n))$. Let

$$s_i - s_j = \int_1^n (X_i^0 - X_j^0)dt \tag{3.9}$$

then when X_i^0 is entirely located above X_j^0, $s_i - s_j \geq 0$; when X_i^0 is entirely underneath X_j^0, $s_i - s_j \leq 0$; and when X_i^0 and X_j^0 alternate their positions, the sign of $s_i - s_j$ is not fixed.

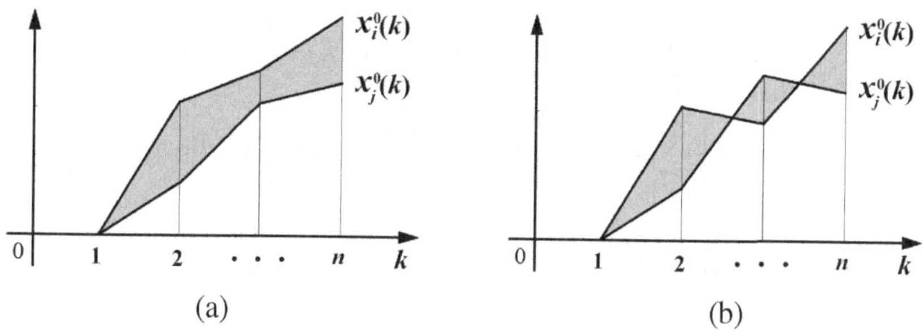

Fig. 3.2 An intuitive description of the relationship between X_i^0 and X_j^0

As shown in Figure 3.2, when X_i^0 is entirely located above X_j^0 (Figure 3.2(a)), the shaded area is positive so that $s_i - s_j \geq 0$. When X_i^0 and X_j^0 alternate their positions (Figure 3.2(b)), the sign of $s_i - s_j$ is not fixed.

Definition 3.2. Let X_i and X_j be two sequences of the same length, which is defined to be the sum of the distances between two consecutive time moments, and s_i and s_j are defined as above. Then,

$$\varepsilon_{ij} = \frac{1 + |s_i| + |s_j|}{1 + |s_i| + |s_j| + |s_i - s_j|} \tag{3.10}$$

is referred to as the *absolute degree of* (grey) incidence between X_i and X_j.

As for sequences of different lengths, the concept of absolute degree of incidence can be defined by either shortening the longer sequence or prolonging the shorter sequence with appropriate methods to make the sequences have the same length. However, by doing so, the ultimate value of the absolute degree of incidence will be affected.

Theorem 3.3. The absolute degree of incidence as defined in equ. (3.1) satisfies the properties of normality, pair symmetry, and closeness.

Proof. Omitted.

Theorem 3.4. Assume that X_i and X_j are two sequences with the same length, same time distances from moment to the next, and equal time moment intervals, then the absolute degree of incidence can also be computed by using

$$\begin{aligned}\varepsilon_{ij} = [1 + \left|\sum_{k=2}^{n-1} x_i^0(k) + \frac{1}{2}x_j^0(n)\right| + \left|\sum_{k=2}^{n-1} x_j^0(k) + \frac{1}{2}x_j^0(n)\right|] \\ \times[1 + \left|\sum_{k=2}^{n-1} x_i^0(k) + \frac{1}{2}x_i^0(n)\right| + \left|\sum_{k=2}^{n-1} x_j^0(k) + \frac{1}{2}x_j^0(n)\right| \\ + \left|\sum_{k=2}^{n-1} (x_i^0(k) - x_j^0(k)) + \frac{1}{2}(x_i^0(n) - x_j^0(n))\right|]^{-1}\end{aligned}$$

Proof. For details, see (Liu and Lin, 2006, pp. 104 – 108).

Theorem 3.5. The absolute degree of incidence ε_{ij} satisfies the following properties:

(1) $0 < \varepsilon_{ij} \le 1$;

(2) ε_{ij} is only related to the geometric shapes of X_i and X_j, and has nothing to do with the spatial positions of these sequences;

(3) Any two sequences are not absolutely unrelated. That is, never equals zero;

(4) The more X_i and X_j are geometrically similar, the greater ε_{ij} is;

(5) If X_i and X_j are parallel or X_i^0 fluctuates around X_j^0 with the area of the parts of X_i^0 located above X_j^0 equal to that of the parts with X_i^0 located underneath X_j^0, then $\varepsilon_{ij} = 1$;

(6) When one of the observed values of X_i and X_j change, ε_{ij} also changes accordingly;

(7) When the lengths of X_i and X_j change, ε_{ij} also changes;

(8) $\varepsilon_{jj} = \varepsilon_{ii} = 1$; and

(9) $\varepsilon_{ij} = \varepsilon_{ji}$.

3.2.2 Relative Degree of Grey Incidence

Let X_i and X_j be sequences of the same length with non-zero initial values and $X_i^{'}$ and $X_j^{'}$ respectively the initial images of X_i and X_j. The absolute degree of grey incidence of $X_i^{'}$ and $X_j^{'}$ is referred to as the *relative degree* of (grey) incidence of X_i and X_j, denoted r_{ij}. This relative degree of incidence is a quantitative representation of the rates of change of the sequences X_i and X_j relative to their initial values. The closer the rates of change of X_i and X_j are the greater r_{ij} is; and vice versa.

Proposition 3.3. Let X_i be a sequence with a non-zero initial value. If $X_j = cX_i$, $c > 0$ is a constant, then $r_{ij} = 1$.

Proof. See (Liu and Lin, 2006, p. 113. QED.

Proposition 3.4. Let X_i and X_j be two sequences of the same length with non-zero initial values. Then

(1) Their relative degree r_{ij} and absolute degree of incidence do not have to have any connections. When ε_{ij} is relatively large, r_{ij} can be very small; when ε_{ij} is very small, r_{ij} can also be very large.

(2) For any non-zero constants *a* and *b*, the relative degree $r_{ij}^{'}$ of incidence between aX_i and bX_j is the same as that r_{ij} of X_i and X_j.

Proof. See (Liu and Lin, 2006, p. 113 – 115, 117). QED.

Theorem 3.6. The relative degree r_{ij} of incidence satisfies the following properties:

(1) $0 < r_{ij} \leq 1$;

(2) The value of r_{ij} has something to do with only the rates of change of the sequences X_i and X_j with respect to their individual initial values and has nothing to do with the magnitudes of other entries. In other words, scalar multiplication does not change the relative degree of incidence;

(3) The rates of change of any two sequences are related somehow. That is, r_{ij} is never zero;

(4) The closer the individual rates of change of X_i and X_j with respect to their initial values, the greater r_{ij} is;

(5) If $X_j = aX_i$, or when the images of zero initial points of the initial images of X_i and X_j satisfy that $X_i^{'0}$ fluctuates around $X_j^{'0}$ and that the area of the parts where $X_i^{'0}$ is located above $X_j^{'0}$ equals that of the parts where $X_i^{'0}$ is located underneath $X_j^{'0}$, $r_{ij} = 1$;

(6) When an entry in X_i or X_j is changed, r_{ij} will change accordingly;

(7) When the length of X_i or X_j changes, r_{ij} also change;

(8) $r_{jj} = r_{ii} = 1$

(9) $r_{ij} = r_{ji}$.

3.2.3 Synthetic Degree of Grey Incidence

Let X_i and X_j be sequences of the same length with non-zero initial entries, ε_{ij} and r_{ij} be respectively the absolute and relative degrees of incidence between X_i and X_j, and $\theta \in [0,1]$. Then

$$\rho_{ij} = \theta\varepsilon_{ij} + (1-\theta)r_{ij} \tag{3.11}$$

is referred to as the *synthetic degree of* (grey) incidence between X_i and X_j. The concept of synthetic degree of incidence reflects the degree of similarity between the zigzagged lines of X_i and X_j as well as the closeness between the rates of changes of X_i and X_j with respect to their individual initial values. It is an index that describes relatively completely the closeness relationship between sequences. In general, we take $\theta = 0.5$. If the interest of study is more about the relationship between the relevant absolute quantities, θ can take a greater value. On the other hand, if the focus is more on comparison between rates of changes, then θ can take a smaller value.

Theorem 3.7. The synthetic degree ρ_{ij} of incidence satisfies the following properties:

(1) $0 < \rho_{ij} \le 1$;

(2) The value of ρ_{ij} is related to the individual observed values of the sequences X_i and X_j as well as the rates of changes of these values with respect to their initial values;

(3) ρ_{ij} will never been zero;

(4) ρ_{ij} changes along with the values in X_i and X_j;

(5) When the lengths of X_i and X_j change, so does ρ_{ij};

(6) With different θ value, ρ_{ij} also varies;

(7) When $\theta = 1$ or 0, $\rho_{ij} = r_{ij}$;

(8) $\rho_{jj} = \rho_{ii} = 1$; and

(9) $\rho_{ij} = \rho_{ji}$.

3.3 Grey Incidence Models Based on Similarity and Closeness

Grey incidence analysis is a very active area of grey systems theory. Its basic idea is to investigate the closeness of connection between sequences using the geometric shapes of the sequences. On the basis of Professor Julong Deng's initial models of grey incidences, many scholars established various different grey incidence models. For instance, in 1992, based on the thinking logic used by Julong Deng in constructing his initial grey incidence models, we proposed the concept of absolute degree of incidence. In the following ten plus years, this model was employed in different applications, leading to the resolutions of a great many practical problems. In particular, J. C. Zhang et al. employed this model in their analysis of cliff demolition; C. J. Zhao et al. applied this model in their analysis of stock markets; C. H. Li employed this model in his causal relationship analysis of accident occurrence in various mining sites; Y. A. Liu and S. L. Chen analyzed on the basis of this model how to track low altitude, small flying targets using multiple radar systems; X. F. Shi et al. applied this model in their analysis of ground-air missile weapon systems; S. L. Tan et al. made use of this model in their study of cleaning targets near and above an airport; X. P. Miao et al. employed this model in their study of vibration control of tapered roller bearings. All these and other scholars had achieved satisfactory results. Based on the recent works by Yong Wei and Naiming Xie, this section focuses on the improvements of our 1992 model of absolute degree of incidence. These new models measure mutual influences and connections between sequences from two different angles: similarity and closeness, while overcoming some of the weaknesses of the original model. These new models have shown to be much easier to be applied in practical situations.

Definition 3.3. Let X_i and X_j be sequences of the same length, and $s_i - s_j$ the same as defined in Proposition 3.2. Then,

$$\varepsilon_{ij} = \frac{1}{1+|s_i - s_j|} \tag{3.12}$$

is referred to as the *similitude degree* of (grey) incidence between X_i and X_j.

When the concept of similitude degree of incidence is employed to measure the geometric similarity of the shapes of sequences X_i and X_j, the more similar the geometric shapes of X_i and X_j, the greater value ε_{ij} takes; and vice versa.

Definition 3.4. Let X_i and X_j be sequences of the same length, and $S_i - S_j$ the same as defined in Proposition 3.2. Then,

$$\rho_{ij} = \frac{1}{1+|S_i - S_j|} \tag{3.13}$$

is referred to as the *closeness degree* of (grey) incidence between X_i and X_j.

When the concept of closeness degree of incidence is employed to measure the spatial closeness of sequence X_i and X_j, the closer X_i and X_j are, the greater value ρ_{ij} takes, and vice versa. This concept is only meaningful when the sequences X_i and X_j possess similar meanings and identical units. Otherwise, it does not stand for any practical significance.

Theorem 3.8. The similitude and closeness degrees (ε_{ij} and ρ_{ij}) of grey incidence both satisfy the axioms of normality, pair symmetry, and closeness of grey incidences.

Proof. We will only look at the argument for the similitude degree ε_{ij} of incidence. For the case of closeness degree ρ_{ij} of incidence, the argument is similar and is omitted.

For normality, from $\varepsilon_{ij} > 0$ and $|s_i - s_j| \geq 0$, it follows that $\varepsilon_{ij} \leq 1$. For pair symmetry, from $|s_i - s_j| = |s_j - s_i|$, it follows that $\varepsilon_{ij} = \varepsilon_{ji}$. The conditions for closeness clearly hold true. QED.

Theorem 3.9 The similitude degree ε_{ij} of incidence satisfies the following properties:

(1) $0 < \varepsilon_{ij} \leq 1$;

(2) The value of ε_{ij} is determined only by the geometric shape of the sequences X_i and X_j without anything to do with their relative spatial positions. In other words, horizontal translations of X_i and X_j will not change the value of ε_{ij};

(3) The more geometrically similar the sequences X_i and X_j are, the great value ε_{ij} takes; and vice versa;

(4) When X_i and X_j are parallel or when X_i^0 fluctuates around X_j^0 and the area of the parts where X_i^0 is located above X_j^0 equals that of the parts where X_i^0 is located beneath X_j^0, $\varepsilon_{ij} = 1$;

(5) $\varepsilon_{ii} = 1$; and

(6) $\varepsilon_{ij} = \varepsilon_{ji}$.

Theorem 3.10. The closeness degree ρ_{ij} of incidence satisfies the following properties:

(1) $0 < \rho_{ij} \leq 1$;

(2) The value of ρ_{ij} is determined not only by the geometric shape of the sequences X_i and X_j but also by their relative spatial positions. In other words, horizontal translations of X_i and X_j will change the value of ρ_{ij};

(3) The closer X_i and X_j are, the greater value ρ_{ij} takes; and vice versa.

(4) If X_i and X_j coincide or X_i fluctuates around X_j with the area of the parts where X_i is located above X_j equals that of the parts where X_i is located beneath X_j, $\rho_{ij} = 1$;

(5) $\rho_{ii} = 1$; and

(6) $\rho_{ij} = \rho_{ji}$.

Example 3.1. Given sequences

$$X_1 = (x_1(1), x_1(2), x_1(3), x_1(4), x_1(5), x_1(7)) = (0.91, 0.97, 0.90, 0.93, 0.91, 0.95)$$
$$X_2 = (x_2(1), x_2(2), x_2(3), x_2(5), x_2(7)) = (0.60, 0.68, 0.61, 0.63, 0.65)$$
$$X_3 = (x_3(1), x_3(3), x_3(7)) = (0.82, 0.90, 0.86)$$

Compute the similitude degrees $\varepsilon_{12}, \varepsilon_{13}$ and the closeness degrees ρ_{12}, ρ_{13} of incidence between X_1 and X_2, X_3, respectively.

Solution. (1) Let us transform both X_2 and X_3 into sequences with the same time intervals as X_1. To this end, let

$$x_2(4) = \frac{1}{2}(x_2(3) + x_2(5)) = \frac{1}{2}(0.61 + 0.63) = 0.62$$

$$x_3(2) = \frac{1}{2}(x_3(1) + x_3(3) = \frac{1}{2}(0.82 + 0.90) = 0.86$$

$$x_3(5) = \frac{1}{2}(x_3(3) + x_3(7)) = \frac{1}{2}(0.90 + 0.86) = 0.88$$

$$x_3(4) = \frac{1}{2}(x_3(3) + x_3(5)) = \frac{1}{2}(0.90 + 0.88) = 0.89$$

So, we have

$$X_2 = (x_2(1), x_2(2), x_2(3), x_2(4), x_2(5), x_2(7)) = (0.60, 0.68, 0.61, 0.62, 0.63, 0.65)$$
$$X_3 = (x_3(1), x_3(2), x_3(3), x_3(4), x_3(5), x_3(7)) = (0.82, 0.86, 0.90, 0.89, 0.88, 0.86)$$

(2) Let us transform X_1, X_2, and X_3 into sequences of equal time distance. To this end, let

$$x_1(6) = \frac{1}{2}(x_1(5) + x_1(7)) = \frac{1}{2}(0.91 + 0.95) = 0.93$$

$$x_2(6) = \frac{1}{2}(x_2(5) + x_2(7)) = \frac{1}{2}(0.63 + 0.65) = 0.64$$

$$x_3(6) = \frac{1}{2}(x_3(5) + x_3(7)) = \frac{1}{2}(0.88 + 0.86) = 0.87$$

So, the sequences

$$X_1 = (x_1(1), x_1(2), x_1(3), x_1(4), x_1(5), x_1(7)) = (0.91, 0.97, 0.90, 0.93, 0.91, 0.93, 0.95)$$
$$X_2 = (x_2(1), x_2(2), x_2(3), x_2(4), x_2(5), x_2(7)) = (0.60, 0.68, 0.61, 0.62, 0.63, 0.64, 0.65)$$
$$X_3 = (x_3(1), x_3(2), x_3(3), x_3(4), x_3(5), x_3(7)) = (0.82, 0.86, 0.90, 0.89, 0.88, 0.87, 0.86)$$

are all 1-time distance, meaning that the time distances between consecutive entries are all 1.

(3) Computing the initial images with zero initial points provides

$$X_1^0 = (x_1^0(1), x_1^0(2), x_1^0(3), x_1^0(4), x_1^0(5), x_1^0(6), x_1^0(7)) = (0, 0.06, -0.01, 0.02, 0, 0.02, 0.04)$$
$$X_2^0 = (x_2^0(1), x_2^0(2), x_2^0(3), x_2^0(4), x_2^0(5), x_2^0(6), x_2^0(7)) = (0, 0.08, 0.01, 0.02, 0.03, 0.04, 0.05)$$
$$X_3^0 = (x_3^0(1), x_3^0(2), x_3^0(3), x_3^0(4), x_3^0(5), x_3^0(6), x_3^0(7)) = (0, 0.04, 0.08, 0.07, 0.06, 0.05, 0.04)$$

(4) Compute $|s_1 - s_2|$, $|s_1 - s_3|$ and $|S_1 - S_2|$, $|S_1 - S_3|$ as follows:

$$|s_1 - s_2| = \left|\sum_{k=2}^{6}(x_1^0(k) - x_2^0(k)) + \frac{1}{2}(x_1^0(7) - x_2^0(7))\right| = 0.095$$

$$|s_1 - s_3| = \left|\sum_{k=2}^{6}(x_1^0(k) - x_3^0(k)) + \frac{1}{2}(x_1^0(7) - x_3^0(7))\right| = 0.21$$

$$|S_1 - S_2| = \left|\sum_{k=2}^{6}(x_1(k) - x_2(k)) + \frac{1}{2}(x_1(7) - x_2(7))\right| = 1.91$$

$$|S_1 - S_3| = \left|\sum_{k=2}^{6}(x_1(k) - x_3(k)) + \frac{1}{2}(x_1(7) - x_3(7))\right| = 0.375$$

(5) Calculate the similitude degrees $\varepsilon_{12}, \varepsilon_{13}$ and closeness degrees ρ_{12}, ρ_{13}.

$$\varepsilon_{12} = \frac{1}{1 + |s_1 - s_2|} = 0.91, \quad \varepsilon_{13} = \frac{1}{1 + |s_1 - s_3|} = 0.83$$

$$\rho_{12} = \frac{1}{1 + |S_1 - S_2|} = 0.34, \quad \rho_{13} = \frac{1}{1 + |S_1 - S_3|} = 0.73$$

Because $\varepsilon_{12} > \varepsilon_{13}$, it follows that X_2 is more similar to X_1 than X_3. Because $\rho_{12} < \rho_{13}$, it follows that X_3 is closer to X_1 than X_2 is.

What needs to be pointed out is that through degrees of incidence grey incidence analysis investigates the mutual influences and connections between sequences. Its focus is about the relevant order relations instead of the specific magnitudes of the values of the degrees of incidence. For instance, when computing the similitude degrees or closeness degrees of incidence based on equs. (3.12) or

(3.13), when the absolute values of the sequence data are relative large, the values of both $\left|s_i - s_j\right|$ and $\left|S_i - S_j\right|$ might be large too, creating the situation that the resultant similitude and closeness degrees of incidence are relatively small. This scenario does not have any substantial impact on the analysis of order relationships. If the situations dealt with demand relatively large numerical magnitudes in the degrees of incidence, one can consider either replacing the number 1 appearing in the numerators or denominators in equs. (3.12) or (3.13) by a relevant constant, or using the grey absolute degree of incidence model or other appropriate models.

3.4 Grey Cluster Evaluations

Grey clustering is a method developed for classifying observation indices or observation objects into definable classes using grey incidence matrices or grey whitenization weight functions. Each cluster can be seen a set consisting of all the observational objects of a same kind. When investigating practical problems, it is often the situation that each observational object possesses quite a few characteristic indices, which are difficult to accurately classify. Based on the objects to be clustered, grey clustering consists of two methods: clustering using grey incidence, and clustering using grey whitenization weight functions. The first method is mainly applied to group same kinds of factors into their individual categories so that a complicated system can be simplified. Though using the lustering method of grey incidence, we can examine whether or not some of the factors under consideration really belong to the same kind so that a synthetic index of these factors or one of these factors can be used to represent all these factors without losing any part of the available information carried by these factors. This problem is about selection of variables to be used in the study of the system of concern. Before conducting a large-scale survey, which generally costs a lot of money and man power, by using the lustering method of grey incidence on a typical sample data, one can reduce the amount of data collection to a minimal level by eliminating the unnecessary variables so that tangible savings can be materialized.

The lustering method of grey whitenization weight functions is mainly applicable for checking whether or not the observational objects belong to predetermined classes so that they can be treated differently.

3.4.1 Grey Incidence Clustering

Assume that there are n observational objects, for each of which m characteristic data points are collected, producing the following sequences: $X_i = \left(x_i(1), x_i(2), \ldots, x_i(n)\right)$, $i = 1,2,\ldots,m$. From these data, the following triangular matrix, referred to as the *incidence matrix* of the characteristic variables:

$$A = \begin{pmatrix} \varepsilon_{11} & \varepsilon_{12} & \cdots & \varepsilon_{11} \\ & \varepsilon_{22} & \cdots & \varepsilon_{2m} \\ & & \ddots & \vdots \\ & & & \varepsilon_{mm} \end{pmatrix}$$

is obtained, where ε_{ij} is the absolute degree of incidence between X_i and X_j, and $\varepsilon_{ii} = 1$, for all $i < j$, $i, j = 1,2,...,m$. For a chosen threshold value $r \in [0,1]$, which in general satisfies $r > 0.5$, if $\varepsilon_{ij} \geq r$, $i \neq j$, the variables X_j and X_i are seen as having the same characteristics. The clustering of the variables with the chosen r value is referred to as the *clustering of r-grey incidence*. When studying a specific problem, the particular r value is determined based on the circumstances involved. The closer the r value is to 1, the finer the clustering is with relatively fewer variables in each cluster. Conversely, the smaller the r value is, the coarser the clustering becomes with relatively more variables in each cluster.

For examples on how to practically employ this clustering method, please consult with (Liu and Lin, 2006, p. 141 – 144).

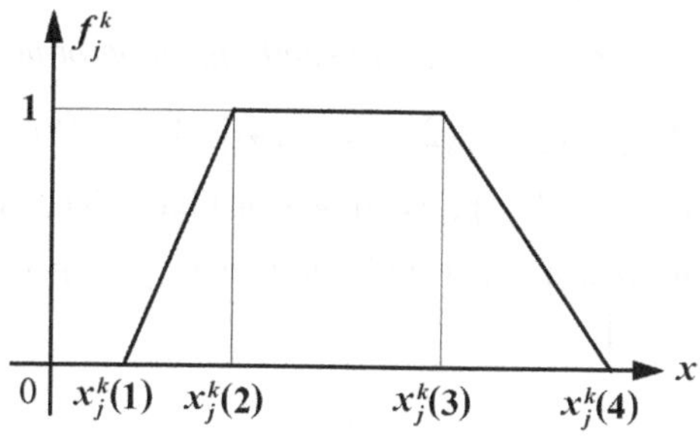

Fig. 3.3 The typical *j*th criterion *k*th subclass whientization weight function

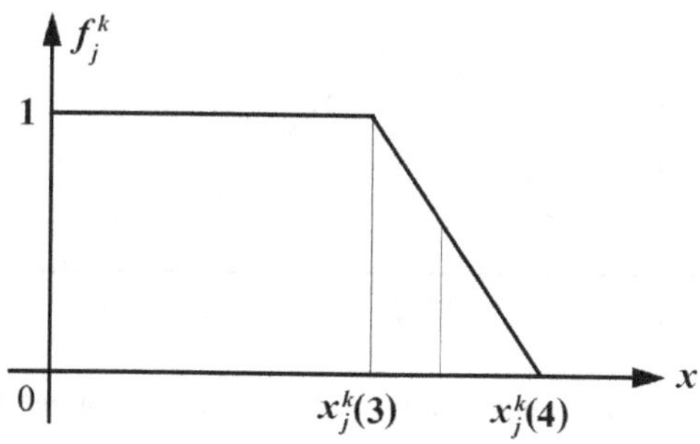

Fig. 3.4 A whitenization weight function of lower measure

3.4.2 Grey Variable Weight Clustering

Assume that there are n objects to be clustered according to m criteria into s different grey classes. Classifying the ith object into the kth grey class according to the observed value x_{ij} of the ith object judged against the jth criterion, $i=1,2,\cdots n,\ j=1,2,\cdots m$, is known as *grey clustering*. Classifying n objects into s grey classes using the jth criterion is referred to as the jth criterion subclassification with the whitenization weight function of the jth criterion kth subclass denoted as $f_j^k(\bullet)$. If the whitenization weight function $f_j^k(\bullet)$ takes the typical form as shown in Figure 3.3, then $x_j^k(1)$, $x_j^k(2)$, $x_j^k(3)$, and $x_j^k(4)$ are referred to as turning points of $f_j^k(\bullet)$. Such a typical whitenization function is written as $f\,_j^k[x_j^k(1),\ x_j^k(2),\ x_j^k(3),\ x_j^k(4)]$. If the whitenization weight function $f_j^k(\bullet)$ does not have the first and second turning points $x_j^k(1)$ and $x_j^k(2)$, as shown in Figure 3.4, then $f_j^k(\bullet)$ is referred to as a *whitenization weight function of lower measure*, and denoted $f\,_j^k[-,-,x_j^k(3),x_j^k(4)]$. If the second and third turning points $x_j^k(2)$ and $x_j^k(3)$ of $f_j^k(\bullet)$ coincide, as shown in Figure 3.5, then $f_j^k(\bullet)$ is referred to as a *whitenization weight function of moderate* (or middle) *measure*, denoted $f\,_j^k[x_j^k(1), x_j^k(2), -, x_j^k(4)]$. If $f_j^k(\bullet)$ does not have the turning points $x_j^k(3)$ and $x_j^k(4)$, as shown in Figure 3.6, then this function is referred to as a *whitenization weight function of upper measure*, denoted $f\,_j^k[x_j^k(1), x_j^k(2), -, -]$.

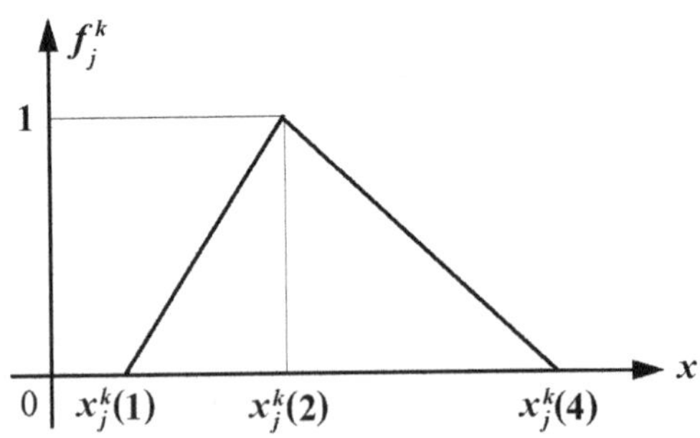

Fig. 3.5 A whitenization weight function of moderate (or middle) measure

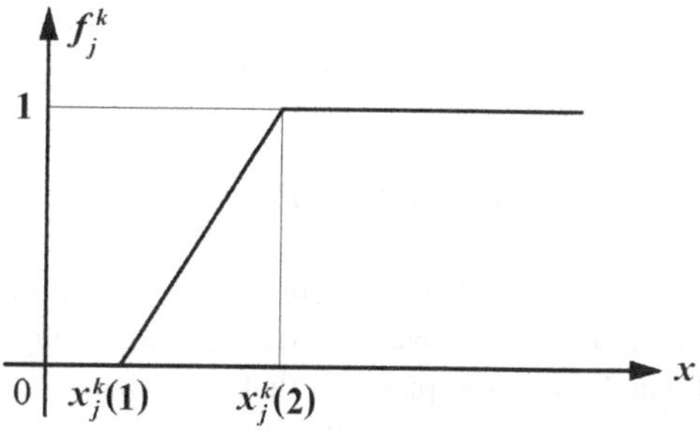

Fig. 3.6 A whitenization weight function of upper measure

For the *k*th subclass of the *j*th criterion, if its whitenization weight function is given as in Figure 3.3, let $\lambda_j^k = \frac{1}{2}(x_j^k(2) + x_j^k(3))$; if its whitenization weight function is given as in Figure 3.4, let $\lambda_j^k = x_j^k(3))$; and if its whitenization weight function is given as in Figure 3.5 or 3.6, let $\lambda_j^k = x_j^k(2)$. This λ_j^k value is referred to as the *critical value* for and $\eta_j^k = \lambda_j^k \Big/ \sum_{j=1}^{m} \lambda_j^k$ the weight of the *k*th subclass of the *j*th criterion.

Assume that x_{ij} is the observed value of object *i* with regard to the *j*th criterion, $f_j^k(\bullet)$ the whitenization weight function and η_j^k the weight of the *k*th subclass of the *j*th criterion, then $\sigma_i^k = \sum_{j=1}^{m} f_j^k(x_{ij}) \cdot \eta_j^k$ is referred to as the (grey) cluster coefficient of variable weight for object *i* to belong to the *k*th grey class, $\sigma_i = (\sigma_i^1, \sigma_i^2, \cdots, \sigma_i^s)$ the cluster coefficient vector of object *i*, and $\sum = (\sigma_i^k)_{n \times s}$ the cluster coefficient matrix. If $\max_{1 \le k \le s}\{\sigma_i^k\} = \sigma_i^{k^*}$, we then say that object *i* belongs to grey class k^*.

Variable weight clustering method is useful to studying problems with such criteria that have the same meanings and units. Otherwise, it is not appropriate to employ this method. Also, if the numbers of observed values of individual criteria are greatly different of each other, this clustering method should not be applied.

In terms of how to determine the whitenization weight function for the *k*th subclass of the *j*th criterion, it generally can be done by using the background information of the problem of concern. When resolving practical problems, one can determine the whitenization weight functions from either the angle of the objects that are to be clustered or by looking at all the same type objects in the whole system, not just the ones involved in the clustering. For more details, please consult with (Liu and Lin, 2006, p. 139 – 153) and the final section of this chapter.

3.4.3 Grey Fixed Weight Clustering

When the criteria for clustering have different meanings, dimensions (units), and drastically different numbers of observed data points, applying variable weight clustering method can lead to the problem that the effects of some of the criteria are very weak. There are two ways to get around this problem: 1) First transform the sample of data values of all the criteria into nondimensional values by applying either the initiating operator or the averaging operator; then cluster the transformed data. When employing this method, all the criteria are treated equally so that no difference played by the criteria in the process of clustering is reflected. 2) Assign each clustering criterion a weight ahead of the clustering process. In this subsection, we will mainly study this second method.

Assume that x_{ij} is the observed value of object i with regard to criterion j, $i=1,2,\cdots n$, $j=1,2,\cdots m$, and $f_j^k(\bullet)$ the whitenization weight function of the kth subclass of the jth criterion, $k=1,2,\cdots,s$, $j=1,2,\cdots m$. If the weight η_j^k of the kth subclass of the jth criterion is not a function of k, $j=1,2,\cdots m$, $k=1,2,\cdots,s$, that is, for any $k_1,k_2 \in \{1,2,\cdots,s\}$, we always have $\eta_j^{k_1} = \eta_j^{k_2}$, then the symbol η_j^k can be written as η_j with the superscript k removed, $j=1,2,\cdots m$. In this case, $\sigma_i^k=\sum_{j=1}^{m} f_j^k(x_{ij})\eta_j$ is referred to as the fixed weight clustering coefficient for object i to belong to the kth grey class. If furthermore $\eta_j=\frac{1}{m}$, for $j=1,2,\cdots,m$, then $\sigma_i^k=\sum_{j=1}^{m} f_j^k(x_{ij})\cdot\eta_j = \frac{1}{m}\sum_{j=1}^{m} f_j^k(x_{ij})$ is referred to as the *equal weight clustering coefficient* for object i to belong to the kth grey class. The method of clustering objects by using grey fixed weight clustering coefficients is known as *grey fixed weight clustering*. The method using grey equal weight clustering coefficients is known as *grey equal weight clustering*.

Grey fixed weight clustering can be carried out according to the following steps:

Step 1: Determine the whitenization weight function $f_j^k(\bullet)$ for the kth subclass of the jth criterion, $j=1,2,\cdots m$, $k=1,2,\cdots,s$.

Step 2: Determine a clustering weight η_j for each criterion $j = 1,2,\cdots m$.

Step 3: Based on the whitenization weight functions $f_j^k(\bullet)$ obtained in step 1, the clustering weights η_j obtained in step 2, and the observed data value x_{ij} of object i with respect to criterion j, calculate the fixed weight clustering coefficients $\sigma_i^k = \sum_{j=1}^{m} f_j^k(x_{ij})$, $i = 1,2,\cdots n$, $j = 1,2,\cdots m$, $k = 1,2,\cdots,s$.

Step 4: If $\max_{1\le k\le s}\{\sigma_i^k\} = \sigma_i^{k^*}$, then object i belongs to grey class k^*.

Example 3.2. To see how grey clustering analysis can be employed to resolve practical problems, let us look at a case study on analyzing different methods of coal mining.

The specific method selected for mining coal is mainly determined by the geological structure of the particular mine location and the available technology for the geological structure. The varied geological mining conditions of different coal mines (or different sections of the same mine) determine the technological levels of the selected mining methods, leading to drastically different economic outcomes. In order to improve the economic return of the coal mining companies, there is a need to select the appropriate mining technology with the optimal economic benefits in mind. To this end, the method of grey fixed weight clustering method provides us a comprehensive way to objectively and quantitatively evaluate various coal mining methods.

For a particular coal mine, four different mining methods are employed. They are fully mechanized mining, top-end conventional machine mining, conventional machine mining, and long-hole mining, which will be our clustering objects with per unit area yield (ten thousand tons/month*area), work efficiency (ton/workday), investment in equipment (ten thousand Yuan), and work cost (Yuan/ton) as the clustering criteria. Each of the criteria is classified into three classes: "good," "acceptable," and "poor." The observed values x_{ij} of the clustering object i with respect to criterion j are given in the following matrix A:

$$A = (x_{ij}) = \begin{bmatrix} 4.34 & 16.37 & 2046 & 10.20 \\ 1.76 & 10.83 & 1096 & 18.67 \\ 1.08 & 6.32 & 523 & 13.72 \\ 1.44 & 4.81 & 250 & 9.43 \end{bmatrix}$$

Solution. Because the meanings of the clustering criteria are different from each other and the magnitudes of the observed data values vary greatly, we employ the method of grey fixed weight clustering.

Through survey questionnaires collected from 20 experts, the whitenization weight functions $f_j^k(\bullet)$ for the kth subclass of the jth criterion, $j=1,2,3,4$, $k=1,2,3$, were respectively given below:

$$f_1^1[2.16,3.24,-,-],\quad f_1^2[1.08,2.16,-,3.24],\quad f_1^3[-,-,1.08,2.16]$$

$$f_2^1[9.6,14.40,-,-],\quad f_2^2[4.80,9.6,-,14.4],\quad f_2^3[-,-,4.8,9.6]$$

$$f_3^1[390,780,-,-],\quad f_3^2[390,780,-,1170],\quad f_3^3[-,-,780,1170]$$

$$f_4^1[-,-,605,13],\quad f_4^2[6.5,13,-,19.5],\quad f_4^3[13,19.5,-,-]$$

Additionally, from the survey we obtained the weights of per unit area yield (ten thousand tons/month*area), work efficiency (ton/workday), investment in equipment (ten thousand Yuan), and work cost (Yuan/ton), respectively, as follows:

$$\eta_1=0.4547,\quad \eta_2=0.2631,\quad \eta_3=0.1411,\quad \eta_4=0.1411$$

From $\sigma_i^k=\sum_{j=1}^m f_j^k(x_{ij})\cdot\eta_j$, it follows that when $i=1$,

$$\sigma_1^1=\sum_{j=1}^4 f_j^1(x_{1j})\eta_j=f_1^1(4.34)\times 0.4547+f_2^1(16.37)\times 0.2631+f_3^1(2046)\times 0.1411+$$
$$+f_4^1(10.20)\times 0.1411=0.8014$$

Similarly, we obtain $\sigma_1^2=0.0803$ and $\sigma_1^3=0$. So, $\sigma_1=(\sigma_1^1,\sigma_1^2,\sigma_1^3)=(0.8014,0.0803,0)$. By going through this procedure repeatedly, we obtain $\sigma_2=(\sigma_2^1,\sigma_2^2,\sigma_2^3)=(0.0674,0.5475,0.4384)$, $\sigma_3=(\sigma_3^1,\sigma_3^2,\sigma_3^3)=(0.0930,0.3743,0.6501)$, and $\sigma_4=(\sigma_4^1,\sigma_4^2,\sigma_4^3)=(0.2186,0.4396,0.6234)$.

Hence, we obtain the coefficient matrix Σ of the grey fixed weight clustering

$$\Sigma=(\sigma_i^k)=\begin{bmatrix}0.8014 & 0.0803 & 0\\ 0.0674 & 0.5475 & 0.4384\\ 0.0930 & 0.3743 & 0.6501\\ 0.2186 & 0.4396 & 0.6234\end{bmatrix}$$

From $\max_{1\le k\le 3}\{\sigma_1^k\}=\sigma_1^1=0.8014$, $\max_{1\le k\le 3}\{\sigma_2^k\}=\sigma_2^2=0.5475$, $\max_{1\le k\le 3}\{\sigma_3^k\}=\sigma_3^3=0.6501$, and $\max_{1\le k\le 3}\{\sigma_4^k\}=\sigma_4^3=0.6234$, it follows that among the four coal mining methods, the comprehensive technological and economic benefits of the method of fully mechanized mining is the best, that of the method of the top-end conventional machine mining is acceptable, and those of conventional and long-hole minings are poor.

3.5 Grey Evaluation Using Triangular Whitenization Functions

This section will study the grey evaluation method based on triangular whitenization weight functions. In 1993, we initially introduced this evaluation model. Since then this model has been widely employed in various practical evaluation scenarios. For instance, X. F. Cai et al. applied this model of triangular whitenization weight functions to evaluate the technology of tunnel construction of subways; X. W. Liu employed this model in his evaluation of the agricultural, environmental ecology of the regions of the Yangtze River delta; J. L Duan, Q. S. Zhang et al. used this model along with the AHP method to evaluate the risk of information systems; Y. Guo, J. J. Liu et al. made use of this model along with TWW functions in their evaluations of highway traffic safety and the quality of highway networks; G. X. Zhang et al. employed this model in their examination and evaluation of research projects on military weaponries; Q. Yang et al. combined this model with the method of AHP in their evaluations of risky commercial investments; S. F. Liu, B. J. Li et al. applied this model in their evaluations of regional signature industries, regional comprehensive strengths of science and technology, and parks of science and industry. Through many years of practical applications and theoretical research on the phenomenon of multiple overlappings of the grey classes of the original model and such problems as the clustering coefficients, selection of endpoints, etc., we established a new class of grey evaluation models on the basis of triangular whitenization weight functions. To perfect these models, we also compared these two classes of methods. In this book, these two classes of grey evaluation models developed on the basis of triangular whitenization weight functions are respectively referred to as endpoint evaluation method and center-point evaluation method.

3.5.1 *Evaluation Model Using Endpoint Triangular Whitenization Functions*

Assume that n objects are to be clustered into s different grey classes according to m criteria. The observed value of object i in terms of criterion j is, $i=1,2,\ldots,n$, $j=1,2,\ldots,m$. We need to evaluate and diagnose object i based on the value x_{ij}. The particular computational steps of the *grey evaluation model* based on endpoint triangular whitenization functions are the following:

Step 1: Based on the predetermined number s of grey classes for the planned evaluation, divide the individual ranges of the criteria into s grey classes. For instance, if the range of values criterion j takes is the interval $[a_1,a_{s+1}]$, then we partition this range into the following s subintervals:

$$[a_1,a_2],\cdots,[a_{k-1},a_k],\cdots,[a_{s-1},a_s],[a_s,a_{s+1}]$$

where $a_k (k=1,2,\cdots,s,s+1)$ can be generally determined based on specific requirements or relevant qualitative analysis.

Step 2: Let the value of the whitenization weight function for $\lambda_k=(a_k+a_{k+1})/2$ to belong to the kth grey class is 1. Connect the point $(\lambda_k,1)$ with the starting point a_{k-1} of the (k – 1)th grey class and the ending point a_{k+2} of the (k + 1)th grey class so that the whitenization weight function $f_j^k(\cdot)$ of the kth grey class in terms of the jth criterion is obtained, $j=1,2,...,m$, $k=1,2,...,s$. As for $f_j^1(\cdot)$ and $f_j^s(\cdot)$, we can expand respectively the value range of criterion j to the left- and right-hand side to a_0 and a_{s+2}, see Figure 3.7.

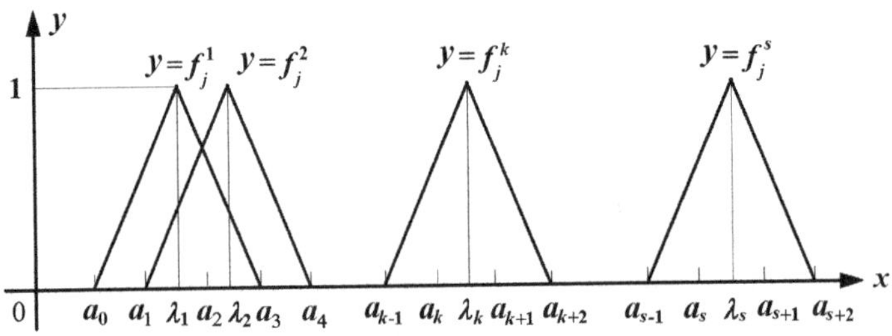

Fig. 3.7 The general endpoint triangular whitenization weight function

For an observed value x of criterion j, its degree $f_j^k(x)$ of membership in the kth grey class, $j=1,2,...,m$, $k=1,2,...,s$, can be computed out of the following formula:

$$f_j^k(x)=\begin{cases}0, & x\notin[a_{k-1},a_{k+2}]\\ \dfrac{x-a_{k-1}}{\lambda_k-a_{k-1}}, & x\in[a_{k-1},\lambda_k]\\ \dfrac{a_{k+2}-x}{a_{k+2}-\lambda_k}, & x\in[\lambda_k,a_{k+2}]\end{cases} \tag{3.14}$$

Step 3: Compute the comprehensive clustering coefficient for object i with respect to grey class k:

$$\sigma_i^k = \sum_{j=1}^{m} f_j^k(x_{ij}) \cdot \eta_j, \quad i=1,2,\ldots,n, \quad k=1,2,\ldots,s \tag{3.15}$$

where $f_j^k(x_{ij})$ is the whitenization weight function of the kth subclass of the jth criterion, and η_j the weight of criterion j in the comprehensive clustering.

Step 4: From $\max_{1\le k\le s}\{\sigma_i^k\} = \sigma_i^{k^*}$, it follows that object i belongs to grey class k^*. When several objects belong to the same k^* grey class, one can further determine the order of preference of these objects in grey class k^* by using the magnitudes of their clustering coefficients.

Example 3.3. Let us look at the evaluation of the ecological environment of a certain geographic region.

The agricultural ecological environment is a large-scale, complex system, made up of social, economic, and natural environments, involving many factors. To evaluate its quality, we must first establish a system of criteria on which to base our scientific evaluation. According to the characteristics of the particular region's agricultural ecological environment, we construct the system of evaluation criteria shown in Figure 3.8. The goal of the first level is to evaluate the quality (x) of the specific agricultural ecological environment. The criterion level contains three areas: the natural ecological background (x_1), degree of natural impact (x_2), and the state of the farm environment (x_3). The next level down contains the following 20 criteria:

x_{11} = number of days of sun shine per year; x_{12} = annual precipitation; x_{13} = actively accumulated temperature; x_{14} = soil quality index; x_{15} = index of forest coverage; x_{16} = index of water shed; x_{17} = occurrence rate of disasters;

x_{21} = index of land plantation; x_{22} = per capita farmland; x_{23} = use rate of water resource; x_{24} = rate of effective irrigation; x_{25} = intensity of fertilizer use; x_{26} = intensity of chemical use; x_{27} = investment rate in environment protection;

x_{31} = deterioration rate of farmland; x_{32} = index of polluted irrigation water; x_{33} = index of polluted farmland; x_{34} = index of farmland air pollution; x_{35} = index of crops polluted by chemicals; x_{36} = index of polluted crops with heavy metals.

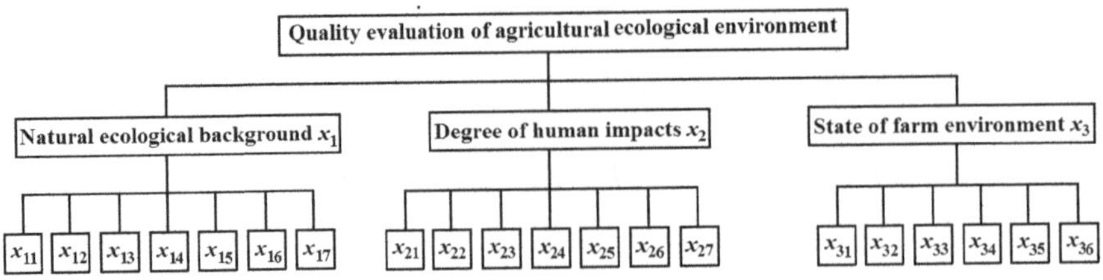

Fig. 3.8 The criteria system for evaluating the quality of an agricultural ecological environment

Step 1: Determine the weights of evaluation criteria. Based on the hierarchical structure of the evaluation criteria system, we conducted a survey with relevant experts in order to construct the matrix of judgments, on which we obtained the weights for the criteria of different levels, see Figures 3.1 – 3.3 for details.

Table 3.1 Weights of the criteria related to natural ecological background

	Natural ecological background x_1 (weight 0.164)						
Criteria	x_{11}	x_{12}	x_{13}	x_{14}	x_{15}	x_{16}	x_{17}
Level weight	0.044	0.070	0.070	0.116	0.209	0.167	0.325
Combined weight	0.007	0.011	0.011	0.019	0.034	0.027	0.053

Table 3.2 Weights of the criteria related to the degree of human impacts

	Degree of human impacts x_2 (weight 0.297)						
Criteria	x_{21}	x_{22}	x_{23}	x_{24}	x_{25}	x_{26}	x_{27}
Level weight	0.070	0.105	0.036	0.048	0.221	0.221	0.300
Combined weight	0.021	0.031	0.011	0.014	0.066	0.066	0.089

Table 3.3 Weights of the criteria related to the state of farm environment

	State of farm environment x_3 (weight 0.539)					
Criteria	x_{31}	x_{32}	x_{33}	x_{34}	x_{35}	x_{36}
Level weight	0.302	0.090	0.134	0.052	0.211	0.211
Combined weight	0.163	0.049	0.072	0.028	0.114	0.114

Step 2: Determination of the evaluation grey classes. To satisfy the need of reflecting the environment quality of the agricultural ecology of the specific region as accurately as possible, we employ five evaluation grey classes, indexed by k = 1, 2, 3, 4, 5. These classes respectively stand for "extremely poor," "poor," "fine," "good," "very good." Considering the experts' opinions, the grey classes of the criteria are determined as shown in Table 3.4.

Table 3.4 Grey classes of evaluation criteria

Criteria	Sub-criteria	Extremely poor	Poor	Fine	Good	Very good
Natural ecological background X_1	x_{11}	$1000 \le x_{11}^1 < 1400$	$1400 \le x_{11}^2 < 2000$	$2000 \le x_{11}^3 < 2600$	$2600 \le x_{11}^4 < 2900$	$2900 \le x_{11}^5 < 3200$
	x_{12}	$400 \le x_{12}^1 < 600$	$600 \le x_{12}^2 < 900$	$900 \le x_{12}^3 < 1000$	$1000 \le x_{12}^4 < 1200$	$1200 \le x_{12}^5 < 1500$
	x_{13}	$2500 \le x_{13}^1 < 3000$	$3000 \le x_{13}^2 < 3500$	$3500 \le x_{13}^3 < 4500$	$4500 \le x_{13}^4 < 5500$	$5500 \le x_{13}^5 < 6500$
	x_{14}	$0.05 \le x_{14}^1 < 0.20$	$0.20 \le x_{14}^2 < 0.40$	$0.40 \le x_{14}^3 < 0.60$	$0.60 \le x_{14}^4 < 0.80$	$0.80 \le x_{14}^5 < 0.95$
	x_{15}	$0.5 \le x_{15}^1 < 5$	$5 \le x_{15}^2 < 15$	$15 \le x_{15}^3 < 20$	$20 \le x_{15}^4 < 30$	$30 \le x_{15}^5 < 35$
	x_{16}	$5 \le x_{16}^1 < 10$	$10 \le x_{16}^2 < 15$	$15 \le x_{16}^3 < 20$	$20 \le x_{16}^4 < 25$	$25 \le x_{16}^5 < 40$
	x_{17}	$35 \le x_{17}^1 < 50$	$20 \le x_{17}^2 < 35$	$10 \le x_{17}^3 < 20$	$5 \le x_{17}^4 < 10$	$0.5 \le x_{17}^5 < 5$
Degree of human impacts X_2	x_{21}	$70 \le x_{21}^1 < 75$	$65 \le x_{21}^2 < 70$	$50 \le x_{21}^3 < 65$	$45 \le x_{21}^4 < 50$	$40 \le x_{21}^5 < 45$
	x_{22}	$0.03 \le x_{22}^1 < 0.05$	$0.05 \le x_{22}^2 < 0.075$	$0.075 \le x_{22}^3 < 0.15$	$0.15 \le x_{22}^4 < 0.25$	$0.25 \le x_{22}^5 < 0.5$
	x_{23}	$85 \le x_{23}^1 < 95$	$75 \le x_{23}^2 < 85$	$60 \le x_{23}^3 < 75$	$50 \le x_{23}^4 < 60$	$40 \le x_{23}^5 < 50$
	x_{24}	$30 \le x_{24}^1 < 40$	$40 \le x_{24}^2 < 50$	$50 \le x_{24}^3 < 65$	$65 \le x_{24}^4 < 80$	$80 \le x_{24}^5 < 98$
	x_{25}	$750 \le x_{25}^1 < 1250$	$550 \le x_{25}^2 < 750$	$450 \le x_{25}^3 < 550$	$350 \le x_{25}^4 < 450$	$300 \le x_{25}^5 < 350$
	x_{26}	$30 \le x_{26}^1 < 45$	$20 \le x_{26}^2 < 30$	$10 \le x_{26}^3 < 20$	$5 \le x_{26}^4 < 10$	$2.5 \le x_{26}^5 < 5$
	x_{27}	$0.5 \le x_{27}^1 < 1$	$1 \le x_{27}^2 < 1.5$	$1.5 \le x_{27}^3 < 2$	$2 \le x_{27}^4 < 2.5$	$2.5 \le x_{27}^5 < 3.5$
State of farm environment X_3	x_{31}	$25 \le x_{31}^1 < 35$	$20 \le x_{31}^2 < 25$	$15 \le x_{31}^3 < 20$	$10 \le x_{31}^4 < 15$	$5 \le x_{31}^5 < 10$
	x_{32}	$28 \le x_{32}^1 < 32$	$24 \le x_{32}^2 < 28$	$18 \le x_{32}^3 < 24$	$14 \le x_{32}^4 < 18$	$10 \le x_{32}^5 < 14$
	x_{33}	$28 \le x_{33}^1 < 32$	$24 \le x_{33}^2 < 28$	$18 \le x_{33}^3 < 24$	$14 \le x_{33}^4 < 18$	$10 \le x_{33}^5 < 14$
	x_{34}	$16 \le x_{34}^1 < 18$	$14 \le x_{34}^2 < 16$	$10 \le x_{34}^3 < 14$	$8 \le x_{34}^4 < 10$	$6 \le x_{34}^5 < 8$
	x_{35}	$32 \le x_{35}^1 < 36$	$26 \le x_{35}^2 < 32$	$20 \le x_{35}^3 < 26$	$16 \le x_{35}^4 < 20$	$12 \le x_{35}^5 < 16$
	x_{36}	$20 \le x_{36}^1 < 22$	$17 \le x_{36}^2 < 20$	$13 \le x_{36}^3 < 17$	$10 \le x_{36}^4 < 13$	$8 \le x_{36}^5 < 10$

Step 3: Expanded values of the fields of the evaluation criteria. Considering the specifics of the geographic region of our study, we expanded the fields of the evaluation criteria. Table 3.5 provides the expanded values and the actually observed values.

Table 3.5 Expanded and actually observed values of evaluation criteria

Labels	x_{11}	x_{12}	x_{13}	x_{14}	x_{15}	x_{16}	x_{17}	x_{21}	x_{22}	x_{23}
x_{ij}^0	750	350	2450	0.02	0.2	2	0.2	35	0.02	35
x_{ij}^7	3400	1600	7000	0.98	40	50	60	80	0.75	98
Actual values	2130.2	1054.0	4916.6	0.73	1.03	27.45	2.21	50.24	0.061	73.2

Labels	x_{24}	x_{25}	x_{26}	x_{27}	x_{31}	x_{32}	x_{33}	x_{34}	x_{35}	x_{36}
x_{ij}^0	25	275	2	0.2	3	8	8	5	10	6
x_{ij}^7	99	1500	55	4	45	34	34	19	38	24
Actual values	96.5	461.62	25.99	2.9	18.3	13	19	7	17	10

Step 4: Determine the whitenization weight clustering coefficients of the criteria. After computing λ_{ij}^k using $\lambda_{ij}^k = \frac{1}{2}(x_{ij}^k + x_{ij}^{k+1})$ and substituting all relevant values into equ. (3.14), we obtain all the whitenization weight clustering coefficients of the criteria, as shown in Table 3.6.

Table 3.6 Whitenization weight clustering coefficients of the criteria

Labels	x_{11}	x_{12}	x_{13}	x_{14}	x_{15}	x_{16}	x_{17}	x_{21}	x_{22}	x_{23}
x_{ij}^1	0	0	0	0	0.326	0	0	0	0.400	0
x_{ij}^2	0.522	0	0	0	0.056	0	0	0	0.954	0.660
x_{ij}^3	0.811	0.584	0.083	0.233	0	0	0	0.400	0.176	0.674
x_{ij}^4	0.174	0.770	0.905	0.880	0	0.717	0.240	0.800	0	0.090
x_{ij}^5	0	0.154	0.333	0.473	0	0.596	0.790	0	0	0

Labels	x_{24}	x_{25}	x_{26}	x_{27}	x_{31}	x_{32}	x_{33}	x_{34}	x_{35}	x_{36}
x_{ij}^1	0	0	0.342	0	0	0	0	0	0	0

Table 3.6 (*continued*)

x_{ij}^2	0	0.059	0.951	0	0.440	0	0.125	0	0	0
x_{ij}^3	0	0.744	0.267	0	0.890	0	0.714	0	0.143	0
x_{ij}^4	0.059	0.589	0	0.480	0.230	0.830	0.625	1	0.833	0.860
x_{ij}^5	0.250	0	0	0.900	0	0.500	0	0.670	0.500	0.500

Step 5: Analysis of the results. From equ. (3.15), we can compute the comprehensive clustering coefficients, see Table 3.7, of the environment quality of the specific regional agricultural ecology for the criteria of the different levels.

Table 3.7 Criteria levels and the regional clustering coefficients

	σ_1	σ_2	σ_3	σ_4	σ_5
x_1	0.068	0.035	0.109	0.426	0.445
x_2	0.118	0.348	0.296	0.339	0.282
x_3	0	0.15	0.396	0.636	0.291
x	0.046	0.19	0.319	0.506	0.313

Based on $\max_{1\le k\le s}\{\sigma_i^k\}=\sigma_i^{k^*}$, it can be seen from Table 3.7 that for the time of our investigation, in terms of the overall environmental quality of the regional agricultural ecology, it is classified as "good." In the area of natural ecological background, the region belongs to the grey class of "very good." In terms of the degree of human impacts, the region belongs to the grey class of "poor." In terms of the state of farm environment, this region belongs to the grey class of "good." These conclusions indicate that the relationship between human activities and the ecological environment should be a main focus of attention. In particular, the community leaders should consider how to control the negative consequences of human activities while actively pursuing those endeavors that produce positive and environmental friendly impacts.

3.5.2 Evaluation Model Using Center-Point Triangular Whitenization Functions

When deciding on the grey classes, the point within a grey class with the maximum degree of greyness is referred to as the center of the class. The specific steps of employing the grey evaluation model based on center-point triangular whitenization weight functions are given below:

Step 1: Based on the number s of grey classes required by the evaluation task, respectively determine the grey classes 1, 2, ..., s 's centers $\lambda_1, \lambda_2, \cdots, \lambda_s$, which stand for the particular points for the observed values to belong to particular grey classes; they could be either their individual intervals centers or not). Also, correspondingly partition the field of each criterion into s grey classes, which are represented respectively by using their centers $\lambda_1, \lambda_2, \cdots, \lambda_s$.

Step 2: Expand the grey classes in two different directions by adding a 0 and $(s+1)$ grey classes with their centers λ_0 and λ_{s+1} determined. So, we have a new sequences of centers: $\lambda_0, \lambda_1, \lambda_2, \cdots, \lambda_s, \lambda_{s+1}$. By respectively connecting the point $(\lambda_k, 1)$ with the center $(\lambda_{k-1}, 0)$ and $(\lambda_{k+1}, 0)$ of the $(k-1)$th and the $(k+1)$th small grey classes, we obtain the triangular whitenization weight function $f_j^k(\cdot)$ for the kth grey class of the jth criterion, $j = 1, 2, \cdots, m$, $k = 1, 2, \cdots, s$, see Figure 3.9 for more details.

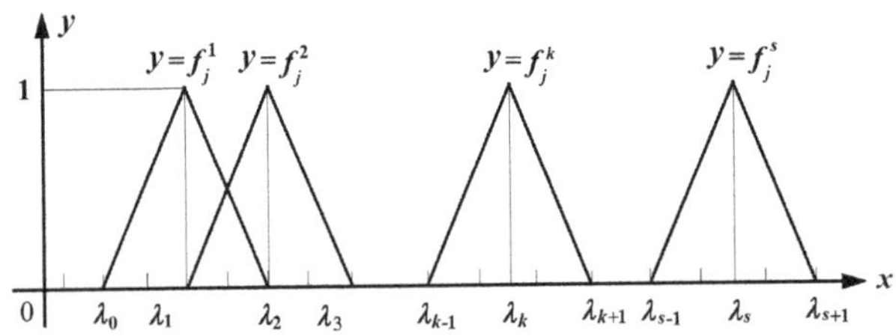

Fig. 3.9 Center-point whitenization weight functions

For an observed value x of criterion j, we can employ the formula

$$f_j^k(x) = \begin{cases} 0, & x \notin [\lambda_{k-1}, \lambda_{k+1}] \\ \dfrac{x - \lambda_{k-1}}{\lambda_k - \lambda_{k-1}}, & x \in (\lambda_{k-1}, \lambda_k] \\ \dfrac{\lambda_{k+1} - x}{\lambda_{k+1} - \lambda_k}, & x \in (\lambda_k, \lambda_{k+1}) \end{cases} \tag{3.16}$$

to compute its degree of membership $f_j^k(x)$ in grey class k, $k = 1, 2, \cdots, s$.

Step 3: Compute the comprehensive clustering coefficient σ_i^k for object i, $i = 1, 2, \cdots, n$, with respect to grey class k, $k = 1, 2, \cdots, s$:

$$\sigma_i^k = \sum_{j=1}^{m} f_j^k(x_{ij}) \cdot \eta_j \tag{3.17}$$

where $f_j^k(x_{ij})$ is the whitenization weight function of the kth subclass of the jth criterion, and η_j the weight of criterion j in the comprehensive clustering.

Step 4: From $\max_{1 \le k \le s}\{\sigma_i^k\} = \sigma_i^{k^*}$, it is decided that object i belongs to grey class k^*. When there are several objects in grey class k^*, these objects can be ordered according to the magnitudes of their comprehensive clustering coefficients.

3.5.3 Comparison between Evaluation Models of Triangular Whitenization Functions

Theorem 3.11. The endpoint triangular whitenization weight functions constructed according to the steps outlined in Subsection 3.5.1 experience multiple overlaps of more than two grey classes.

Proof. Assume that in the clustering using endpoint triangular whitenization weight functions, the intersection of the grey class $k-1$ and grey class k is D_1, and the intersection of grey class k and grey class $k+1$ is D_2. From how the endpoint triangular whitenization functions are constructed, the fields of the grey classes $k-1$, k, and $k+1$ are respectively: $[a_{k-2}, a_{k+1}], [a_{k-1}, a_{k+2}], [a_k, a_{k+3}]$. So, we have

$$D_1 = [a_{k-2}, a_{k+1}] \cap [a_{k-1}, a_{k+2}] = [a_{k-1}, a_{k+1}]$$
$$D_2 = [a_{k-1}, a_{k+2}] \cap [a_k, a_{k+3}] = [a_k, a_{k+2}]$$

That is, $D_1 \cap D_2 = [a_k, a_{k+1}] \neq \varnothing$. QED.

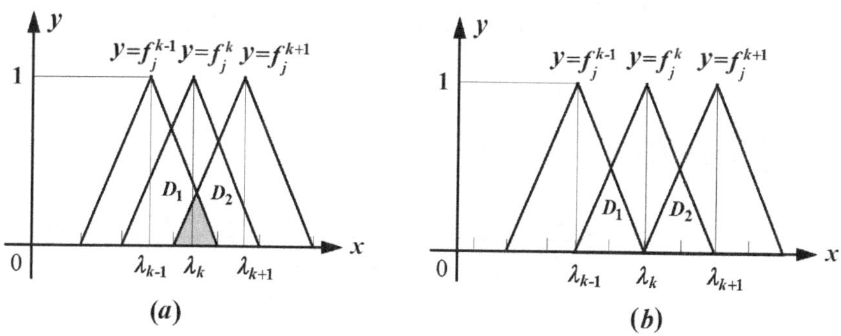

Fig. 3.10 Intersections of grey classes of triangular whitenization weight functions

Theorem 3.12. For the center-point triangular weight functions constructed according to the steps outlined in Subsection 3.5.2, there is not the phenomenon of multiple overlaps.

Proof. Assume that in the clustering using center-point triangular whitenization weight functions the intersection of grey class $k-1$ and grey class k is D_1, and the intersection of grey class k and grey class $k+1$ is D_2. According to the construction of the center-point triangular whitenization weight functions, the value fields of the grey classes $k-1$, k, and $k+1$ are respectively $[\lambda_{k-2},\lambda_k],[\lambda_{k-1},\lambda_{k+1}],[\lambda_k,\lambda_{k+2}]$. Therefore,

$$D_1=[\lambda_{k-2},\lambda_k]\cap[\lambda_{k-1},\lambda_{k+1}]=[\lambda_{k-1},\lambda_k]$$
$$D_2=[\lambda_{k-1},\lambda_{k+1}]\cap[\lambda_k,\lambda_{k+2}]=[\lambda_k,\lambda_{k+1}]$$

So, it is obvious that $D_1\cap D_2=\varnothing$. QED.

The intersections of the grey classes of the endpoint and center-point triangular whitenization weight functions are shown in Figure 3.10. By combining the results of Theorems 3.11 and 3.12, it can be seen that the grey evaluation models based on center-point triangular whitenization weight functions are more superior than those based on end-point triangular whitenization weight functions.

Theorem 3.13. For any evaluation model based on end-point triangular whitenization functions, let x be the observed value of criterion *j*, when $x\in[\lambda_{k-1},\lambda_k]$, $f_j^1(x)+f_j^2(x)+\cdots+f_j^s(x)\neq 1$.

Proof. For convenience without losing the generality, let us look at the case of $s=2$. For the endpoint triangular whitenization weight functions as shown in Subsection 3.5.1, assume $x\in[\lambda_{k-1},\lambda_k]$. Then we have

$$f_j^{k-1}(x)=\frac{a_{k+1}-x}{a_{k+1}-\lambda_{k-1}},\quad f_j^k(x)=\frac{x-a_{k-1}}{\lambda_k-a_{k-1}}$$

therefore,

$$f_j^{k-1}(x)+f_j^k(x)=\frac{a_{k+1}-x}{a_{k+1}-\lambda_{k-1}}+\frac{x-a_{k-1}}{\lambda_k-a_{k-1}}$$
$$>\frac{a_{k+1}-x}{\lambda_k-\lambda_{k-1}}+\frac{x-a_{k-1}}{\lambda_k-\lambda_{k-1}}>\frac{\lambda_k-x}{\lambda_k-\lambda_{k-1}}+\frac{x-\lambda_{k-1}}{\lambda_k-\lambda_{k-1}}=1$$

QED.

Theorem 3.14. For any grey evaluation model based on center-point triangular whitenization weight functions, let x be the observed value of criterion *j*, then when $x \in [\lambda_{k-1}, \lambda_k]$, $f_j^1(x) + f_j^2(x) + \cdots + f_j^s(x) = 1$.

Proof. Let us prove this result for the case of $s = 2$. For the center-point triangular whitenization weight functions as given in Subsection 3.5.2, assume $x \in [\lambda_{k-1}, \lambda_k]$. Then, we have

$$f_j^{k-1}(x) = \frac{\lambda_k - x}{\lambda_k - \lambda_{k-1}},\ f_j^k(x) = \frac{x - \lambda_{k-1}}{\lambda_k - \lambda_{k-1}}$$

so that

$$f_j^{k-1}(x) + f_j^k(x) = \frac{\lambda_k - x}{\lambda_k - \lambda_{k-1}} + \frac{x - \lambda_{k-1}}{\lambda_k - \lambda_{k-1}} = 1 .\ \text{QED.}$$

Theorems 3.13 and 3.14 imply that the grey cluster evaluation models based on center-point triangular whitenization weight functions satisfy the property of normality, while those based on endpoint triangular whitenization weight functions do not. Additionally, for the endpoint triangular whitenization weight functions, there is not any formal procedure for choosing their endpoints $a_0, a_1, a_2, \cdots, a_s, a_{s+1}, a_{s+2}$; and insufficient scientific basis exists for the determination of the grey classes. On the other hand, for center-point triangular whitenization weight functions, the center point of grey class *k* can be determined by simply using λ_k that represents the most likely point to belong to class *k*. Then, based on the points $\lambda_0, \lambda_1, \lambda_2, \cdots, \lambda_s, \lambda_{s+1}$, it is relatively an easy task to construct the triangular whitenization weight functions for all the grey classes.

Example 3.4. In this case study, we evaluate the achievements and effectiveness of establishing disciplinary, educational programs of a certain university, where the name of the university is omitted for the purpose of privacy.

Developing quality programs is one of the most fundamental constructions universities heavily invest in. Differences in quality between institutions of higher learning are essentially those of characteristics and levels of sophistication in their educational programs. Evaluations of educational programs offered by universities can provide decision-makers with more accurate information on the effectiveness of the already implemented program constructions and the achievements and the levels of development of various disciplinary programs offered by various colleges. So, successful exercises can be repeated in other program developments and existing problems can be addressed accordingly so that the university resources can be optimized while the efficacy of the university administration is manifested.

On the basis of referencing the available literature on evaluating educational programs, large-scale surveys, and in-depth discussions, an initial system of criteria was established for evaluating the achievements and effectiveness of disciplinary constructions. On the basis of this initial system, survey questionnaires were developed and the eventual survey was conducted and administrated through the university office of program development. This specific survey collected data from university administrators, disciplinary leaders, and other levels of administration and teaching professionals of major universities from across China. A total of 244 questionnaires were distributed with 234 retrieved. After removing incomplete and non-usable questionnaires, we obtained 226 valid questionnaires, representing a 92.6% of return rate. Among those valid questionnaires, 65% were from professors (including three who were also administrators), 1% from research professors, 12% from associate professors, 1% from research associate professors, 20% from lecturers, and 1% from assistant professors.

Based on the valid survey data, the research task force further adjusted and eventually finalized the system of criteria for evaluating the achievements and effectiveness of educational programs. This finalized system contained 6 first level criteria and 57 second level criteria. To make this presentation not too tedious, we will only select three of the items to show our comprehensive clustering evaluation using the results of the first level criteria.

The six first level criteria used for our evaluation of the achievements and effectiveness of educational programs include: teaching faculty, scholarly works, production of talents, construction of disciplinary platform, structure of working environment, and scholarly exchange, which are respectively denoted by using x_1, x_2, x_3, x_4, x_5, and x_6. The respective weights of criteria are $\eta_1 = 0.21$, $\eta_2 = 0.24$, $\eta_3 = 0.23$, $\eta_4 = 0.14$, $\eta_5 = 0.1$, and $\eta_6 = 0.08$. For the convenience of our analysis, we first converted all the scores of the criteria into percentage points. Then, the survey results of the three items of each of the six first level criteria are provided in Table 3.8.

Table 3.8 The actual observations of three educational programs in a certain university

Program #	Faculty	Scholarly works	Talent Production	Disciplinary platform	Working environment	Scholarly exchange
1	83	89	93	78	74	63
2	79	80	82	66	72	47
3	83	40	51	56	70	45

According to the given evaluation requirements, we divided the comprehensive evaluation results into four grey classes: excellent, good, fine, and poor. Because the scores of all the educational programs of the said university were within the range of 40 to 95, we took the interval [40,95] as the uniform field for each criterion to take value from. Then, we respectively employed the grey evaluation models based on endpoint and center-point triangular whitenization weight functions to conduct our comprehensive evaluation of the university's achievements and effectiveness of its programs and made our comparison of these evaluation models.

(1) Let us first do our comprehensive evaluation using the grey evaluation model based on end-point triangular whitenization weight functions.

Step 1: According to the four grey classes as required by the evaluation work, let us also partition the value field of each criterion into 4 grey classes. Since all the six first level criteria had been converted to percentage points with their observations falling within the grey interval [40,95], let us partition this interval into the following four subintervals corresponding to "excellent," "good,", "fine," and "poor": $[40,60),[60,75),[75,85),[85,95)$.

Step 2: Let us expand the given grey interval [40,95] to 30 on the left and 100 on the right so that the corresponding sequence of subintervals is $[30,40),[40,60),[60,75),[75,85),[85,95),[95,100]$. Let the whitenization weight function value for $\lambda_k=(a_k+a_{k+1})/2$ to belong to the kth grey class is 1 so that we obtained the triangular whitenization weight function $f_j^k(\cdot)$ for the kth grey class with respect to the jth criterion by connecting point $(\lambda_k,1)$ with the starting point of the $(k-1)$th grey class and the ending point of the $(k+1)$th class, $k=1,2,3,4$, $j=1,2,\ldots,6$, as follows:

$$f_j^1(x)=\begin{cases}0, & x\notin[75,100]\\ \dfrac{x-75}{15}, & x\in(75,90]\\ \dfrac{100-x}{10}, & x\in(90,100]\end{cases},\quad f_j^2(x)=\begin{cases}0, & x\notin[60,95]\\ \dfrac{x-60}{20}, & x\in(60,80]\\ \dfrac{95-x}{15}, & x\in(80,95)\end{cases}$$

$$f_j^3(x)=\begin{cases}0, & x\notin[40,85]\\ \dfrac{x-40}{27.5}, & x\in(40,67.5]\\ \dfrac{85-x}{17.5}, & x\in(67.5,85)\end{cases},\quad f_j^4(x)=\begin{cases}0, & x\notin[30,75]\\ \dfrac{x-30}{20}, & x\in(30,50]\\ \dfrac{75-x}{25}, & x\in[50,75)\end{cases}$$

Step 3: Construct the comprehensive clustering coefficient σ_i^k for program i with respect to grey class k, $i=1,2,3$, $k=1,2,3,4$. See Tables 3.9 – 3.11 for details.

Table 3.9 The comprehensive clustering coefficient σ_i^k for program 1

Grey class	x_1	x_2	x_3	x_4	x_5	x_6	x
1	0.5333	0.9333	0.8667	0. 2	0	0	0.5633
2	0.8	0.4	0.1333	0.9	0.7	0. 15	0.5027
3	0.1143	0	0	0.4	0.6286	0.8364	0.2098
4	0	0	0	0	0.04	0.48	0.0424

Table 3.10 The comprehensive clustering coefficient σ_i^k for program 2

Grey class	x_1	x_2	x_3	x_4	x_5	x_6	x
1	0.2667	0.3333	0.4667	0	0	0	0.2433
2	0.95	1	0.8667	0.3	0.6	0	0.7408
3	0.3429	0.2857	0.1714	0.9455	0.7429	0.2545	0.407
4	0	0	0	0.36	0.12	0.85	0.1304

Table 3.11 The comprehensive clustering coefficient σ_i^k for program 3

Grey class	x_1	x_2	x_3	x_4	x_5	x_6	x
1	0.5333	0	0	0	0	0	0.112
2	0.8	0	0	0	0.5	0	0.218
3	0.1143	0	0.4	0.5818	0.8571	0.1818	0.2977
4	0	0.5	0.96	0.76	0.2	0.75	0.5272

Step 4: From $\max_{1\le k\le 4}\{\sigma_1^k\}=\sigma_1^1=$ 0.5633, $\max_{1\le k\le 4}\{\sigma_2^k\}=\sigma_2^2=$ 0.7408, and $\max_{1\le k\le 4}\{\sigma_3^k\}=\sigma_3^4=$ 0.5272, it follows that the evaluation outcome for program 1 is "excellent," that for program 2 is "good," and that for program 3 is "poor." Because σ_1^2 = 0.5027 is quite close to σ_1^1 = 0.5633, it can be further concluded that the achievements and effectiveness of program 1 is really somewhere between "excellent" and "good" instead of a satisfactory "excellent."

(2) Secondly, let us repeat our comprehensive evaluation by using the grey evaluation model based on center-point triangular whitenization weight functions.

Step 1: For the required four grey classes "excellent," "good," "fine," and "poor," we first determine the center points as follows: $\lambda_1=90, \lambda_2=80, \lambda_3=70, \lambda_4=50$.

Step 2: Expand the required grey classes by adding a class named "extra-excellent" and another class named "very poor." Assume that the centers of these additional classes are $\lambda_0 = 100, \lambda_5 = 30$ so that we obtain a new sequences of class centers: $\lambda_0 = 100, \lambda_1 = 90, \lambda_2 = 80, \lambda_3 = 70, \lambda_4 = 50, \lambda_5 = 30$. By connecting point $(\lambda_k, 1)$ and the center points $(\lambda_{k-1}, 0)$ and $(\lambda_{k+1}, 0)$ of the $(k-1)$th grey class and the $(k+1)$th grey class, we produce the triangular whitenization weight function $f_j^k(\cdot)$ for the kth grey class with respect to the jth criterion, $i = 1,2,3,4$, $k = 1,2,3,\ldots,6$, as follows:

$$f_j^1(x) = \begin{cases} 0, & x \notin [80,100] \\ \frac{x-80}{10}, & x \in (80,90] \\ \frac{100-x}{10}, & x \in (90,100) \end{cases}, \quad f_j^2(x) = \begin{cases} 0, & x \notin [70,90] \\ \frac{x-70}{10}, & x \in (70,80] \\ \frac{90-x}{10}, & x \in (80,90) \end{cases}$$

$$f_j^3(x) = \begin{cases} 0, & x \notin [50,80] \\ \frac{x-50}{20}, & x \in (50,70] \\ \frac{80-x}{10}, & x \in (70,80] \end{cases}, \quad f_j^4(x) = \begin{cases} 0, & x \notin [30,70] \\ \frac{x-30}{20}, & x \in (30,50] \\ \frac{70-x}{20}, & x \in (50,70) \end{cases}$$

Step 3: Compute the comprehensive clustering coefficient σ_i^k for program i, $i = 1,2,3$, $k = 1,2,3,4$. For details see Tables 3.12 – 3.14.

Table 3.12 The comprehensive clustering coefficient σ_i^k for program 1

Grey class	x_1	x_2	x_3	x_4	x_5	x_6	x
1	0.3	0.9	0.7	0	0	0	0.44
2	0.7	0.1	0	0.8	0.4	0	0.323
3	0	0	0	0.2	0.6	0.65	0.14
4	0	0	0	0	0	0.35	0.028

Table 3.13 The comprehensive clustering coefficient σ_i^k for program 2

Grey class	x_1	x_2	x_3	x_4	x_5	x_6	x
1	0	0	0.2	0	0	0	0.046
2	0.9	1	0.8	0	0.2	0	0.817
3	0.1	0	0	0.8	0.8	0	0.213
4	0	0	0	0.2	0	0.85	0.096

Table 3.14 The comprehensive clustering coefficient σ_i^k for program 3

Grey class	x_1	x_2	x_3	x_4	x_5	x_6	x
1	0.3	0	0	0	0	0	0.063
2	0.7	0	0	0	0	0	0.147
3	0	0	0.05	0.3	1	0	0.1535
4	0	0. 5	0.95	0.7	0	0.75	0.4965

Step 4: From $\max_{1\le k\le 4}\{\sigma_1^k\}=\sigma_1^1=$ 0.44, $\max_{1\le k\le 4}\{\sigma_2^k\}=\sigma_2^2=$ 0.817, and $\max_{1\le k\le 4}\{\sigma_3^k\}=\sigma_3^4=$ 0.4965, it follows that the achievements and effectiveness of program 1 are in the grey class of "excellent," the comprehensive evaluation of program 2 is "good," and that of program 3 is "poor." These conclusions are exactly the same as those obtained earlier using the grey evaluation model based on endpoint triangular whitenization weight functions.

By comparing the clustering results presented in these tables, it can be seen that the most sums of the clustering coefficients for belonging to the grey classes "excellent," "good," "fine," and "poor," of the criteria in Tables 3.9 – 3.11 are greater than 1, while quite a few of the criteria have three of their clustering coefficients great than 0. This fact shows something not quite right. On the other hand, in Tables 3.12 – 14, there is not problem with multiple overlaps. And if we also computed the clustering coefficients of the expanded grey classes, it can be verified that the clustering coefficients of each criterion satisfy $f_j^1(x)+f_j^2(x)+\cdots+f_j^s(x)=1$.

In particular, the center-point triangular whitenization weight functions use the points λ_k that most likely belong to the kth grey class as the center points so that these weight functions can be quite easily constructed using $\lambda_0,\lambda_1,\lambda_2,\cdots,\lambda_s,\lambda_{s+1}$. Based on the ordinary ways of thinking, people tend to have more assurance about the center points of grey classes than the grey classes. So, the conclusions derived based on this cognitive assurance are more scientific and more reliable.

3.6 Applications

3.6.1 *Order of Grey Incidences*

All the degrees of grey incidence studied earlier are about how closely related sequences are. For any chosen grey incidence operator, see Section 3.1 for details, the values of the absolute and relative degrees of grey incidence are uniquely determined. When the grey incidence operator and the θ-value are fixed, the synthetic degree of grey incidence is also unique. These kinds of conditional uniqueness do not affect how we analyze our problems. As a matter of fact, when analyzing a system, we investigate the relationship between the system's characteristic of interest and relevant factors. Our main focus is the ordering of the degrees of incidence between the characteristic sequence and the individual factor sequences without much attention given to the specific magnitudes of the degrees of incidence.

Let X_0 be a system's characteristic sequence, $X=\{X_1, X_2, \cdots, X_m\}$ a set of relevant factor sequences, and γ the degree of grey incidence. If $\gamma_{0i} \geq \gamma_{0j}$, we say that factor X_i is superior to factor X_j, written $X_i \succ X_j$. The order relation "$\succ$" is referred to as the *grey incidence order* derived from degrees of grey incidence. That is, the general grey incidence order contains such special cases as *grey absolute incidence order, grey relative incidence order*, and *grey synthetic incidence order*. It can be shown (Liu and Lin, 2006, p. 119 – 120) that grey incidence order, absolute incidence order, relative incidence order, and synthetic incidence order are all partial ordering on the set *X*. Furthermore, if the factor sequences in *X* all have the same length as X_0, then both grey incidence order and grey absolute incidence order are linear orderings on *X*. If the initial values of all the sequences are non-zero, then both grey relative incidence order and grey synthetic incidence order are also linear orders on *X*.

3.6.2 Preference Analysis

Assume that $Y_1, Y_2, \cdots, Y_s$ are a system's characteristic behavioral sequences and $X_1, X_2, \cdots, X_m$ relevant factor behavioral sequences with the same length. Let γ_{ij} be the degree of grey incidence between Y_i and X_j, $i=1,2,\cdots,s$, $j=1,2,\cdots,m$. Then $\Gamma=(\gamma_{ij})_{s\times m}$ is referred to as the *grey incidence matrix* of the system, where the *i*th row is made up of the degrees of grey incidence between the characteristic sequence $Y_i (i=1,2,\cdots,s)$ and each of the factor sequences $X_1, X_2, \cdots, X_m$; and the *j*th column the degrees of grey incidences between each of the characteristic sequences $Y_1, Y_2, \cdots, Y_s$ and $X_j (j=1,2,\cdots,m)$.

Similarly, we can define general grey incidence matrices, such as the grey *absolute incidence matrix* $A=(\varepsilon_{ij})_{s\times m}$, the grey *relative incidence matrix* $B=(r_{ij})_{s\times m}$, and the grey *synthetic incidence matrix* $C=(\rho_{ij})_{s\times m}$.

Now, for the grey incidence matrix $\Gamma=(\gamma_{ij})_{s\times m}$, if there are $k, i \in \{1,2,\cdots,s\}$ such that $\gamma_{kj} \geq \gamma_{ij}$, $j=1,2,\cdots,m$, then the system's characteristic Y_k is said to be more favorable than the system's characteristic Y_i, written $Y_k \succ Y_i$. If $\forall i = 1,2,\cdots,s, i \neq k$, $Y_k \succ Y_i$ always holds true, then Y_k is said to be the *most favorable characteristic*. Additionally, if there are $l, j \in \{1,2,\cdots,m\}$ such that $\gamma_{il} \geq \gamma_{ij}$, $i = 1,2,\cdots,s$, then we say that the system's factor X_l is more favorable than the factor X_j, written $X_l \succ X_j$. If $\forall j = 1,2,\cdots,m, j \neq l$, $X_l \succ X_j$ always holds true, then is said to be the *most favorable factor*.

If there are $k, i \in \{1,2,\cdots,s\}$ satisfying $\sum_{j=1}^{m} \gamma_{kj} \geq \sum_{j=1}^{m} \gamma_{ij}$, then the system's characteristic Y_k is said to be more quasi-favorable than Y_i, denoted $Y_k \succeq Y_i$. If there are $l, j \in \{1,2,\cdots,m\}$ satisfying $\sum_{i=1}^{m} \gamma_{il} \geq \sum_{i=1}^{m} \gamma_{ij}$, then the system's factor X_l is said to be more *quasi-favorable* than X_j, denoted $X_l \succeq X_j$.

If there is $k \in \{1,2,\cdots,s\}$ such that $\forall\ i = 1,2,\cdots,s, i \neq k$, $Y_k \succeq Y_i$, then the system's characteristic is said to be quasi-preferred. If there is $l \in \{1,2,\cdots,m\}$ such that $\forall\ j = 1,2,\cdots,m, j \neq l$, $X_l \succeq X_j$, then the system's factor is said to be *quasi-preferred*.

Proposition 3.5. In a system of s characteristics and m relevant factors, there may not be any most favorable characteristic and most favorable factor. However, there must be quasi-preferred characteristic and factor.

When investigating practical problems, the analyses of the three incidence orders may not provide cohesive conclusions. It is because the absolute incidence order looks at the relationship between absolute quantities, the relative incidence order focuses on the rates of change with respect to the initial values of the observed sequences, while the synthetic incidence order combines both the relationships between absolute quantities and between the rates of change. Considering the background of the problem of concern, we can choose one of the incidence orders. For the purpose of simplicity, after a particular grey incidence operator is applied on the system's characteristic behavioral sequences and relevant factor sequences, one only needs to employ the absolute incidence order on the processed data.

3.6.3 Practical Applications

Example 3.5. In this example, we look at how to apply grey incidence models in analyzing the time difference of economic indices.

In order to effectively monitor the performance of macro-economic systems and provide timely warning, there is a need to investigate the time relationship of various economic indices with respect to the economic cycles in terms of the peaks and valleys by addressing such questions as: Which indices can provide warning ahead of time? Which ones would be synchronic with the evolution of the economic systems? And which ones tend to be lag behind the economic development? In other words, we need to divide economic indices into three classes: leading indicators, synchronic indices, and stagnant representations. To this end, the method of grey incidence analysis provides an effective tool for classifying economic indices.

Through careful research and analysis, we selected the following 8 major classes and 17 criteria as indices for economic performance:

(1) The class of energy and raw materials: the total production of energy;
(2) The class of investment: the total investment in real estate;
(3) The class of production: increase in industry output, increase in light industry output, increase in heavy industry output;
(4) The class of revenue: national income, national expenditure;
(5) The class of currency and credit: currency in circulation, savings at various financial institutions, amount of loans issued by financial institutions, cash payout in the form of salary and wages, net amount of currency in circulation;
(6) The class of consumption: the gross retail amount of the society;
(7) The class of foreign trade: gross amount of imports, gross amount of exports, direct investments by foreign entities;
(8) The class of commodity prices: the consumer price index.

By applying the following standards, we classify the previous criteria into three classes: leading indicators, synchronic indices, and stagnant representations.

The standards for determining leading indicators:

(1) The indicated appearance of economic cyclic peaks needs to be at least three months ahead of the actual occurrence. Such leading relationship needs to be relatively stable with few exceptions;
(2) The indicated cycles and the historical cycles are nearly one-to-one corresponded to each other. And, for the most recent three economic cycles, at least two times the indicated cycles are ahead of the actual occurrences with at least 3 months of lead time;
(3) The economic characteristics of the indices provide relatively definite and clear leading relationships with respect to the background economic cycles.

The standards for determining both synchronic indices and stagnant representations are similar to what are outlined above with the exception that for synchronic indices, the time differences between the indicated appearances and the actual occurrences of economic cycles need to be within plus and minus 3 months, while for stagnant representations, the indicated appearances of economic cycles are behind the actual occurrences by at least 3 months.

As expected, in practical selections it is almost impossible to actually find an index that meets all the stated standards. So, what can be done is that based on the recorded reference cycles, we look for the statistical indices that meet the previously stated standards as closely as possible. Even so, the actual situation is that an identified leading indicator can also sometimes lag behind the actual economic development, while an identified stagnant representation can also provide nice lead-time in its forecast of a specific economic evolution. Similar scenarios also occur with regard to synchronic indices. However, theoretically, if among the one-to-one correspondences between an index and the actually recorded cycles over 2/3 of times the index is leading the actual occurrences, then we treat such an index as leading. Similar treatments are applied to synchronic indices and stagnant representations.

Because for Chinese economy the increase in industry output has played a very important role, as a synchronic index it has a quite high quality. So, it can be employed as the basic index in our grey incidence analysis. We will compute not only the absolute degree of grey incidence between each criterion and the increase in industry output, but also the absolute degrees of grey incidence of other 16 criteria with their data translated 1 – 12 months along the time axis either left or right. When the data are translated to the left, the months will take negative values, when translated to the right, the months will take positive values. The amount of horizontal translation is denoted by L. That is, we compute the absolute degrees of grey incidence between all the 16 individual criteria, not including that of increase in industry output, and that of the increase in industry output for $L = -12, \ldots, 12$. For each L-value, we order the obtained absolute degrees of grey incidence from the smallest to the largest with the criterion listed in the front chosen as candidate criterion for that specific L-value. For instance, when $L = 0$, the absolute degrees of grey incidence of the criteria are listed in Table 3.15.

Table 3.15 The absolute degrees of grey incidence of the criteria when $L = 0$

Index	Absolute degree of grey incidence	Index	Absolute degree of grey incidence
increase in heavy industry output	0.979810	national income	0.559540
increase in light industry output	0.972655	gross amount of exports	0.544870
gross retail amount of the society	0.862105	total production of energy	0.541044
cash payout as salaries	0.789278	net amount of currency in circulation	0.525936
currency in circulation	0.753681	loans issued by financial institutions	0.507958
total investment in real estate	0.726366	savings at financial institutions	0.505226
gross amount of imports	0.598248	consumer price index	0.500173
national expenditure	0.566914	direct investments by foreign entities	0.500002

Synchronic indices should be selected from those with larger absolute degrees of grey incidence, because the large degrees of incidence indicate that these criteria have greater similarities in comparison with that of increase in industry output, which we employ as the basic standard of the Chinese economic cycles. However, we still do not have theoretical evidence to support that an index with large absolute degree of grey incidence must be synchronic. To this end, we also need to consider whether or not the related absolute degrees of grey incidence will be even greater when $L \neq 0$. If when $L = 0$ the value of the absolute degree of grey incidence of a certain index is ranked quite in the front, and when $L = -4$, its value is even greater, it means that after this index is translated to four months earlier, it is more similar to the pattern of the increase in industry output. So, in this case, this specific index can be seen as one leading the economic cycle by as much as about four months. By using these two standards, we can not only classify indices as synchronic, leading, or stagnant, but also the specific amount of leading or staggering amount of time.

When $L = 0$, the index of "cash payout as salaries" is ranked relatively in the front. So, it is a natural candidate for being a synchronic indicator. When L–value changes, the relevant changes in its absolute degrees of grey incidence are given in Table 3.16.

Table 3.16 The absolute degrees of "cash payout as salaries" when $L \neq 0$

L	Absolute degree of grey incidence	L	Absolute degree of grey incidence
-12	0.664615	1	0.877090
-11	0.705983	2	0.867859
-10	0.733564	3	0.857366
-9	0.752740	4	0.832260
-8	0.753598	5	0.825027
-7	0.732221	6	0.806787
-6	0.723942	7	0.806782
-5	0.731232	8	0.820384
-4	0.742249	9	0.803771
-3	0.752628	10	0.806649
-2	0.770216	11	0.805679
-1	0.800838	12	0.836308
0	0.789278		

From Table 3.16, it follows that when $L = 1$, the absolute degree of grey incidence reaches its maximum. So, speaking strictly, this specific index should be seen as one that is lagging the economic cycle by as much as one month. According to the custom, leading or lagging no more than two months is seen as synchronic, while more than this range of time will it be treated as either a leading or staggering index.

As a second example, for the index of "gross retail amount of the society," the computational results are provided in Table 3.17.

Table 3.17 The absolute degrees of "gross retail amount of the society"

L	Absolute degree of grey incidence	L	Absolute degree of grey incidence
-12	0.914466	1	0.856944
-11	0.915117	2	0.866789
-10	0.918527	3	0.876758
-9	0.887243	4	0.882430
-8	0.888258	5	0.889590
-7	0.928151	6	0.895899
-6	0.948684	7	0.900899
-5	0.939351	8	0.900130
-4	0.923900	9	0.895977
-3	0.909621	10	0.894374
-2	0.884610	11	0.892662
-1	0.846814	12	0.889532
0	0.862105		

From Table 3.17, it can be seen that when $L = -6$ the absolute degree of grey incidence of the particular index reaches its maximum. Therefore, it can be seen as a leading indicator.

By using this method, we can compute the L–values corresponding to the maximum absolute degrees of grey incidence of each of the indices of our interest. The results are listed in Table 3.18.

Table 3.18 L–values corresponding to maximum absolute degrees of incidence of the indices of our interest

Index	L	Absolute degree	Index	L	Absolute degree
currency in circulation	−6	0.983452	national income	+12	0.718998
increase in heavy industry output	0	0.979810	gross amount of imports	−9	0.606556
increase in light industry output	0	0.972655	gross amount of exports	+10	0.560054
gross retail amount of the society	−6	0.948684	total production of energy	−6	0.555035
cash payout as salaries	+1	0.877090	direct investments by foreign entities	−11	0.510016
national expenditure	+12	0.800533	loans issued by financial institutions	−5	0.508375
net amount of currency in circulation	+8	0.796688	savings at financial institutions	−6	0.505588
total investment in real estate	−11	0.769778	consumer price index	+11	0.503235

Table 3.18 indicates that we can classify the 16 indices of the eight major classes into three classes as leading, synchronic, and stagnant indices as shown in Table 3.19.

Table 3.19 Classifications of leading, synchronic, and stagnant indices

	Leading index	Synchronic index	Stagnant index
Energy & raw materials	total production of energy (-6)*		
Investment	total investment in real estate (-11)		
production		increase in light industry output (0) increase in heavy industry output (0)	
Finance			national income (+12) national expenditure (+12)
Currency and credit	currency in circulation (-6) savings at financial institutions (-6) loans issued by financial institutions (-5)	cash payout as salaries (+1)	net amount of currency in circulation (+8)
Consumption	gross retail amount of the society (-6)		
Foreign trade	gross amount of imports (-9) direct investments by foreign entities (-11)		gross amount of exports (+10)
Commodity price			consumer price index (+11)

* Numbers in parentheses stand for the time difference between indicated cycles and reference cycles.

Example 3.6. In this case study, we will look at the system of evaluation criteria for dominant regional industries and an actual selection of dominant industries for Jiangsu province, PR China.

By a dominant industry, it means that during a special time period of economic development of either a country or geographic region, the particular industry plays a leading role in the regional structure of industries and dominates the overall development of the regional economy. Such an industry possesses a promising market perspective and a capability for advancing its own level of technology. Historically, each process for the regional structure of industries to rise to a higher level of sophistication has always been signaled by the development of new dominant industries and changes of dominance from one group of industrial leaders to another group. Japanese scholar Shinohara Miyohei proposed two theoretical standards for planning for Japan's industrial structure: (1) The income elasticity of the demand; and (2) the rising productivity. Economist Albert Otto Hirschman studied the problem of dominant industries from the angle of mutual connection of different industries. He believed that some complementary relationships between certain industries are stronger than those between other industries. The government locates the "strategic" or dominant industries in the economic system at places where the chain connections of input-output are much stronger and much more intimate than those at other places. So, in the development of a regional economy, individual industries play different roles and functions in the regional industrial system, in which one or several industries control the leading position. These industries constitute the leading industries and form the group of dominant industries of the region. Each development of modern regional economies in fact is represented by the development of dominant industries. The impacts of dominant industries on regional economic growth can be seen in three aspects. One is their forward looking influence. That is the impacts of the dominant industries on other industries that use the products of the dominant industries. The second is the backward looking influence. That is the impacts the dominant industries exert on those industries from which they receive their production materials. And, the third is the side influence. That is the impacts the dominant industries place on the development and prosperity of the regional economy. Through these three areas of impacts, dominant industries project their industrial superiority over onto other industries of the region so that the overall regional economy is positively affected. So, there is the need for first locating the dominant industries of a specific time period; through further developing these dominant industries, coordinated development of various economic sectors of the region can be expected.

Based on the theory of dominant industries, each regional industrial system can be decomposed into dominant industries, the serving industries that are established to provide support for the smooth development of the dominant industries, and other potential leaders existing among the other general industries.

In short, the basic idea for adjusting and optimizing regional structure of industries is to correctly select and optimize the development of the dominant industries so that they can effectively carry out the task of dividing and forming specialized economic zones across the region of their coverage. In terms of increasing the dominant industries' capability to bring along a healthy development of other regional industries, help develop the relevant industries, especially those forwardly impacted by the dominant industries, by purposefully prolonging the product chain as much as possible, by maintaining and strengthening the existing supporting economic sectors while increasing these sectors' quality, by actively developing basic economic infrastructures, while removing the "bottlenecks" that place constraints on the development of the regional economy, by supporting other general business activities so that when the current dominant industries are weakened due to the changing environment, new leading industries can emerge in a timely fashion in order to keep the regional economy in its normal cycle of development without interruption.

Because regional dominant industries determine the formation and evolution of the region's economic system, clearly identifying these industries and well understanding their relationships with other relevant industries, supporting economic sectors, and the more general economic areas are extremely important in terms of promoting health growth of the regional economy. On the other hand, the formation and evolution of regional dominant industries are influenced to a degree by the region's economic structure, the market preference, and the available resources. So, encouraging and promoting the appearance of dominant industries are both the start and the ultimate goal of introducing regional economic policies.

When selecting dominant industries, we should comply with the following: principle of resource superiority, principle of market demand-supply, principle of economic benefits, principle of related benefits, principle of industrial superiority, principle of technological advancement, principle of sustainable development, principle of sufficient employment, principle of giving priority to the development of "bottleneck" areas, etc.

The characteristics of regional dominant industries and the principles of how to select regional dominant industries provide us a clear direction on how to actually pinpoint out particular regional industries as the leaders. Considering the specific circumstances of Jiangsu Province, People's Republic of China, we choose the following as the system of evaluation criteria for determining regional dominant industries of the province.

(1) The degrees of industrial incidence

The degrees of industrial incidence contain two parts: degree of forward incidence and degree of backward incidence. Both of these concepts represent the degrees of influence, also known as the spreading effect, on the changes of the input-output of other industries as directly or indirectly affected by changes in the demand and supply of one industry. The most commonly employed are the sensitivity coefficient and the influence coefficient. Assume that b_{ij} is the (i,j)-entry in the complete consumption coefficient matrix B studied in the input-output analysis, n the number of defined industries. Then, the sensitivity coefficient μ_i of the ith industry is defined by

$$\mu_i = \frac{\sum_{j=1}^{n} b_{ij}}{\frac{1}{n}\sum_{i=1}^{n}\sum_{j=1}^{n} b_{ij}}$$

The magnitude of μ_i reflects the degree of influence of an increase in the output of each industry on the output of the ith industry. It is a degree of forward incidence. The greater the sensitivity coefficient μ_i is, the more forward incidence capability the ith industry possesses.

The influence coefficient γ_j of the jth industry is defined by

$$\gamma_j = \frac{\sum_{i=1}^{n} b_{ij}}{\frac{1}{n}\sum_{i=1}^{n}\sum_{j=1}^{n} b_{ij}}$$

The magnitude of γ_j reflects the influence of an increase in the demand of the *j*th industry on the output of other industries. It is a degree of backward incidence. The greater the influence coefficient γ_j is, the stronger capability of backward incidence the *j*th industry has. If $\gamma_j > 1$, it means that the effect of this industry is above the average among all industries. If $\gamma_j = 1$, it means that the effect of this industry is right at the mean level of all industries. If $\gamma_j < 1$, it means that the effect of this industry is below the mean level. Similar explanations can be given to μ_i.

(2) The Income Elasticity of Demand

Assume *Y* is the domestic GDP and x_i the market demand for the products of the *i*th industry. Then the income elasticity ε_i of demand of the *i*th industry is given by

$$\varepsilon_i = \frac{y}{x_i}\frac{\partial x_i}{\partial y}$$

which represents the effect of changes in the domestic GDP on the demand of the products of the *i*th industry. Generally speaking, the industries with large income elasticity of demand have strong capability of market expansion. When $\varepsilon_i < 1$, it means that the expanding rate of demand of the *i*th industry is smaller than the rate of increase of the national GDP. When $\varepsilon_i > 1$, it means that the expanding rate of demand of the *i*th industry is greater than the rate of increase of the national GDP. When ε_i is equal to or close to 1, it implies that the expanding rate of demand of the *i*th industry is about the national average of increase. A regional dominant industry should be one with relatively large income elasticity of demand.

(3) The Growth

Assume that x_i^0 is the initial demand for input of the *i*th industry and r_i the average rate of increase. Then the demand for input of the *i*th industry is given by

$$x_i^t = x_i^0 (1 + r_i)^t$$

The greater the r_i value is, the faster the *i*th industry grows. and the more important position and role it occupies and plays in the regional economic system.

(4) The Employment Rate

Because current China experiences a huge amount of surplus of labor, the problem of employment has become the number one most important matter of all levels of the government. In the governmental effort of adjusting the industrial structure, how to create a scenario of sufficient employment has been one of the main contents of the adjustment. To this end, let us employ the integrated employment coefficient to evaluate the absorption capability of each industry. Assume that v_i stands for the wage of labor of the ith industry, q_i the total output of the ith industry, and c_{ij} the Wassily Leontief inverse matrix coefficient as listed in the input-output table. Then the integrated employment coefficient β_i of the ith industry is given as follows:

$$\beta_i = \sum_{j=1}^{n} \frac{v_i}{q_i} c_{ij}$$

The magnitude of β_i reflects the amount of labor needed to produce 1 unit of the final product directly and indirectly by the ith industry and its relevant industries.

In the rest of this case study, we look at the actual selection of dominant industries in Jiangsu Province, the People's Republic of China. Based on the input-output table of Jiangsu Province of 1997 and the yearly statistical almanacs, we classified all industries in Jiangshu Province into three classes: dominant, supporting, and general industries, respectively, using the grey evaluation method of fixed weights, where

(1) Based on the 1997 input-output table, we calculated the sensitivity coefficient, influence coefficient, and integrated employment coefficient, for each industry;
(2) By using the annual statistical almanacs, we computed the income elasticity of demand and rates of growths.

Because the industries listed in the annual almanacs are different of those used in the input-output table, we combined the 40 industries as shown in the input-output table into 34 industries. The specific data values of the criteria are provided in Table 3.20.

According to the grey fixed weight clustering model and the observed values of the criteria, let the whitenization weight function of the kth subclass of the jth criterion be $f_j^k(*)$. Through using Delphi postal mail survey, we determined the types of whitenization weight functions and the weights of the individual criteria in the comprehensive evaluation. In particular, the whitenization weight functions

Table 3.20 Computational results of all the criteria for the 34 industries

Name	Sensitivity coefficient μ_i	Influence coefficient γ_i	Income elasticity of demand ε_i	Growth rate r_i	Overall employment rate β_i
Agriculture	0.9633	0.6756	0.4488	1.8615	0.7286
Coal mining	0.8889	0.6876	-0.10524	-1.13421	0.5936
Extraction of petroleum & natural gas	0.8199	0.3589	2.173455	17.73033	0.3528
Metal ore mining	0.2161	0.7366	-0.05618	-0.60063	0.6717
Non-metal ore mining	0.6359	0.7649	-0.30407	-3.39012	0.4226
Food and tobacco processing	0.8249	1.0321	0.178727	1.842096	0.5096
Textile	1.2055	1.2583	0.385003	3.850923	0.3982
Clothing, leather, down, & fabrics	0.2598	1.1266	0.474869	4.690773	0.6244
Lumber processing, furniture making	0.2896	1.0386	0.961863	8.923825	0.4542
Paper making, printing, office supply	0.9162	1.1368	1.099007	10.0312	0.4746
Petroleum refinery, coking	1.2970	0.9357	1.239179	11.1295	0.3630
Chemicals	3.3553	1.0795	1.724493	14.70029	0.3834
Manufactures using non-metal minerals	1.1903	1.0556	1.220044	10.98148	0.4596
Metal smelting, rolling processing	4.7383	1.4733	0.142253	1.474251	0.4607
Metal products	0.8404	1.3129	1.875618	15.74695	0.4002
Machinery	1.2090	1.2996	0.867199	8.139289	0.4296
Transportation equipment	0.6622	1.5949	0.70133	6.721973	0.4370
Electric machinery, equipment	0.8473	1.4695	0.822944	7.766583	0.4154
Electronic, communication equipment	1.1501	1.2383	1.078045	9.864103	0.3957
Instruments, meters, & office appliances	0.1256	1.2873	3.107475	23.37564	0.4362
Other manufacturing	1.1311	1.0108	1.921372	16.0583	0.5896
Supply of Electric, steam hot water	1.6183	0.7644	1.323333	11.77354	0.4471
Production, supply of coal gas	0.0946	1.4203	2.832106	21.79184	0.6892
Production, supply of tap water	0.1208	0.7659	-0.0766	-0.82168	0.5050
Construction	0.3474	1.1201	1.621324	13.96896	0.5311
Transportation, postal service	1.8850	0.6446	1.75804	14.9351	0.4722
Food services	3.1201	0.7710	0.935783	8.709375	0.5482
Financial, insurance	1.1328	0.5166	0.468103	4.628241	0.2747
Real estate	0.2238	0.9941	1.606129	13.86004	0.4061
Societal Services	1.1264	0.7829	2.027072	16.76815	0.5008
Health, sports, welfare	0.2152	1.2420	1.624586	13.99229	0.5205
Education, arts, radio, TV, movie	0.1110	0.7258	2.370946	18.99219	0.7162
Service to scientific research, technology	0.1969	0.8553	1.451781	12.73562	0.5796
Government & others	0.2412	0.9139	1.72827	14.7268	0.5478

of the sensitivity coefficient, influence coefficient, income elasticity coefficient, growth rate, and integrated employment coefficient are respectively given below:

$$f_1^1[0.9,1.3,—],\ f_1^2[0.6,0.9,-,1.2],\ f_1^3[-,-,0.5,1];$$
$$f_2^1[0.9,1.3,-,-],\ f_2^2[0.5,0.9,-,1.3],\ f_2^3[-,-,0.4,0.9];$$
$$f_3^1[1,1.6,-,-],\ f_3^2[0.5,1-,1.5],\ f_3^3[-,-,0.4,1];$$
$$f_4^1[6,12,-,-],\ f_4^2[3,7,-,11],\ f_4^3[-,-,4,10];$$
$$f_5^1[0.3,0.5,-,-],\ f_5^2[0.25,0.45,-,0.7],\ f_5^3[-,-0.25,0.4].$$

The corresponding weights for these criteria are respectively: 0.14, 0.14, 0.24, 0.24, and 0.24. The computed comprehensive evaluation results of all the industries are given in Table 3.21.

Table 3.21 The comprehensive evaluation results of all the 34 industries

code	Name	Dominant	Supporting	General	MAX	Clustering results
1	Agriculture	0.2621	0.1719	0.5335	0.5335	General
2	Coal mining	0.24	0.3026	0.5705	0.5705	General
3	Extraction of petroleum & natural gas	0.5433	0.2259	0.2659	0.5433	Dominant
4	Metal ore mining	0.24	0.1099	0.6657	0.6657	General
5	Non-metal ore mining	0.1471	0.3165	0.6197	0.6197	General
6	Food and tobacco processing	0.2862	0.3815	0.529	0.529	General
7	Textile	0.3501	0.2434	0.4828	0.4828	General
8	Clothing, leather, down, & fabrics	0.3193	0.2347	0.5624	0.5624	General
9	Lumber processing, furniture making	0.3505	0.6737	0.1983	0.6737	Supporting
10	Paper making, printing, office supply	0.4989	0.6565	0.0234	0.6565	Supporting
11	Petroleum refinery, coking	0.5278	0.3882	0.0592	0.5278	Dominant
12	Chemicals	0.7829	0.2372	0.0265	0.7829	Dominant
13	Manufactures using non-metal minerals	0.6348	0.4563	0	0.6348	Dominant
14	Metal smelting, rolling processing	0.4728	0.2297	0.48	0.4728	Dominant
15	Metal products	0.7402	0.2924	0.0446	0.7402	Dominant
16	Machinery	0.4891	0.5635	0.1275	0.5635	Supporting
7	Transportation equipment	0.3332	0.5733	0.3451	0.5733	Supporting
18	Electric machinery, equipment	0.34914	0.6629	0.2029	0.6629	Supporting
19	Electronic, communication equipment	0.5065	0.4904	0.012	0.5065	Dominant
20	Instruments, meters, & office appliances	0.7789	0.2278	0.14	0.7789	Dominant
21	Other manufacturing	0.8396	0.2393	0	0.8396	Dominant
22	Supply of Electric, steam hot water	0.6767	0.4138	0.0379	0.6767	Dominant

Table 3.21 (*continued*)

23	Production, supply of coal gas	0.86	0.001	0.14	0.86	Dominant
24	Production, supply of tap water	0.24	0.2802	0.6575	0.6575	General
25	Construction	0.797	0.2251	0.14	0.797	Dominant
26	Transportation, postal service	0.8266	0.2692	0.0715	0.8266	Dominant
27	Food services	0.4883	0.5871	0.1134	0.5871	Supporting
28	Financial, insurance	0.0814	0.1645	0.7354	0.7354	General
29	Real estate	0.6402	0.2943	0.14	0.6402	Dominant
30	Societal Services	0.7992	0.3245	0.0327	0.7992	Dominant
31	Health, sports, welfare	0.8397	0.1926	0.14	0.8397	Dominant
32	Education, arts, radio, TV, movie	0.72	0.079	0.1887	0.72	Dominant
33	Service to scientific research, technology	0.6607	0.263	0.1525	0.6607	Dominant
34	Government & others	0.7248	0.2812	0.14	0.7248	Dominant

By analyzing the previous comprehensive evaluation results, we can obtain the following conclusions: Firstly, the industries that can be possibly considered as dominant include: Extraction of petroleum & natural gas, Petroleum refinery, coking, Chemicals, Manufactures using non-metal minerals, Metal smelting, rolling processing, Metal products, Electronic & communication equipment, Instruments, meters, & office appliances, Other manufacturing, Supply of Electric & steam hot water, Production & supply of coal gas, Construction, Transportation & postal service, Real estate, Societal Services, Health, sports, and welfare, Education, arts, radio, TV, and movie, Service to scientific research & technology, and Government & others.

Secondly, the industries that can be possibly considered as supporting include: Lumber processing & furniture making, Paper making, printing & office supply, Machinery, Transportation equipment, Electric machinery and equipment, and Food services.

Thirdly, the class of general industries include: Agriculture, Coal mining, Metal ore mining, Non-metal ore mining, Food and tobacco processing, Textile, Clothing, leather, down, & fabrics, and Financial & insurance.

One of the main tasks facing the economic development of Jiangsu Province is to materialize the desired economic growth through a transformation from the current extensive type to a more concentrated and constrained type, and through an elevation of the current industrial structure. One key of success is to correctly select and actively promote the development of dominant industries. By taking advantage of the relative strong capability to attract and attain advanced technologies of these industries, their high degrees of inter-connections, and superior ability of penetration, the overall consumptions of material and energy resources can be reduced while product qualities and attached commercial values are increased. Therefore, elevations of the entire industrial structure can appear in a sustainable fashion while creating great employment opportunities in large numbers. Each elevation of the industrial structure also represents a process of replacing the obsolete leading industries with new ones. The existence of dominant industries is variable; it changes and evolves along with development of the regional economy. When some old branded leaders are replaced by some newcomers, the existing industrial structure also changes accordingly. Therefore, at the same time when dominant industries are fostered and promoted, the government officers have to have an eye on the overall planning, correctly deal with the development relationships between various industries, and maintain a coordinated development of the regional economy.

For additional examples of practical applications, please consult with (Liu and Lin, 2006, p. 132 – 139; 174 – 190).

Chapter 4
Grey Systems Modeling

4.1 The GM(1,1) Model

4.1.1 The Basic Form of GM(1,1) Model

Let $X^{(0)} = (x^{(0)}(1), x^{(0)}(2), \cdots, x^{(0)}(n))$ be a sequence of raw data. Denote its accumulation generated sequence by $X^{(1)} = (x^{(1)}(1), x^{(1)}(2), \cdots, x^{(1)}(n))$. Then

$$x^{(0)}(k) + ax^{(1)}(k) = b \tag{4.1}$$

is referred to as the *original form* of the GM(1,1) model, where the symbol GM(1,1) stands for "first order grey model in one variable."

Let $Z^{(1)} = (z^{(1)}(2), z^{(1)}(3), \cdots, z^{(1)}(n))$ be the sequence generated from $X^{(1)}$ by adjacent neighbor means. That is, $z^{(1)}(k) = \frac{1}{2}(x^{(1)}(k) + x^{(1)}(k-1))$, $k = 2,3,...,n$. Then,

$$x^{(0)}(k) + az^{(1)}(k) = b \tag{4.2}$$

is referred to as the *basic form* of the GM(1,1) model.

Theorem 4.1. Let $X^{(0)}$, $X^{(1)}$, and $Z^{(1)}$ be the same as above except that $X^{(0)}$ is non-negative. If $\hat{a} = (a,b)^T$ is a sequence of parameters, and

$$Y = \begin{bmatrix} x^{(0)}(2) \\ x^{(0)}(3) \\ \vdots \\ x^{(0)}(n) \end{bmatrix}, \quad B = \begin{bmatrix} -z^{(1)}(2) & 1 \\ -z^{(1)}(3) & 1 \\ \vdots & \vdots \\ -z^{(1)}(n) & 1 \end{bmatrix} \tag{4.3}$$

then the least squares estimate sequence of the GM(1,1) model equ. (4.2) satisfies $\hat{a} = (B^T B)^{-1} B^T Y$.

Proof. See (Liu and Lin, 2006, p. 199 – 202) for details. QED.

Continuing all the notations from Theorem 4.1, if $[a,b]^T = (B^T B)^{-1} B^T Y$, then

$$\frac{dx^{(1)}}{dt} + ax^{(1)} = b$$

is referred to as a *whitenization* (or *image*) *equation* of the GM(1,1) model in equ. (4.2).

Theorem 4.2. Let $B, Y, \hat{a}$ be the same as in Theorem 4.1. If $\hat{a} = [a,b]^T = (B^T B)^{-1} B^T Y$, then

(1) The solution, also known as time response function, of the whitenization equation $\frac{dx^{(1)}}{dt} + ax^{(1)} = b$ is given by

$$x^{(1)}(t) = \left(x^{(1)}(1) - \frac{b}{a} \right) e^{-at} + \frac{b}{a} \tag{4.4}$$

(2) The time response sequence of the GM(1,1) model in equ. (4.2) is given below:

$$\hat{x}^{(1)}(k+1) = \left(x^{(0)}(1) - \frac{b}{a} \right) e^{-ak} + \frac{b}{a},\ k = 1,2,\cdots n \tag{4.5}$$

(3) The restored values of $x^{(0)}(k)$'s are given as follows:

$$\begin{aligned} \hat{x}^{(0)}(k+1) &= \alpha^{(1)} \hat{x}^{(1)}(k+1) = \hat{x}^{(1)}(k+1) - \hat{x}^{(1)}(k) \\ &= \left(1 - e^{a}\right)\left(x^{(0)}(1) - \frac{b}{a} \right) e^{-ak},\ k = 1,2,\cdots n \end{aligned} \tag{4.6}$$

The parameters ($-a$) and b of the GM(1,1) model are referred to as the development coefficient and grey action quantity, respectively. The former reflects the development states of $\hat{x}^{(1)}$ and $\hat{x}^{(0)}$. In general, the variables that act upon the system of interest should be external or pre-defined. Because GM(1,1) is a model constructed on a single sequence, it uses only the behavioral sequence (or referred to as output sequence or background values) of the system without considering any externally acting sequences (or referred to as input sequences, or driving quantities). The grey action quantity in the GM(1,1) model is a value derived from the background values. It reflects changes contained in the data and its exact intension is grey. This quantity realizes the extension of the relevant intension. Its existence distinguishes grey systems modeling from the general input-output (or black-box) modeling. It is also an important test stone of separating the thoughts of grey systems and those of grey boxes.

4.1.2 Expanded Forms of GM(1,1) Model

Theorem 4.3. The GM(1,1) model in equ. (4.2) can be transformed into

$$x^{(0)}(k) = \beta - \alpha x^{(1)}(k-1) \tag{4.7}$$

where $\beta = \dfrac{b}{1+0.5a}$ and $\alpha = \dfrac{a}{1+0.5a}$.

Proof. See (Liu and Lin, 2006, p. 204) for details. QED.

Theorem 4.4. If β and α are the same as defined in Theorem 4.3, and $\hat{X}^{(1)} = (\hat{x}^{(1)}(1), \hat{x}^{(1)}(2), \cdots, \hat{x}^{(1)}(n))$ the time response sequence of the GM(1,1) model such that

$$\hat{x}^{(1)}(k) = (x^{(0)}(1) - \frac{b}{a})e^{-a(k-1)} + \frac{b}{a},\ k = 1,2,\cdots n$$

then $x^{(0)}(k) = (\beta - \alpha x^{(0)}(1))e^{-\alpha(k-2)}$.

Proof. See (Liu and Lin, 2006, p. 204 – 205) for details. QED.

Theorem 4.5. The GM(1,1) model in equ. (4.2) can be transformed into $\hat{x}^{(0)}(k) = (1-\alpha)x^{(0)}(k-1)$, $k = 3,4,\cdots,n$.

Proof. From Theorem 4.3, it follows that

$$\begin{aligned} x^{(0)}(k) &= \beta - \alpha x^{(1)}(k-1) \\ &= \beta - \alpha[x^{(1)}(k-2) + x^{(0)}(k-1)] \\ &= [\beta - \alpha x^{(1)}(k-2)] - \alpha x^{(0)}(k-1) \\ &= x^{(0)}(k-1) - \alpha x^{(0)}(k-1) = (1-\alpha)x^{(0)}(k-1). \end{aligned}$$ QED.

Theorem 4.6. The GM(1,1) model in equ. (4.2) can be transformed into

$$\hat{x}^{(0)}(k) = \frac{1-0.5a}{1+0.5a}x^{(0)}(k-1); k = 3,4\cdots n \tag{4.8}$$

Proof. From Theorem 4.5, it follows that $\hat{x}^{(0)}(k) = (1-\alpha)x^{(0)}(k-1)$. Considering the fact that

$$1-\alpha = 1-\frac{a}{1+0.5a} = \frac{1-0.5a}{1+0.5a}$$

we have

$$\hat{x}^{(0)}(k) = \frac{1-0.5a}{1+0.5a}x^{(0)}(k-1); k = 3,4\cdots n\text{. QED.}$$

Theorem 4.7. The GM(1,1) model in equ. (4.2) can be transformed into

$$\hat{x}^{(0)}(k) = \frac{x^{(1)}(k)-0.5b}{x^{(1)}(k-1)+0.5b}x^{(0)}(k-1); k = 3,4\cdots n \tag{4.9}$$

Proof. Equ. (4.2) implies that

$$a = \frac{b-x^{(0)}(k)}{z^{(1)}(k)}.$$

Substituting this expression into $\frac{1-0.5a}{1+0.5a}$ provides

$$\frac{1-0.5a}{1+0.5a} = \frac{1-0.5(\frac{b-x^{(0)}(k)}{z^{(1)}(k)})}{1+0.5(\frac{b-x^{(0)}(k)}{z^{(1)}(k)})} = \frac{z^{(1)}(k)-0.5b+0.5x^{(0)}(k)}{z^{(1)}(k)+0.5b-0.5x^{(0)}(k)}$$

By noticing $z^{(1)}(k) = 0.5x^{(1)}(k)+0.5x^{(1)}(k-1) = x^{(1)}(k-1)+0.5x^{(0)}(k)$, we have

$$\frac{1-0.5a}{1+0.5a} = \frac{x^{(1)}(k-1)+0.5x^{(0)}(k)-0.5b+0.5x^{(0)}(k)}{x^{(1)}(k-1)+0.5x^{(0)}(k)+0.5b-0.5x^{(0)}(k)}$$

$$= \frac{x^{(1)}(k-1)+x^{(0)}(k)-0.5b}{x^{(1)}(k-1)+0.5b}$$

$$= \frac{x^{(1)}(k)-0.5b}{x^{(1)}(k-1)+0.5b}$$

So, it is ready to obtain $\hat{x}^{(0)}(k)=\dfrac{x^{(1)}(k)-0.5b}{x^{(1)}(k-1)+0.5b}x^{(0)}(k-1); k=3,4\cdots n$. QED.

Theorem 4.8. The GM(1,1) model in equ. (4.2) can be transformed into

$$\hat{x}^{(0)}(k)=\frac{b-ax^{(1)}(k-1)}{1+0.5a} \tag{4.10}$$

Proof. From equ. (4.2) and $z^{(1)}(k)=0.5x^{(1)}(k)+0.5x^{(1)}(k-1)=x^{(1)}(k-1)+0.5x^{(0)}(k)$, it follows that

$$\hat{x}^{(0)}(k)=b-az^{(1)}(k)=b-a[\ x^{(1)}(k-1)+0.5x^{(0)}(k)]$$

$$=b-ax^{(1)}(k-1)-0.5ax^{(0)}(k)$$

That is, $\hat{x}^{(0)}(k)+0.5a\hat{x}^{(0)}(k)=(1+0.5a)\hat{x}^{(0)}(k)=b-ax^{(1)}(k-1)$ so that we obtain

$$\hat{x}^{(0)}(k)=\frac{b-ax^{(1)}(k-1)}{1+0.5a}.\ \text{QED.}$$

Theorem 4.9. The GM(1,1) model in equ. (4.2) can be transformed into

$$\hat{x}^{(0)}(k)=\left(\frac{1-0.5a}{1+0.5a}\right)^{k-2}\left(\frac{b-ax^{(0)}(1)}{1+0.5a}\right); k=2,3,\cdots,n \tag{4.11}$$

Proof. From Theorem 4.6, it follows that $\hat{x}^{(0)}(k)=\dfrac{1-0.5a}{1+0.5a}x^{(0)}(k-1)$. By using $\hat{x}^{(0)}(k-1)$, $\hat{x}^{(0)}(k-2)$, $\hat{x}^{(0)}(k-3)$, …, $\hat{x}^{(0)}(3)$ to substitute for $x^{(0)}(k-1)$, $x^{(0)}(k-2)$, $x^{(0)}(k-3)$, …, $x^{(0)}(3)$, respectively, leads to

$$\hat{x}^{(0)}(k)=\frac{1-0.5a}{1+0.5a}x^{(0)}(k-1)=\left(\frac{1-0.5a}{1+0.5a}\right)^{2}x^{(0)}(k-2)=\cdots=\left(\frac{1-0.5a}{1+0.5a}\right)^{k-2}x^{(0)}(2)$$

Letting $k=2,$, equ. (4.10) provides $x^{(0)}(2)=\dfrac{b-ax^{(1)}(1)}{1+0.5a}$ so that we have

$$\hat{x}^{(0)}(k)=\left(\frac{1-0.5a}{1+0.5a}\right)^{k-2}\left(\frac{b-ax^{(0)}(1)}{1+0.5a}\right);k=2,3,\cdots,n\text{. QED.}$$

Theorem 4.10. The GM(1,1) model in equ. (4.2) can be transformed into

$$\hat{x}^{(0)}(k)=\hat{x}^{(0)}(3)e^{(k-3)\ln(1-\alpha)},\ k=3,4,\cdots,n \tag{4.12}$$

Proof. Since the whitenization equation of the GM(1,1) model is of the exponential form, without loss of generality, let us assume $\hat{x}^{(0)}(k)=Ce^{\lambda k}$. From Theorem 4.5, it follows that $\hat{x}^{(0)}(k)=(1-\alpha)x^{(0)}(k-1)$. So, we have

$$\frac{1}{1-\alpha}=\frac{x^{(0)}(k)}{x^{(0)}(k-1)}=e^{\lambda}$$

So, $\lambda=\ln(1-\alpha)$. By letting $k=3$ in $\hat{x}^{(0)}(k)=Ce^{\lambda k}$, we obtain

$$\hat{x}^{(0)}(3)=Ce^{3\lambda}=Ce^{3\ln(1-\alpha)}$$

Solving this equation produces

$$C=\hat{x}^{(0)}(3)e^{-3\ln(1-\alpha)}$$

Therefore, we have

$$\hat{x}^{(0)}(k)=\hat{x}^{(0)}(3)e^{(k-3)\ln(1-\alpha)}\text{. QED.}$$

Theorem 4.11. If $X^{(0)}$ is quasi-smooth, then the development coefficient ($-a$) of the GM(1,1) model can be written as follows:

$$a=\frac{\dfrac{b}{x^{(1)}(k-1)}-\rho(k)}{1+0.5\rho(k)} \tag{4.13}$$

where $\rho(k)=\dfrac{x^{(0)}(k)}{x^{(1)}(k-1)}$.

Proof. See (Liu and Lin, 2006, p. 211 – 212) for details. QED.

Example 4.1. Given a sequence of raw data $X^{(0)}=(x^{(0)}(1),x^{(0)}(2),x^{(0)}(3),x^{(0)}(4),x^{(0)}(5))$ = (2.874, 3.278, 3.337, 3.390, 3.679), simulate this sequence $X^{(0)}$ by respectively using the following three GM(1,1) models and compare the simulation accuracy: 1) $x^{(0)}(k)+az^{(1)}(k)=b$; 2) $x^{(0)}(k)=\beta-\alpha x^{(1)}(k-1)$; and 3) $x^{(0)}(k)=(\beta-\alpha x^{(0)}(1))e^{-a(k-2)}$.

Solution. 1) Let us look at the GM(1,1) model $x^{(0)}(k)+az^{(1)}(k)=b$. Step 1: Compute the accumulation generation of $X^{(0)}$ as follows:

$X^{(1)}=(x^{(1)}(1),x^{(1)}(2),x^{(1)}(3),x^{(1)}(4),x^{(1)}(5))=(2.874,6.152,9.489,12.897,16.558)$

Step 2: Check the quasi-smoothness of $X^{(0)}$. From $\rho(k)=\dfrac{x^{(0)}(k)}{x^{(1)}(k-1)}$, it follows that $\rho(3)\approx 0.54$, $\rho(4)\approx 0.36$ < 0.5, and $\rho(5)\approx 0.29$ < 0.5. So, for $k>3$, the condition of quasi-smoothness is satisfied.

Step 3: Determine whether or not complies with the law of quasi-exponentiality. From $\sigma^{(1)}(k)=\dfrac{x^{(1)}(k)}{x^{(1)}(k-1)}$, it follows that $\sigma^{(1)}(3)\approx 1.54, \sigma^{(1)}(4)\approx 1.36, \sigma^{(1)}(5)\approx 1.29$. So, for $k>3$, $\sigma^{(1)}(k)\in[1,1.5]$ with $\delta=0.5$. That is, the law of quasi-exponentiality is satisfied. So, we can establish a GM(1,1) model for $X^{(1)}$.

Step 4: By using the adjacent neighbors of $X^{(1)}$, we obtain the neighbor means sequence $Z^{(1)}=(z^{(1)}(2),z^{(1)}(3),z^{(1)}(4),z^{(1)}(5))$ = $(4.513,7.820,11.184,14.718)$. So,

$$B=\begin{bmatrix}-z^{(1)}(2) & 1\\ -z^{(1)}(3) & 1\\ -z^{(1)}(4) & 1\\ -z^{(1)}(5) & 1\end{bmatrix}=\begin{bmatrix}-4.513 & 1\\ -7.820 & 1\\ -11.184 & 1\\ -14.718 & 1\end{bmatrix},\quad Y=\begin{bmatrix}x^{(0)}(2)\\ x^{(0)}(3)\\ x^{(0)}(4)\\ x^{(0)}(5)\end{bmatrix}=\begin{bmatrix}3.278\\ 3.337\\ 3.390\\ 3.679\end{bmatrix}$$

Step 5: By using the least squares estimate, we obtain the sequence of parameters $\hat{a}=[a,b]^T$ as follows

$$\hat{a}=(B^TB)^{-1}B^TY=\begin{bmatrix}-0.03720\\ 3.06536\end{bmatrix}$$

Step 6: We establish the following model

$$\frac{dx^{(1)}}{dt}-0.0372x^{(1)}=3.06536$$

and its time response formula:

$$\hat{x}^{(1)}(k)=(x^{(0)}(1)-\frac{b}{a})e^{-a(k-1)}+\frac{b}{a}=85.276151e^{0.0372k}-82.402151$$

Step 7: Compute the simulated values:

$$\hat{X}^{(1)}=(\hat{x}^{(1)}(1),\hat{x}^{(1)}(2),\hat{x}^{(1)}(3),\hat{x}^{(1)}(4),\hat{x}^{(1)}(5))=(2.8704,6.1060,9.4605,12.9422,16.5558)$$

Step 8: Compute the simulated values of $X^{(0)}$ by using $\hat{x}^{(0)}(k)=\alpha^{(1)}\hat{x}^{(1)}(k)=\hat{x}^{(1)}(k)-\hat{x}^{(1)}(k-1)$:

$$\hat{X}^{(0)}=(2.8740,3.2320,3.3545,3.4817,3.6136)$$

Step 9: Check errors. From Table 4.1, we can compute the sum of squared errors

$$s=\varepsilon^T\varepsilon=[\varepsilon(2),\varepsilon(3),\varepsilon(4),\varepsilon(5)]\begin{bmatrix}\varepsilon(2)\\\varepsilon(3)\\\varepsilon(4)\\\varepsilon(5)\end{bmatrix}=0.01511$$

and the mean relative error:

$$\Delta=\frac{1}{4}\sum_{k=2}^{5}\Delta_k=1.6025\%$$

Table 4.1 Error checks

Ordinality	Actual data $x^{(0)}(k)$	Simulated data $\hat{x}^{(0)}(k)$	Error $\varepsilon(k)=x^{(0)}(k)-\hat{x}^{(0)}(k)$	Relative error $\Delta_k=\lvert\varepsilon(k)\rvert/x^{(0)}(k)$
2	3.278	3.230	0.0460	1.40%
3	3.337	3.3545	−0.0175	0.52%
4	3.390	3.4817	−0.0917	2.71%
5	3.679	3.6136	0.0654	1.78%

2) Let us look at the GM(1,1) model $x^{(0)}(k)=\beta-\alpha x^{(1)}(k-1)$. From the previous work in 1), it follows that $a=-0.03720, b=3.06536$. So, we have

$$\alpha=\frac{a}{1+0.5a}=\frac{-0.03720}{1+0.5\cdot(-0.03720)}=0.0379$$

$$\beta=\frac{b}{1+0.5a}=\frac{3.06536}{1+0.5\cdot(-0.03720)}=3.1235$$

Therefore, we obtain the specific model $x^{(0)}(k)=\beta-\alpha x^{(1)}(k-1)$ = 3.1235 + 0.0379 $x^{(1)}(k-1)$. So, the simulated values are given by $\hat{X}^{(0)}=(\hat{x}^{(0)}(1),\hat{x}^{(0)}(2),\hat{x}^{(0)}(3),\hat{x}^{(0)}(4),\hat{x}^{(0)}(5))=$ (3.2324,3.2324,3.3567,3.4820,3.6105). For the error checking, from Table 4.2, we can compute the sum of squared errors as follows: $s=\varepsilon^T\varepsilon=0.0156$.

Table 4.2 Error checks

Ordinality	Actual data $x^{(0)}(k)$	Simulated data $\hat{x}^{(0)}(k)$	Error $\varepsilon(k)=x^{(0)}(k)-\hat{x}^{(0)}(k)$	Relative error $\Delta_k=\lvert\varepsilon(k)\rvert/x^{(0)}(k)$
2	3.278	3.2324	0.0456	1.39%
3	3.337	3.3567	−0.0197	0.59%
4	3.390	3.4820	−0.092	2.71%
5	3.679	3.6105	0.0685	1.86%

The average relative error is

$$\Delta=\frac{1}{4}\sum_{k=2}^{5}\Delta_k=1.6375\%\,.$$

3) Let us look at the GM(1,1) model $x^{(0)}(k)=(\beta-\alpha x^{(0)}(1))e^{-a(k-2)}$. From both 1) and 2) above, it follows that $a=-0.0372,\alpha=0.0379,\beta=3.1235$. So, we have

$$\begin{aligned}x^{(0)}(k)&=(\beta-\alpha x^{(0)}(1))e^{-a(k-2)}=(3.1235+0.0379\times2.874)e^{0.0372(k-2)}\\&=3.2324246e^{0.0372(k-2)}\end{aligned}$$

and $\hat{X}^{(0)}=(3.1144,3.2324,3.3549,3.4821,3.6141)$. From Table 4.3, we compute the sum of squared errors $s=\varepsilon^T\varepsilon$ = 0.01509 and the average relative error $\Delta=\frac{1}{4}\sum_{k=2}^{5}\Delta_k=1.6025\%$.

4) From the sums of squared errors and the average relative errors of these three models, it can be seen that for sequence $X^{(0)}$= (2.874,3.278,3.337,3.390,3.679), both

$$\begin{cases} \hat{x}^{(1)}(k) = (x^{(0)}(1) - \frac{b}{a})e^{-a(k-1)} + \frac{b}{a} \\ \hat{x}^{(0)}(k) = \hat{x}^{(1)}(k) - \hat{x}^{(1)}(k-1) \end{cases}$$

and

$$\hat{x}^{(0)}(k) = (\beta - \alpha x^{(0)}(1))e^{-a(k-2)}$$

provide better better simulation accuracies than the difference model $\hat{x}^{(0)}(k) = \beta - \alpha x^{(1)}(k-1)$.

Table 4.3 Error checks

Ordinality	Actual data $x^{(0)}(k)$	Simulated data $\hat{x}^{(0)}(k)$	Error $\varepsilon(k) = x^{(0)}(k) - \hat{x}^{(0)}(k)$	Relative error $\Delta_k = \lvert\varepsilon(k)\rvert / x^{(0)}(k)$
2	3.278	3.2324	0.0456	1.39%
3	3.337	3.3549	−0.0179	0.54%
4	3.390	3.4821	−0.0921	2.72%
5	3.679	3.6141	0.0649	1.76%

4.2 Improvements on GM(1,1) Models

4.2.1 Remnant GM(1,1) Model

When the accuracy of a GM(1,1) model does not meet the predetermined requirement, one can establish another GM(1,1) model using the error sequence to remedy the original model to improve the accuracy. In particular, let $X^{(0)}$ be a sequence of raw data, $X^{(1)}$ the accumulation generated sequence based on $X^{(0)}$, and the time response formula of the GM(1,1) model is

$$\hat{x}^{(1)}(k+1) = (x^{(0)}(1) - \frac{b}{a})e^{-ak} + \frac{b}{a}$$

then

$$d\hat{x}^{(1)}(k+1) = (-a)(x^{(0)}(1) - \frac{b}{a})e^{-ak} \tag{4.14}$$

is referred to as the *restored value* through derivatives. Then, in general, $d\hat{x}^{(1)}(k+1) \neq \hat{x}^{(0)}(k+1)$, where $\hat{x}^{(0)}(k+1) = \hat{x}^{(1)}(k+1) - \hat{x}^{(1)}(k)$ stands for the restored value through inverse accumulation. This very fact implies that the GM(1,1) is neither a differential equation nor a difference equation. However, when $|a|$ is sufficiently small, from $1 - e^{a} \approx -a$, it follows that $d\hat{x}^{(1)}(k+1) \approx \hat{x}^{(0)}(k+1)$, meaning that the results of differentiation and difference are quite close. Therefore, the GM(1,1) model in this case can be seen as both differential equation and a difference equation.

Because the restored values through derivatives and through inverse accumulation are different, to reduce possible errors caused by reciprocating operators, the errors of $X^{(1)}$ are often used to improve the simulated values $\hat{x}^{(1)}(k+1)$ of $X^{(1)}$. In particular, assume that $\varepsilon^{(0)} = (\varepsilon^{(0)}(1), \varepsilon^{(0)}(2), \cdots, \varepsilon^{(0)}(n))$, where $\varepsilon^{(0)}(k) = x^{(1)}(k) - \hat{x}^{(1)}(k)$, is the error sequence of $X^{(1)}$. If there a k_0 satisfying that $n - k_0 \geq 4$ and $\forall k \geq k_0$, the signs of $\varepsilon^{(0)}(k)$ stay the same, then $(|\varepsilon^{(0)}(k_0)|, |\varepsilon^{(0)}(k_0+1)|, , \cdots, |\varepsilon^{(0)}(n)|)$ is known as the *error sequence of modelability*, is and still denoted $\varepsilon^{(0)} = (\varepsilon^{(0)}(k_0), \varepsilon^{(0)}(k_0+1), \cdots, \varepsilon^{(0)}(n))$. In this case, let the sequence $\varepsilon^{(1)} = (\varepsilon^{(1)}(k_0), \varepsilon^{(1)}(k_0+1), \cdots, \varepsilon^{(1)}(n))$ be accumulation generated on $\varepsilon^{(0)}$ with the following GM(1,1) time response formula

$$\hat{\varepsilon}^{(1)}(k+1) = \left(\varepsilon^{(0)}(k_0) - \frac{b_\varepsilon}{a_\varepsilon}\right)\exp[-a_\varepsilon(k-k_0)] + \frac{b_\varepsilon}{a_\varepsilon}, \quad k \geq k_0$$

then the simulation sequence of $\varepsilon^{(0)}$ is given by $\hat{\varepsilon}^{(0)} = (\hat{\varepsilon}^{(0)}(k_0), \hat{\varepsilon}^{(0)}(k_0+1), \cdots, \hat{\varepsilon}^{(0)}(n))$, where

$$\hat{\varepsilon}^{(0)}(k+1) = (-a_\varepsilon)\left(\varepsilon^{(0)}(k_0) - \frac{b_\varepsilon}{a_\varepsilon}\right)\exp[-a_\varepsilon(k-k_0)], \quad k \geq k_0.$$

If $\hat{\varepsilon}^{(0)}$ is used to improve $\hat{X}^{(1)}$, the modified time response formula

$$\hat{x}^{(1)}(k+1) = \begin{cases} \left(x^{(0)}(1) - \frac{b}{a}\right)e^{-ak} + \frac{b}{a}, & k < k_0 \\ \left(x^{(0)}(1) - \frac{b}{a}\right)e^{-ak} + \frac{b}{a} \pm a_\varepsilon(\varepsilon^{(0)}(k_0) - \frac{b_\varepsilon}{a_\varepsilon})e^{-a_\varepsilon(k-k_0)}, & k \geq k_0 \end{cases} \tag{4.15}$$

is referred to as the GM(1,1) *model with error modification*, or simply *remnant* GM(1,1) for short, where the sign of the error modification value

$$\hat{\varepsilon}^{(0)}(k+1) = a_\varepsilon \times \left(\varepsilon^{(0)}(k_0) - \frac{b_\varepsilon}{a_\varepsilon}\right)\exp[-a_\varepsilon(k-k_0)]$$

needs to stay the same as those in $\varepsilon^{(0)}$.

If a modeling of the error sequence $\varepsilon^{(0)}=(\varepsilon^{(0)}(k_0),\varepsilon^{(0)}(k_0+1),\cdots,\varepsilon^{(0)}(n))$ of $X^{(0)}$ and $\hat{X}^{(0)}$ is used to modify the simulation value $\hat{X}^{(0)}$, then different methods of restoration from $\hat{X}^{(1)}$ to $\hat{X}^{(0)}$ can produce different time response sequences of error modification. In particular, if

$$\hat{x}^{(0)}(k)=\hat{x}^{(1)}(k)-\hat{x}^{(1)}(k-1)=\left(1-e^{a}\right)\left(x^{(0)}(1)-\frac{b}{a}\right)e^{-a(k-1)}$$

then the corresponding time response sequence of error modification

$$\hat{x}^{(0)}(k+1)=\begin{cases}\left(1-e^{a}\right)\left(x^{(0)}(1)-\dfrac{b}{a}\right)e^{-ak}, & k<k_0\\ \left(1-e^{a}\right)\left(x^{(0)}(1)-\dfrac{b}{a}\right)e^{-ak}\pm a_\varepsilon\left(\varepsilon^{(0)}(k_0)-\dfrac{b_\varepsilon}{a_\varepsilon}\right)e^{-a_\varepsilon(k-k_0)}, & k\geq k_0\end{cases}$$

is known as the *error modification model* of inverse accumulation restoration.

If $\hat{x}^{(0)}(k+1)=(-a)\left(x^{(0)}(1)-\frac{b}{a}\right)e^{-ak}$, the corresponding time response sequence of error modification

$$\hat{x}^{(0)}(k+1)=\begin{cases}(-a)\left(x^{(0)}(1)-\dfrac{b}{a}\right)e^{-ak}, & k<k_0\\ (-a)\left(x^{(0)}(1)-\dfrac{b}{a}\right)e^{-ak}\pm a_\varepsilon\left(\varepsilon^{(0)}(k_0)-\dfrac{b_\varepsilon}{a_\varepsilon}\right)e^{-a_\varepsilon(k-k_0)}, & k\geq k_0\end{cases}$$

is known as the *error modification model* of derivative restoration.

In the previous discussion, all the error simulation terms in remnant GM(1,1) have been taken as the derivative restoration. Of course, they can be taken as inverse accumulation restoration. That is, one can take

$$\hat{\varepsilon}^{(0)}(k+1)=\left(1-e^{a_\varepsilon}\right)\left(\varepsilon^{(0)}(k_0)-\frac{b_\varepsilon}{a_\varepsilon}\right)e^{-a_\varepsilon(k-k_0)},\quad k\geq k_0$$

As long as $|a_\varepsilon|$ is sufficiently small, the effects of different error restoration methods on the modified $\hat{x}^{(0)}(k+1)$ are almost the same.

4.2.2 Groups of GM(1,1) Models

In practice, one does not have to use all the available data in his modeling. Each subsequence of the original data can be employed to establish a model. Generally speaking, different subsequences lead to different models. Even though the same kind of the GM(1,1) is applied, different subsequences lead to different a,b values. These changes reflect the fact that varied circumstances and conditions have different effect on the system of our concern.

For a given sequence $X^{(0)}=(x^{(0)}(1),x^{(0)}(2),\cdots,x^{(0)}(n))$, if we take $x^{(0)}(n)$ as the origin of the time axis, then $t < n$ is seen as the past, $t = n$ the present, and $t > n$ the future. If

$$\hat{x}^{(0)}(k+1)=(1-e^{a})\left(x^{(0)}(1)-\frac{b}{a}\right)e^{-ak}$$

is the restored values of inverse accumulation of the GM(1,1) time responses of $X^{(0)}$, then for $t \le n$, $\hat{x}^{(0)}(t)$ is referred to as the *simulated value* out of the model; and when $t > n$, $\hat{x}^{(0)}(t)$ is known as the *prediction* of the model. The main purpose of modeling is to make predictions. To improve the prediction accuracy, one first needs to guarantee sufficiently high accuracy in his simulation, especially for the simulation of the time moment $t = n$. Therefore, in general, the data, including $x^{(0)}(n)$, used for modeling should be an *equal-time-interval sequence*.

For the given sequence $X^{(0)}$, the GM(1,1) model established using the entire sequence $X^{(0)}$ is known as the *all-data* GM(1,1). For $\forall k_0 > 1$, the GM(1,1) model established on the tail sequence $X^{(0)} = (x^{(0)}(k_0),x^{(0)}(k_0+1),\cdots,x^{(0)}(n))$ is known as a *partial-data* GM(1,1). If $x^{(0)}(n+1)$ stands for a piece of new information, then the GM(1,1) model established on the prolonged sequence $X^{(0)} = (x^{(0)}(1),x^{(0)}(2),\cdots,x^{(0)}(n),x^{(0)}(n+1))$ is known as a *new-information* GM(1,1); the GM(1,1) model established on $X^{(0)}=((x^{(0)}(2),\cdots,x^{(0)}(n),x^{(0)}(n+1))$ with the new information added and the oldest piece $x^{(0)}(1)$ of information removed is known as a *metabolic* GM(1,1).

For detailed empirical analysis of these GM(1,1) models, please consult with (Liu and Lin, 2006, pp. 224 – 228).

4.3 Applicable Ranges of GM(1,1) Models

Proposition 4.1. When $(n-1)\sum_{k=2}^{n}\left[z^{(1)}(k)\right]^2 \to \left[\sum_{k=2}^{n} z^{(1)}(k)\right]^2$, the GM(1,1) becomes invalid.

Proof. By using the model parameters obtained by the least squared estimate, we have

$$\hat{a}=\frac{\sum_{k=2}^{n} z^{(1)}(k)\sum_{k=2}^{n} x^{(0)}(k)-(n-1)\sum_{k=2}^{n} z^{(1)}(k)x^{(0)}(k)}{(n-1)\sum_{k=2}^{n}\left[z^{(1)}(k)\right]^2-\left[\sum_{k=2}^{n} z^{(1)}(k)\right]^2}$$

$$\hat{b}=\frac{\sum_{k=2}^{n} x^{(0)}(k)\sum_{k=2}^{n}\left[z^{(1)}(k)\right]^2-\sum_{k=2}^{n} z^{(1)}(k)\sum_{k=2}^{n} z^{(1)}(k)x^{(0)}(k)}{(n-1)\sum_{k=2}^{n}\left[z^{(1)}(k)\right]^2-\left[\sum_{k=2}^{n} z^{(1)}(k)\right]^2}$$

When $(n-1)\sum_{k=2}^{n}\left[z^{(1)}(k)\right]^2\to\left[\sum_{k=2}^{n} z^{(1)}(k)\right]^2$, $\hat{a}\to\infty,\hat{b}\to\infty$, so that the model parameters cannot be determined. Hence, the GM(1,1) becomes invalid. QED.

Proposition 4.2. When the development coefficient a of the GM(1,1) model satisfies $|a|\geq 2$, the GM(1,1) model becomes invalid.

Proof. From the following expression of the GM(1,1) model

$$x^{(0)}(k)=\left(\frac{1-0.5a}{1+0.5a}\right)^{k-2}\left(\frac{b-ax^{(0)}(1)}{1+0.5a}\right);k=2,3,\cdots,n$$

it can be seen that when $a=-2$, $x^{(0)}(k)\to\infty$; when $a=2$, $x^{(0)}(k)=0$; and when $|a|>2$, $\frac{b-ax^{(0)}(1)}{1+0.5a}$ becomes a constant, while the sign of $\left(\frac{1-0.5a}{1+0.5a}\right)^{k-2}$ changes with k being even or odd. So, the sign of $x^{(0)}(k)$ flips with k being even or odd. QED.

The discussion above indicates that $(-\infty,-2]\cup[2,\infty)$ is the forbidden area for the development coefficient $(-a)$ of the GM(1,1) model. When $a\in(-\infty,-2]\cup[2,\infty)$, the GM(1,1) model loses its validity. In general, when $|a|<2$, the GM(1,1) model is meaningful. However, for different values of a, the prediction effect of the model is different. For the case of $-2<a<0$, let us respectively take $-a=0.1$,

0.2, 0.3, 0.4, 0.5, 0.6, 0.8, 1.5, 1.8 to conduct a simulation analysis. By taking $k = 0$, 1, 2, 3, 4, 5, from $x_i^{(0)}(k+1) = e^{-ak}$, we obtain the following sequences:

If $-a$ =0.1, $X_1^{(0)} = (x_1^{(0)}(1), x_1^{(0)}(2), x_1^{(0)}(3), x_1^{(0)}(4), x_1^{(0)}(5), x_1^{(0)}(6))$
= (1, 1.1051, 1.2214, 1.3499, 1.4918, 1.6487).

If $-a$ =0.2, $X_2^{(0)}$ = (1, 1.2214, 1.4918, 1.8221, 2.2255, 2.7183).

If $-a$ =0.3, $X_3^{(0)}$ = (1, 1.3499, 1.8221, 2.4596, 3.3201, 4.4817).

If $-a$ =0.4, $X_4^{(0)}$ = (1, 1.4918, 2.225, 3.3201, 4.9530, 7.3890).

If $-a$ =0.5, $X_5^{(0)}$ = (1, 1.6487, 2.7183, 4.4817, 7.3890, 12.1825).

If $-a$ =0.6, $X_6^{(0)}$ = (1, 1.8821, 3.3201, 6.0496, 11.0232, 20.0855).

If $-a$ =0.8, $X_7^{(0)}$ = (1, 2.2255, 4.9530, 11.0232, 24.5325, 54.5982).

If $-a$ =1, $X_8^{(0)}$ = (1, 2.7183, 7.3890, 20.0855, 54.5982, 148.4132).

If $-a$ =1.5, $X_9^{(0)}$ = (1, 4.4817, 20.0855, 90.0171, 403.4288, 1808.0424).

If $-a$ =1.8, $X_{10}^{(0)}$ = (1, 6.0496, 36.5982, 221.4064, 1339.4308, 8103.0839).

Let us respectively apply $X_1^{(0)}$, $X_2^{(0)}$, …, and $X_9^{(0)}$ to establish a GM(1,1) model and obtain the following time response sequences:

$$\hat{x}_1^{(1)}(k+1) = 10.50754e^{0.09992182k} - 9.507541,$$
$$\hat{x}_2^{(1)}(k+1) = 5.516431e^{0.1993401k} - 4.516431,$$
$$\hat{x}_3^{(1)}(k+1) = 3.85832e^{0.297769k} - 2.858321,$$
$$\hat{x}_4^{(1)}(k+1) = 3.033199e^{0.394752k} - 2.033199,$$
$$\hat{x}_5^{(1)}(k+1) = 2.541474e^{0.4898382k} - 1.541474,$$
$$\hat{x}_6^{(1)}(k+1) = 2.216363e^{0.5826263k} - 1.216362,$$
$$\hat{x}_7^{(1)}(k+1) = 1.815972e^{0.7598991k} - 0.8159718,$$
$$\hat{x}_8^{(1)}(k+1) = 1.581973e^{0.9242348k} - 0.5819733,$$
$$\hat{x}_9^{(1)}(k+1) = 1.287182e^{1.270298k} - 0.2871823,$$
$$\hat{x}_{10}^{(1)}(k+1) = 0.198197e^{1.432596k} - 0.1981966.$$

From $\hat{x}_i^{(0)}(k+1) = \hat{x}_i^{(1)}(k+1) - \hat{x}_i^{(1)}(k)$, $i = 1,2,\cdots,10$, we obtain ,

$$\hat{x}_1^{(0)}(k+1) = 0.99918e^{0.09992182k}, \quad \hat{x}_2^{(0)}(k+1) = 0.99698e^{0.1993401k},$$
$$\hat{x}_3^{(0)}(k+1) = 0.99362e^{0.297769k}, \quad \hat{x}_4^{(0)}(k+1) = 0.989287e^{0.394752k},$$

$$\hat{x}_5^{(0)}(k+1) = 0.984248e^{0.4898382k},\ \hat{x}_6^{(0)}(k+1) = 0.97868e^{0.5826263k},$$
$$\hat{x}_7^{(0)}(k+1) = 0.966617e^{0.7598991k},\ \hat{x}_8^{(0)}(k+1) = 0.95419e^{0.9242348k},$$
$$\hat{x}_9^{(0)}(k+1) = 0.925808e^{1.270298k},\ \hat{x}_{10}^{(0)}(k+1) = 0.91220e^{1.432596k}.$$

From the mean generation of $z^{(1)}(k) = \frac{1}{2}(x^{(1)}(k) + x^{(1)}(k-1))$ of the GM(1,1) model $x^{(0)}(k) + az^{(1)}(k) = b$, for increasing sequences, it has the effect of weakening the growth. For an exponential sequence, the established GM(1,1) has a small development coefficient.

Let us compare the errors between the original sequence $X_i^{(0)}$ and the simulation sequence $\hat{X}_i^{(0)}$, Table 4.4.

Table 4.4 Simulation errors

Development coefficient $(-a)$	$\frac{1}{5}\sum_{i=2}^{6}[\hat{x}^{(0)}(k) - x^{(0)}(k)]$	Mean relative error $\frac{1}{5}\sum_{k=2}^{6}\Delta_k$
0.1	0.004	0.104%
0.2	0.010	0.499%
0.3	0.038	1.300%
0.4	0.116	2.613%
0.5	0.307	4.520%
0.6	0.741	7.074%
0.8	3.603	14.156%
1	14.807	23.544%
1.5	317.867	51.033%
1.8	1632.240	65.454%

It can be seen that as the development coefficient increases, the simulation error grows drastically. When the development coefficient is smaller than or equal to 0.3, the simulation accuracy can reach above 98%. When the coefficient is smaller than or equal 0.5, the simulation accuracy can reach above 95%. When the coefficient is greater than 1, the simulation accuracy is lower than 70%. When the coefficient is greater than 1.5, the simulation accuracy is lower than 50%.

Let us now further focus on the first step, second step, fifth step, and 10th step prediction errors. Table 4.5.

Table 4.5 Prediction errors

$-a$	0.1	0.2	0.3	0.4	0.5	0.6	0.8	1	1.5	1.8
Step 1 error	0.129%	0.701%	1.998%	4.317%	7.988%	13.405%	31.595%	65.117%	—	—
Step 2 error	0.137%	0.768%	2.226%	4.865%	9.091%	15.392%	36.979%	78.113%	—	—
Step 5 error	0.160%	0.967%	2.912%	6.529%	12.468%	21.566%	54.491%	—	—	—
Step 5 error	0.855%	1.301%	4.067%	9.362%	18.330%	32.599%	88.790%	—	—	—

It can be seen that when the development coefficient is smaller than 0.3, the step 1 prediction accuracy is above 98%, with both steps 2 and 5 accuracies above 97%. When $0.3 < -a \le 0.5$, the steps 1 and 2 prediction accuracies are all above 90%; and the step 10 prediction accuracy also above 80%. When the development coefficient is greater than 0.8, the step 1 prediction accuracy is below 70%. The horizontal bars in Table 4.5 represent that the relevant errors are greater than 100%.

From this analysis, we can draw the following conclusions: When $-a \le 0.3$, GM(1,1) can be applied to make mid-term to long-term predictions; when $0.3 < -a \le 0.5$, GM(1,1) can be applied to make short-term and mid-term predictions with caution; when $0.5 < -a \le 0.8$ and GM(1,1) is used to make short-term predictions, one needs to be very cautious about the prediction results; when $0.8 < -a \le 1$, one should employ the remnant GM(1,1) model; and when $-a > 1$, the GM(1,1) should not be applied.

4.4 The GM(*r*,*h*) Models

4.4.1 The GM(1, N) Model

Let $X_i^{(0)} = (x_i^{(0)}(1), x_i^{(0)}(2), \cdots, x_i^{(0)}(n))$, $i = 1,2,...,N$, be given, where $X_1^{(0)}$ stands for a sequence of a system's characteristic data, $X_i^{(0)}$ data sequences of relevant factors, $X_i^{(1)}$ the accumulated generation of $X_i^{(0)}$, $i = 1,2,...,N$, and $Z_1^{(1)}$ the adjacent neighbor mean generated sequence of $X_1^{(1)}$. Then,

$$x_1^{(0)}(k) + az_1^{(1)}(k) = \sum_{i=2}^{N} b_i x_i^{(1)}(k) \tag{4.16}$$

is known as the GM(1,*N*) *model*. The constant $(-a)$ is known as the system's development coefficient, $b_i x_i^{(1)}(k)$ the *driving term*, b_i the *driving coefficient*, and $\hat{a} = [a, b_1, b_2, \cdots, b_N]^T$ the sequence of parameters.

Theorem 4.12. For the previously defined terms $X_i^{(0)}$, $X_i^{(1)}$, and $Z_1^{(1)}$, $i = 1,2,...,N$, let

$$B = \begin{bmatrix} -z_1^{(1)}(2) & x_2^{(1)}(2) & \cdots & x_N^{(1)}(2) \\ -z_1^{(1)}(3) & x_2^{(1)}(3) & \cdots & x_N^{(1)}(3) \\ \cdots & \cdots & \cdots & \cdots \\ -z_1^{(1)}(n) & x_2^{(1)}(n) & \ddots & x_N^{(1)}(n) \end{bmatrix}, Y = \begin{bmatrix} x_1^{(0)}(2) \\ x_1^{(0)}(3) \\ \vdots \\ x_1^{(0)}(n) \end{bmatrix}$$

then the least squares estimate of the sequence $\hat{a}=[a,b_1,b_2,\cdots,b_N]^T$ of parameters satisfies

$$\hat{a}=(B^TB)^{-1}B^TY\,.$$

For $\hat{a}=[a,b_1,b_2,\cdots,b_N]^T$,

$$\frac{dx^{(1)}}{dt}+ax_1^{(1)}=b_2x_2^{(1)}+b_3x_3^{(1)}+\cdots+b_Nx_N^{(1)} \tag{4.17}$$

is known as the *whitenization* (or *image*) *equation* of the GM(1,*N*) model in equ. (4.16).

Theorem 4.13. Let $X_i^{(0)}$, $X_i^{(1)}$, $Z_1^{(1)}$, *B*, and *Y*, $i=1,2,\cdots,N$, are defined as in Theorem 4.12, and

$$\hat{a}=[a,b_1,b_2,\cdots,b_N]^T=(B^TB)^{-1}B^TY\,.$$

Then,

(1) The solution of the whitenization equation $\frac{dx_1^{(1)}}{dt}+ax_1^{(1)}=\sum_{i=2}^{N}b_ix_i^{(1)}$ is:

$$\begin{aligned}x^{(1)}(t)&=e^{-at}[\sum_{i=2}^{N}\int b_ix_i^{(1)}(t)e^{at}dt+x^{(1)}(0)-\sum_{i=2}^{N}\int b_ix_i^{(1)}(0)dt]\\&=e^{-at}[x_1^{(1)}(0)-t\sum_{i=2}^{N}b_ix_i^{(1)}(0)+\sum_{i=2}^{N}\int b_ix_i^{(1)}(t)e^{at}dt\end{aligned} \tag{4.18}$$

(2) When $X_i^{(1)}$, $i=1,2,\cdots,N$, vibrate in small magnitudes, $\sum_{i=2}^{N}b_ix_i^{(1)}(k)$ can be seen as a grey constant. In this case, an approximate time response sequence of the GM(1,*N*) model in equ. (4.16) is given by

$$\hat{x}_1^{(1)}(k+1)=(x_1^{(1)}(0)-\frac{1}{a}\sum_{i=2}^{N}b_ix_i^{(1)}(k+1))e^{-ak}+\frac{1}{a}\sum_{i=2}^{N}b_ix_i^{(1)}(k+1) \tag{4.19}$$

where $x_1^{(1)}(0)$ is taken as $x_1^{(0)}(1)$

(3) The restored sequence of inverse accumulation is

$$\hat{x}_1^{(0)}(k+1)=\alpha^{(1)}\hat{x}_1^{(1)}(k+1)=\hat{x}_1^{(1)}(k+1)-\hat{x}_1^{(1)}(k)$$

(4) The GM(1.*N*) difference model is

$$\hat{x}_1^{(0)}(k)=-az_1^{(1)}(k)+\sum_{i=2}^{N}b_i\hat{x}_i^{(1)}(k)\,.$$

4.4.2 The GM(0,N) Model

Let be $X_1^{(0)}$ a data sequence of a system's characteristic, $X_i^{(0)}$ the data sequences of relevant factors, and $X_i^{(1)}$ the accumulation generated sequence of $X_i^{(0)}$, $i = 2,3,\cdots,N$, then

$$x_1^{(1)}(k) = a + b_2 x_2^{(1)}(k) + b_3 x_3^{(1)}(k) + \cdots + b_N x_N^{(1)}(k) \tag{4.20}$$

is known as the GM(0,*N*) *model*. Because this model does not contain any derivative, it is a static model. Although its form looks like a multivariate linear regression model, it is essentially different from any of the statistical models. In particular, the general multivariate linear regression model is established on the basis of the original data sequences, while the GM(0,*N*) model is constructed on the accumulation generation of the original data.

Theorem 4.14. With the same sequences $X_i^{(0)}$ and $X_i^{(1)}$ as above given, let

$$B = \begin{bmatrix} 1 & x_2^{(1)}(2) & x_3^{(1)}(2) & \cdots & x_N^{(1)}(2) \\ 1 & x_2^{(1)}(3) & x_3^{(1)}(3) & \cdots & x_N^{(1)}(3) \\ \cdots & \cdots & \cdots & \cdots & \cdots \\ 1 & x_2^{(1)}(n) & x_3^{(1)}(n) & \cdots & x_N^{(1)}(n) \end{bmatrix}, \ Y = \begin{bmatrix} x_1^{(1)}(2) \\ x_1^{(1)}(3) \\ \vdots \\ x_1^{(1)}(n) \end{bmatrix}$$

then the least squares estimate of the parametric sequence $\hat{a} = [a, b_1, b_2, \cdots, b_N]^T$ is given by

$$\hat{a} = (B^T B)^{-1} B^T Y .$$

4.4.3 The GM(2,1) and Verhulst Models

The GM(1,1) model is suitable for sequences that show an obvious exponential pattern and can be used to describe monotonic changes. As for non-monotonic wavelike development sequences, or saturated sigmoid sequences, one can consider establishing GM(2,1) and Verhulst models.

4.4.3.1 The GM(2,1) Model

For a given sequence of raw data $X^{(0)} = (x^{(0)}(1), x^{(0)}(2), \cdots, x^{(0)}(n))$, let its sequences of accumulation generation and inverse accumulation generation be $X^{(1)} = (x^{(1)}(1), x^{(1)}(2), \cdots, x^{(1)}(n))$ and $\alpha^{(1)} X^{(0)} = (\alpha^{(1)} x^{(0)}(2), \cdots, \alpha^{(1)} x^{(0)}(n))$, where $\alpha^{(1)} x^{(0)}(k) = x^{(0)}(k) - x^{(0)}(k-1)$, $k = 2,3,\cdots,n$, and the sequence of adjacent neighbor mean generation of $X^{(1)}$ be $Z^{(1)} = (z^{(1)}(2), z^{(1)}(3), \cdots z^{(1)}(n))$. Then,

$$\alpha^{(1)}x^{(0)}(k) + a_1 x^{(0)}(k) + a_2 z^{(1)}(k) = b \tag{4.21}$$

is known as the GM(2,1) *model*; and

$$\frac{d^2 x^{(1)}}{dt^2} + \alpha_1 \frac{dx^{(1)}}{dt} + \alpha_2 x^{(1)} = b \tag{4.22}$$

the *whitenization equation* of the GM(2,1) model.

Theorem 4.15. For the sequences $X^{(0)}$, $X^{(1)}$, $Z^{(1)}$, and $\alpha^{(1)}X^{(0)}$, as defined above, let

$$B = \begin{bmatrix} -x^{(0)}(2) & -z^{(1)}(2) & 1 \\ -x^{(0)}(3) & -z^{(1)}(3) & 1 \\ \cdots & \cdots & \cdots \\ -x^{(0)}(n) & -z^{(1)}(n) & 1 \end{bmatrix},$$

$$Y = \begin{bmatrix} \alpha^{(1)}x^{(0)}(2) \\ \alpha^{(1)}x^{(0)}(3) \\ \cdots \\ \alpha^{(1)}x^{(0)}(n) \end{bmatrix} = \begin{bmatrix} x^{(0)}(2) - x^{(0)}(1) \\ x^{(0)}(3) - x^{(0)}(2) \\ \cdots \\ x^{(0)}(n) - x^{(0)}(n-1) \end{bmatrix}$$

then the least squares estimate of the parametric sequence $\hat{a} = [a_1, a_2, b]^T$ of the GM(2,1) is given as follows:

$$\hat{a} = (B^T B)^{-1} B^T Y .$$

Theorem 4.16. For the solution of the GM(2,1) whitenization equation, the following hold true:

(1) If $X^{(1)^*}$ is a special solution of $\frac{d^2 x^{(1)}}{dt^2} + \alpha_1 \frac{dx^{(1)}}{dt} + \alpha_2 x^{(1)} = b$ and $\overline{X}^{(1)}$ the general solution of the corresponding homogeneous equation $\frac{d^2 x^{(1)}}{dt^2} + \alpha_1 \frac{dx^{(1)}}{dt} + \alpha_2 x^{(1)} = 0$, then $X^{(1)^*} + \overline{X}^{(1)}$ is the general solution of the GM(2,1) whitenization equation.

(2) There are the following three cases for the general solution of the homogeneous equation above: (i) when the characteristic equation $r^2+\alpha_1 r+\alpha_2=0$ has two distinct real roots r_1, r_2,

$$\overline{X}^{(1)}=c_1e^{r_1t}+c_2e^{r_2t}; \tag{4.23}$$

(ii) when the characteristic equation has a repeated root r,

$$\overline{X}^{(1)}=e^{rt}(c_1+c_2t); \tag{4.24}$$

(iii) when the characteristic equation has two complex conjugate roots $r_1=\alpha+i\beta$ and $r_2=\alpha-i\beta$,

$$\overline{X}^{(1)}=e^{\alpha t}(c_1\cos\beta t+c_2\sin\beta t). \tag{4.25}$$

(3) A special solution of the whitenization equation may take of the three possibilities: (i) When 0 is not a root of the characteristic equation, $X^{(1)^*}=C$; (ii) when 0 is one of the two distinct roots of the characteristic equation, $X^{(1)^*}=Cx$; and (iii) when 0 is the only root of the characteristic equation, $X^{(1)^*}=Cx^2$.

4.4.3.2 The VerhulstModel

Assume that $X^{(0)}$ is a sequence of raw data, $X^{(1)}$ the sequence of accumulation generation of $X^{(0)}$, and $Z^{(1)}$ the adjacent neighbor mean generation of $X^{(1)}$. Then,

$$x^{(0)}(k)+az^{(1)}(k)=b(z^{(1)}(k))^{\alpha} \tag{4.26}$$

is known as the GM(1,1) *power model*. And,

$$\frac{dx^{(1)}}{dt}+ax^{(1)}=b(x^{(1)})^{\alpha} \tag{4.27}$$

is known as the *whitenization equation* of the GM(1,1) power model.

Theorem 4.17. The solution of the whitenization equation of the GM(1,1) power model is

$$x^{(1)}(t)=\left\{e^{-(1-a)at}\left[(1-a)\int be^{(1-a)at}dt+c\right]\right\}^{\frac{1}{1-a}} \tag{4.28}$$

Theorem 4.18. Let $X^{(0)}$, $X^{(1)}$, and $Z^{(1)}$ be defined above. Let

$$B=\begin{bmatrix}-z^{(1)}(2) & (z^{(1)}(2))^{\alpha}\\ -z^{(1)}(3) & (z^{(1)}(3))^{\alpha}\\ \vdots & \vdots\\ -z^{(1)}(n) & (z^{(1)}(n))^{\alpha}\end{bmatrix},\ Y=\begin{bmatrix}x^{(0)}(2)\\ x^{(0)}(3)\\ \vdots\\ x^{(0)}(n)\end{bmatrix}$$

then the least squares estimate of the parametric sequence $\hat{a}=[a,b]^T$ of the GM(1,1) power model is

$$\hat{a}=(B^TB)^{-1}B^TY\ .$$

When the power in the GM(1,1) power model $\alpha=2$, the resultant model

$$x^{(0)}(k)+az^{(1)}(k)=b(z^{(1)}(k))^2 \tag{4.29}$$

is known as the *grey Verhulst model*; and

$$\frac{dx^{(1)}}{dt}+ax^{(1)}=b(x^{(1)})^2 \tag{4.30}$$

is known as the *whitenization equation* of the grey Verhulst model.

Theorem 4.19. (1) The solution of the Verhulst whitenization equation is

$$\begin{aligned}x^{(1)}(t)&=\frac{1}{e^{at}\left[\dfrac{1}{x^{(1)}(0)}-\dfrac{b}{a}(1-e^{-at})\right]}=\frac{ax^{(1)}(0)}{e^{at}\left[a-bx^{(1)}(0)(1-e^{-at})\right]}\\&=\frac{ax^{(1)}(0)}{bx^{(1)}(0)+(a-bx^{(1)}(0))e^{at}}\end{aligned} \tag{4.31}$$

(2) The time response sequence of the grey Verhulst model is

$$\hat{x}^{(1)}(k+1)=\frac{ax^{(1)}(0)}{bx^{(1)}(0)+(a-bx^{(1)}(0))e^{ak}}\,. \tag{4.32}$$

The Verhulst model is mainly used to describe and to study processes with saturated states (or sigmoid processes). For instance, this model is often used in the prediction of human populations, biological growth, reproduction, and economic life span of consumable products, etc. From the solution of the Verhulst equation, it can be seen that when $t\to\infty$, if $a>0$, then $x^{(1)}(t)\to 0$; if $a<0$, then $x^{(1)}(t)\to\dfrac{a}{b}$. That is, there is a sufficiently large t such that for any $k>t$, both $x^{(1)}(k+1)$ and

$x^{(1)}(k)$ will be sufficiently close to each other. In this case, $x^{(0)}(k+1)=x^{(1)}(k+1)-x^{(1)}(k)\approx 0$, meaning that the system approaches distinction.

In practice, one often faces with sigmoid processes in the original data sequences. When such an instance appears, we can simply take the original sequence as $X^{(1)}$ with its accumulation generation as $X^{(0)}$ to establish a Verhulst model to directly simulate $X^{(1)}$.

Example 4.2. Assume that the expenditures on the research of a certain kind of torpedo are given in Table 4.6. Try to employ the Verhulst model to simulate the data and make predictions.

Table 4.6 Expenditures on the research of a certain kind of torpedo (in 10 thousand Yuan)

Year	1995	1996	1997	1998	1999	2000	2001	2002	2003	2004
Expenditure	496	779	1187	1025	488	255	157	110	87	79

The accumulated expenditures are given in Table 4.7.

Table 4.7 Accumulated expenditures (in ten thousand Yuan)

Year	1995	1996	1997	1998	1999	2000	2001	2002	2003	2004
Expenditure	496	1275	2462	3487	3975	4230	4387	4497	4584	4663

From theorem 4.18, we compute the parameters as follows:

$$\hat{a}=[a,b]^T=\begin{bmatrix}-0.98079\\-0.00021576\end{bmatrix}$$

so that the whitenization equation is

$$\frac{dx^{(1)}}{dt}-0.98079x^{(1)}=-0.00021576(x^{(1)})^2$$

By taking $x^{(1)}(0)=x^{(0)}(1)=496$, we obtain the time response sequence

$$\hat{x}^{(1)}(k+1)=\frac{ax^{(1)}(0)}{bx^{(1)}(0)+(a-bx^{(1)}(0))e^{ak}}=\frac{-486.47}{-0.10702-0.87378e^{-0.98079k}}$$

On the basis of this formula, we produce the simulated values $\hat{x}^{(1)}(k)$ as shown in Table 4.8.

Table 4.8 Error checks

Ordinality	Actual data $x^{(0)}(k)$	Simulated data $\hat{x}^{(0)}(k)$	Error $\varepsilon(k)=x^{(0)}(k)-\hat{x}^{(0)}(k)$	Relative error $\Delta_k=\frac{\lvert\varepsilon(k)\rvert}{x^{(0)}(k)}$
2	1275	1119.1	155.9	0.12226
3	2462	2116	346	0.14053
4	3487	3177.5	309.5	0.08876
5	3975	3913.7	61.3	0.01541
6	4230	4286.2	-56.2	0.01328
7	4387	4444.8	-57.8	0.01318
8	4497	4507.4	-10.4	0.0023
9	4584	4531.3	52.7	0.0115
10	4663	4540.3	122.7	0.02631

From Table 4.8, we can obtain the mean relative error

$$\Delta=\frac{1}{9}\sum_{k=2}^{10}\Delta_k=4.3354\%$$

and predict the research expenditure for the year of 2005 on the special kind of torpedo as

$$\hat{x}_1^{(0)}(11)=\hat{x}_1^{(1)}(11)-\hat{x}_1^{(1)}(10)=9.0342\,.$$

This value indicates that the research work on the torpedo is nearing its conclusion.

4.4.4 The GM(r,h) Models

In this subsection, we focus on the investigation of the structure of the GM(r,h) model, its relationships with the GM(1,1) model, GM(1,N) model, GM(0,N) model, GM(2,1) model, and other models.

Definition 4.1. Assume that $X_i^{(0)}=(x_i^{(0)}(1),x_i^{(0)}(2),\cdots,x_i^{(0)}(n))$, $i=1,2,...,h$, be given, where $X_1^{(0)}$ stands for a data sequence of a system's characteristic, and $X_i^{(0)}$, $i=1,2,...,h$, data sequences of relevant factors. Let

$$\alpha^{(1)}\hat{x}_1^{(1)}(k)=\hat{x}_1^{(1)}(k)-\hat{x}_1^{(1)}(k-1)=\hat{x}_1^{(0)}(k)$$
$$\alpha^{(2)}\hat{x}_1^{(1)}(k)=\alpha^{(1)}\hat{x}_1^{(1)}(k)-\alpha^{(1)}\hat{x}_1^{(1)}(k-1)=\hat{x}_1^{(0)}(k)-\hat{x}_1^{(0)}(k-1)$$
$$\cdots\ \cdots\ \cdots\ \cdots\ \cdots\ \cdots$$
$$\alpha^{(r)}\hat{x}_1^{(1)}(k)=\alpha^{(r-1)}\hat{x}_1^{(1)}(k)-\alpha^{(r-1)}\hat{x}_1^{(1)}(k-1)=\alpha^{(r-2)}\hat{x}_1^{(0)}(k)-\alpha^{(r-2)}\hat{x}_1^{(0)}(k-1)$$

and $z^{(1)}(k)=\frac{1}{2}(x^{(1)}(k)+x^{(1)}(k-1))$, then

$$\alpha^{(r)}x_1^{(1)}(k)+\sum_{i=1}^{r-1}a_i\alpha^{(r-i)}x_1^{(1)}(k)+a_r z_1^{(1)}(k)=\sum_{j=1}^{h-1}b_j x_{j+1}^{(1)}(k)+b_h \tag{4.33}$$

is referred to as the GM(r,h) *model.*

The GM(r,h) model is a rth order grey model in h variables.

Definition 4.2. In the GM(r,h) model, $-\hat{a}=[-a_1,-a_2,\cdots,-a_h]^T$ is referred to as the *development coefficient vector,* $\sum_{j=1}^{h-1}b_j x_{j+1}^{(1)}(k)$ the *driving term,* and $\hat{b}=[b_1,b_2,\cdots,b_h]^T$ the *vector of driving coefficients.*

Theorem 4.20. Let $X_1^{(0)}$ be a data sequence of a system's characteristic, $X_i^{(0)}$, $i=2,3,\cdots,N$, the data sequences of relevant factors, $X_i^{(1)}$ the accumulation generated sequence of $X_i^{(0)}$, $Z_1^{(1)}$ the adjacent neighbor mean generated sequence from $X_1^{(1)}$, and $\alpha^{(r-i)}X_1^{(1)}$ the ($r-i$)th order inverse accumulation sequence of $X_1^{(1)}$. Define

$$B=\begin{bmatrix} -\alpha^{(r-1)}x_1^{(1)}(2) & -\alpha^{(r-2)}x_1^{(1)}(2) & \cdots & -\alpha^{(1)}x_1^{(1)}(2) & -z_1^{(1)}(2) & x_2^{(1)}(2) & \cdots & x_h^{(1)}(2) & 1 \\ -\alpha^{(r-1)}x_1^{(1)}(3) & -\alpha^{(r-2)}x_1^{(1)}(3) & \cdots & -\alpha^{(1)}x_1^{(1)}(3) & -z_1^{(1)}(3) & x_2^{(1)}(3) & \cdots & x_h^{(1)}(3) & 1 \\ \cdots & \cdots & \cdots & \cdots & \cdots & \cdots & \cdots & \cdots & \cdots \\ -\alpha^{(r-1)}x_1^{(1)}(n) & -\alpha^{(r-2)}x_1^{(1)}(n) & \cdots & -\alpha^{(1)}x_1^{(1)}(n) & -z_1^{(1)}(n) & x_2^{(1)}(n) & \cdots & x_h^{(1)}(n) & 1 \end{bmatrix},$$

$$Y=\begin{bmatrix} \alpha^{(r)}x_1^{(1)}(2) \\ \alpha^{(r)}x_1^{(1)}(3) \\ \vdots \\ \alpha^{(r)}x_1^{(1)}(n) \end{bmatrix}$$

then the parametric sequence $\hat{a}=[a,b_1,b_2,\cdots,b_N]^T$ of the least squares estimate satisfies

$$\hat{a}=(B^T B)^{-1}B^T Y\ .$$

Definition 4.3. Let $-\hat{a}=[-a_1,-a_2,\cdots,-a_h]^T$ be the development coefficient vector and $\hat{b}=[b_1,b_2,\cdots,b_h]^T$ the driving coefficient vector, then

$$\frac{d^r x_1^{(1)}}{dt^r}+a_1\frac{d^{r-1}x_1^{(1)}}{dt^{r-1}}+a_2\frac{d^{r-2}x_1^{(1)}}{dt^{r-2}}+\cdots+a_{r-1}\frac{dx_1^{(1)}}{dt}+a_r x_1^{(1)}=b_1x_2^{(1)}+b_2x_3^{(1)}+\cdots+b_{h-1}x_h^{(1)}+b_h \tag{4.34}$$

is referred to as the *whitenization* (or *image*) *equation* of the GM(r,h) model

$$\alpha^{(r)}x_1^{(1)}(k)+\sum_{i=1}^{r-1}a_i\alpha^{(r-i)}x_1^{(1)}(k)+a_r z_1^{(1)}(k)=\sum_{j=1}^{h-1}b_j x_{j+1}^{(1)}(k)+b_h\cdot$$

The transformation between the GM(r,h) model and its whitenization equation is realized by either differentiation or difference. That is,

$$\frac{dx_1^{(1)}}{dt}=\frac{x_1^{(1)}(k)-x_1^{(1)}(k-1)}{k-(k-1)}=x_1^{(1)}(k)-x_1^{(1)}(k-1)=\alpha^{(1)}x_1^{(1)}(k)=x_1^{(0)}(k)$$

$$\frac{d^2x_1^{(1)}}{dt^2}=\frac{x_1^{(0)}(k)-x_1^{(0)}(k-1)}{k-(k-1)}=x_1^{(0)}(k)-x_1^{(0)}(k-1)=\alpha^{(1)}x_1^{(0)}(k)$$

....

$$\frac{d^{(r-i)}x_1^{(1)}}{dt^{(r-i)}}=\frac{\alpha^{(r-i-1)}x_1^{(1)}(k)-\alpha^{(r-i-1)}x_1^{(1)}(k-1)}{k-(k-1)}=\alpha^{(r-i-1)}x_1^{(1)}(k)-\alpha^{(r-i-1)}x_1^{(1)}(k-1)=\alpha^{(r-i-1)}x_1^{(0)}(k)$$

....

$$\frac{d^{r}x_1^{(1)}}{dt^{r}}=\frac{\alpha^{(r-1)}x_1^{(1)}(k)-\alpha^{(r-1)}x_1^{(1)}(k-1)}{k-(k-1)}=\alpha^{(r-1)}x_1^{(1)}(k)-\alpha^{(r-1)}x_1^{(1)}(k-1)=\alpha^{(r-1)}x_1^{(0)}(k)$$

The GM(r,h) model is the general form of grey systems models. In particular,

(1) When $r = 1$ and $h = 1$, the previous becomes:

$$\frac{dx_1^{(1)}}{dt}+a_1x_1^{(1)}=b_1 \text{ and } \alpha^{(1)}x_1^{(1)}(k)+a_1z_1^{(1)}(k)=b_1$$

which is the GM(1,1) model.

(2) When $r = 1$ and $h = N$, the previous model takes the form of

$$\frac{dx_1^{(1)}}{dt}+a_1x_1^{(1)}=b_1x_2^{(1)}+b_2x_3^{(1)}+\cdots+b_{N-1}x_N^{(1)}+b_N$$

$$\alpha^{(1)}x_1^{(1)}(k)+a_1z_1^{(1)}(k)=b_1x_2^{(1)}(k)+b_2x_3^{(1)}(k)+\cdots+b_{N-1}x_N^{(1)}(k)+b_N$$

which is the GM(1,N) model.

(3) When $r = 0$ and $h =$ N, the previous model is

$$x_1^{(1)}(k)=b_1x_2^{(1)}(k)+b_2x_3^{(1)}(k)+\cdots+b_{N-1}x_N^{(1)}(k)+b_N$$

which is the GM(0,N) model.

(4) When $r = 2$ and $h = 1$, the previous model reduces to

$$\frac{d^2x_1^{(1)}}{dt^2}+a_1\frac{dx_1^{(1)}}{dt}+a_2x_1^{(1)}=b_1$$

$$\alpha^{(2)}x_1^{(1)}(k)+a_1\alpha^{(1)}x_1^{(1)}(k)+a_2z_1^{(1)}(k)=b_1$$

which is the GM(2,1) model.

Based on this discussion, it can be seen that such models as GM(1,1), GM(1,N), GM(0,N), GM(2,1), etc., are all special cases of the GM(r,h) model. So, it is very important to further the study of the GM(r,h) model.

4.5 Grey Systems Predictions

No matter what needs to be done, one should always get familiar with the situation, think through the details, make educated predictions, and lay out a detailed plan before he could potentially arrive at his desired successful conclusions. For matters as great as international affairs, national events and citizens livings, the developments of regional or business entities, and for matters as small as personal daily works or living details, scientifically sound predictions are needed everywhere. Otherwise, dealing with situations according to personal wishes without knowing any of the related particular details would generally run into difficulties.

The so-called *prediction* is about foretelling the possible course of development of societal events, political matters, economic ups and downs, etc., using scientific methods and techniques based on attainable historical and present data and information so that appropriate actions can be planned and carried out. In short, prediction is about making scientific inferences about the evolutions of materials and events ahead of time. The general prediction includes not only the *static inference* about the unknown matters based on what is known within the same time frame, but also the *dynamic inference* about the future based on the history and the present state of affairs of a certain matter. A specific prediction is only about a dynamic forecast within which a scientific inference about the future evolution of a certain event is given.

Grey prediction is about making scientific, quantitative forecasts about the future states of systems based on understandings of unascertained characteristics of the systems by making use of sequence operators on the original data sequences in order to generate, treat, and excavate the hidden laws of systems evolution so that grey systems models can be established and the future predicted. All the methods of the grey systems theory studied so far can be employed to make predictions. For a given problem, the appropriate prediction model is chosen by making use of the conclusions of a sufficiently and carefully done qualitative analysis. Also, the choice of models should vary along with the changing conditions. Each model chosen has to be tested through many different methods in order to decide its appropriateness and effectiveness. Only the models that pass various tests can be meaningfully employed to make predictions.

Let $X^{(0)} = (x^{(0)}(1), x^{(0)}(2), \cdots, x^{(0)}(n))$ be some raw data, $\hat{X}^{(0)} = (\hat{x}^{(0)}(1), \hat{x}^{(0)}(2), \cdots, \hat{x}^{(0)}(n))$ the simulated data out of a chosen prediction model, $\varepsilon^{(0)} = \left(\varepsilon(1), \varepsilon(2), \ldots \varepsilon(n)\right) = \left(x^{(0)}(1) - \hat{x}^{(0)}(1),\ x^{(0)}(2) - \hat{x}^{(0)}(2),\ \ldots,\ x^{(0)}(n) - \hat{x}^{(0)}(n)\right)$ the error sequence, and

$$\Delta = \left(\left| \frac{\varepsilon(1)}{x^{(0)}(1)} \right|, \left| \frac{\varepsilon(2)}{x^{(0)}(2)} \right|, \cdots, \left| \frac{\varepsilon(n)}{x^{(0)}(n)} \right| \right) = \{\Delta_k\}_1^n$$

the relative error sequence. For $k \le n$, $\Delta_k = \left| \frac{\varepsilon(k)}{x^{(0)}(k)} \right|$ is known as *relative error* of the simulation at point k, and $\bar{\Delta} = \frac{1}{n}\sum_{k=1}^{n} \Delta_k$ the *mean relative error*; $1-\bar{\Delta}$ is known as the *mean relative accuracy*, and $1-\Delta_k$ the *simulation accuracy* at point k, $k = 1,2,\cdots,n$; for a given α, when $\bar{\Delta} < \alpha$ and $\Delta_n < \alpha$ hold true, the prediction model is said to be *error-satisfactory*.

Let ε stands for the absolute degree of incidence between the raw data $X^{(0)}$ and the simulated values $\hat{X}^{(0)}$. If for a given $\varepsilon_0 > 0$ the absolute degree ε of incidence satisfies $\varepsilon > \varepsilon_0$, then the simulation model is said to be *incidence satisfactory*.

Based on the sequences $X^{(0)}$, $\hat{X}^{(0)}$, and $\varepsilon^{(0)}$, consider the relevant means and variances

$$\bar{x} = \frac{1}{n}\sum_{k=1}^{n} x^{(0)}(k), \quad S_1^2 = \frac{1}{n}\sum_{k=1}^{n} (x^{(0)}(k) - \bar{x})^2$$

and

$$\bar{\varepsilon} = \frac{1}{n}\sum_{k=1}^{n} \varepsilon(k), \quad S_2^2 = \frac{1}{n}\sum_{k=1}^{n} (\varepsilon(k) - \bar{\varepsilon})^2.$$

If for a given $C_0 > 0$, the ratio of variances $C = \frac{S_2}{S_1} < C_0$, then the model is said to be *variance ratio satisfactory*. If $p = P(\left| \varepsilon(k) - \bar{\varepsilon} \right| < 0.6745 S_1)$ is seen as a small error probability and for a given $p_0 > 0$, when $p > p_0$, then the model is said to be *small-error probability satisfactory*.

What discussed above are three different ways to test a chosen model. Each of them is based on observations of the error to determine the accuracy of the model. For both the mean relative error $\bar{\Delta}$ and the simulation error, the smaller they are, the better. For the degree of incidence ε, the greater it is the better. For the variance ratio C, the smaller the value is the better, it is because a small C indicates that S_2 is relatively small while S_1 is relatively large, meaning that the error

Table 4.9 Commonly used scales of accuracy for model testing

Threshold / Accuracy scale	Relative error α	Degree of incidence ε_0	Variance ratio C_0	Small error probability p_0
1st level	0.01	0.90	0.35	0.95
2nd level	0.05	0.80	0.50	0.80
3th level	0.10	0.70	0.65	0.70
4th level	0.20	0.60	0.80	0.60

variance is small while the variance of the original data is large so that the errors are relatively more concentrated with little fluctuation than the original data. Therefore, for better simulation results, the smaller S_2 is when compared to S_1 the better. For the case of small error probability p , as soon as a set of α , ε_0 , C_0 , and p_0 values are chosen, a scale of accuracy for testing models is determined. The most commonly used scales of accuracy for testing models are listed in Table 4.9.

In most applications published so far in the area of grey systems, the most commonly used is the criterion of relative errors.

4.5.1 Sequence Predictions

A *sequence prediction* is an attempt of foretelling the future behaviors of a system's variables. To this end, the GM(1,1) model is commonly employed. Based on the practical situation involved, other models have also been considered. On the basis of qualitative analysis, appropriate sequence operators are used on the available data sequence. Then, an suitable GM(1,1) model is established. After making sure that the model passes the accuracy test, it is applied to make predictions. For related practical applications, please consult with (Liu and Lin, 2006, p. 277 – 281).

4.5.2 Interval Predictions

If the given sequence of raw data is chaotic and it is difficult for any model to pass the accuracy test, the researcher will then have trouble producing accurate quantitative predictions. In this case, one can consider providing a range for the future values to fall within. In particular, let $X(t)$ be a zigzagged line. If there are smooth and continuous curves $f_u(t)$ and $f_s(t)$, satisfying that for any t , $f_u(t) < X(t) < f_s(t)$, then $f_u(t)$ is known as the *lower bound function* of $X(t)$ and $f_s(t)$ the *upper bound function*, and $S = \{(t, X(t)) | X(t) \in [f_u(t), f_s(t)]\}$ the *value band* of $X(t)$. If the upper and lower bound of $X(t)$ are the same kind functions, then S is known as a uniform band. When S is a uniform band with exponential functions as its upper and lower bounds $f_u(t)$ and $f_s(t)$, then S is known as a *uniform exponential band*. If a uniform band S has linear upper and lower bound functions $f_u(t)$ and $f_s(t)$, then S is known as a *uniform linear band* or a *straight band* for short. If for $t_1 < t_2$, $f_s(t_1) - f_u(t_1) < f_s(t_2) - f_u(t_2)$ always holds true, then S is known as a *trumpet-like band*.

Let $X^{(0)} = (x^{(0)}(1), x^{(0)}(2), \cdots, x^{(0)}(n))$ be a sequence of raw data, and its accumulation generation be $X^{(1)} = (x^{(1)}(1), x^{(1)}(2), \cdots, x^{(1)}(n))$. Define

$$\sigma_{\max} = \max_{1 \le k \le n}\{x^{(0)}(k)\}, \sigma_{\min} = \min_{1 \le k \le n}\{x^{(0)}(k)\}$$

and respectively take the upper and lower bound functions $f_u(n+t)$ and $f_s(n+t)$ of $X^{(1)}$ as follows

$$f_u(n+t) = x^{(1)}(n) + t\sigma_{\min}, \ f_s(n+t) = x^{(1)}(n) + t\sigma_{\max}.$$

Then $S = \{(t, X(t)) | t > n, X(t) \in [f_u(t), f_s(t)]\}$ is known as the *proportional band*.

Proposition 4.3. Each proportional band is also a straight trumpet-like band.

In fact, both the upper and lower bound functions of a proportional band are increasing straight lines of time with slopes $\sigma_{\min}$ and $\sigma_{\max}$, respectively.

For a sequence $X^{(0)}$ of raw data, let $X_u^{(0)}$ be the sequence corresponding to the curve that connects all the low points of $X^{(0)}$ and $X_s^{(0)}$ the sequence corresponding to the curve of all the upper points of $X^{(0)}$. Assume that

$$\hat{x}_u^{(1)}(k+1) = \left(x_u^{(0)}(1) - \frac{b_u}{a_u} \right) \exp(-a_u k) + \frac{b_u}{a_u}$$

and

$$\hat{x}_s^{(1)}(k+1) = \left(x_s^{(0)}(1) - \frac{b_s}{a_s} \right) \exp(-a_s k) + \frac{b_s}{a_s}$$

are respectively the GM(1,1) time response sequences of $X_u^{(0)}$ and $X_s^{(0)}$. Then

$$S = \{(t, X(t)) | X(t) \in [\hat{X}_u^{(1)}(t), \hat{X}_s^{(1)}(t)]\}$$

is known as a *wrapping band*, Figure 4.1.

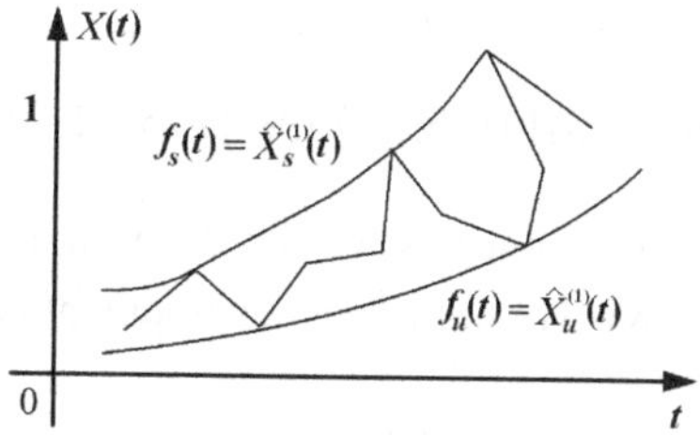

Fig. 4.1 A wrapping band

For a given sequence $X^{(0)}$ of raw data, let us take m different subsequences to establish m GM(1,1) models with the corresponding parameters $\hat{a}_i=[a_i,b_i]^T;i=1,2,\cdots,m$. Let

$$-a_{\max}=\max_{1\le i\le m}\{-a_i\},\ -a_{\min}=\min_{1\le i\le m}\{-a_i\}$$

$$\hat{x}_u^{(1)}(k+1)=\left(x_u^{(0)}(1)-\frac{b_{\min}}{a_{\min}}\right)\exp(-a_{\min}k)+\frac{b_{\min}}{a_{\min}}$$

$$\hat{x}_s^{(1)}(k+1)=\left(x_s^{(0)}(1)-\frac{b_{\max}}{a_{\max}}\right)\exp(-a_{\max}k)+\frac{b_{\max}}{a_{\max}}$$

then $S=\{(t,X(t))\big|X(t)\in[\hat{X}_u^{(1)}(t),\hat{X}_s^{(1)}(t)]\}$ is known as a *development band.*

Proposition 4.4. All wrapping bands and development bands are exponential bands.

For a sequence $X^{(0)}=(x^{(0)}(1),x^{(0)}(2),\cdots,x^{(0)}(n))$ of raw data, let $f_u(t)$ and $f_s(t)$ be a upper and a lower bound function of the accumulation generation $X^{(1)}$ of $X^{(0)}$. For any $k>0$,

$$\hat{x}^{(0)}(n+k)=\frac{1}{2}[f_u(n+k)+f_s(n+k)]$$

is known as *basic prediction value*, and $\hat{x}_u^{(0)}(n+k)=f_u(n+k)$ and $\hat{x}_s^{(0)}(n+k)=f_s(n+k)$ respectively the *lowest and highest predicted values.*

Example 4.3. The data (in ten thousand) for the bicycle retails in a certain city are given as follows:

$$X^{(0)}=(x^{(0)}(1),x^{(0)}(2),x^{(0)}(3),x^{(0)}(4),x^{(0)}(5),x^{(0)}(6))$$
$$=(5.0810,4.6110,5.1177,9.3775,11.0574,11.3524)$$

where $x^{(0)}(1)=$ 5.0810 is the annual sales for the year of 1978, …, and $x^{(0)}(6)=$ 11.3524for the year of 1983. Try to make a development band prediction.

Solution. Take the following subsequences

$$X_1^{(0)}=(x^{(0)}(1),x^{(0)}(2),x^{(0)}(3),x^{(0)}(4),x^{(0)}(5),x^{(0)}(6))$$
$$=(5.0810,4.6110,5.1177,9.3775,11.0574,11.3524)$$
$$X_2^{(0)}=(x^{(0)}(1),x^{(0)}(2),x^{(0)}(3),x^{(0)}(4),x^{(0)}(5))$$
$$=(5.0810,4.6110,5.1177,9.3775,11.0574)$$
$$X_3^{(0)}=(x^{(0)}(2),x^{(0)}(3),x^{(0)}(4),x^{(0)}(5),x^{(0)}(6))$$
$$=(4.6110,5.1177,9.3775,11.0574,11.3524)$$
$$X_4^{(0)}=(x^{(0)}(3),x^{(0)}(4),x^{(0)}(5),x^{(0)}(6))$$
$$=(5.1177,9.3775,11.0574,11.3524)$$

Based on each of these subsequences, let us establish the corresponding GM(1,1) models:

$$\frac{dx^{(1)}}{dt}+a_ix^{(1)}=b_i,\ i=1,2,3,4$$

with their individual parameters $\hat{a}_i=[a_i,b_i]^T, i=1,2,3,4$, given below:

$$\hat{a}_1=[a_1,b_1]^T=[-0.2202,3.4689]^T,$$
$$\hat{a}_2=[a_2,b_2]^T=[-0.3147,2.1237]^T$$
$$\hat{a}_3=[a_3,b_3]^T=[-0.2013,5.0961]^T,$$
$$\hat{a}_4=[a_4,b_4]^T=[-0.0911,8.7410]^T$$

Because

$$-a_{\min}=\min_{1\le i\le 4}\{-a_i\}=\min\{0.2202,0.3147,0.2013,0.0911\}=0.0911=-a_4$$
$$-a_{\max}=\max_{1\le i\le 4}\{-a_i\}=\max\{0.2202,0.3147,0.2013,0.0911\}=0.3147=-a_2$$

the upper bound time response sequence of the development band is

$$\begin{cases} \hat{x}_s^{(1)}(k+1) = \left(x^{(0)}(1) - \dfrac{b_2}{a_2}\right)e^{-a_2 k} + \dfrac{b_2}{a_2} = 11.8293e^{0.3147k} - 6.7483 \\ \hat{x}_s^{(0)}(k+1) = \hat{x}_s^{(1)}(k+1) - \hat{x}_s^{(1)}(k) \end{cases}$$

That is, $\hat{x}_s^{(0)}(k+1) = 11.8293e^{0.3147k} - 11.8293e^{0.3147k-0.3147} = 3.1938e^{0.3147k}$. So, the highest predicted values are $\hat{x}_s^{(0)}(7)$ = 21.1029, $\hat{x}_s^{(0)}(8)$ = 28.9078, and $\hat{x}_s^{(0)}(9)$=39.5993. Because the starting value of $X_4^{(0)}$ is $x^{(0)}(3)$, the lower bound time response sequence of the development band is

$$\begin{cases} \hat{x}_u^{(1)}(k+3) = \left(x^{(0)}(3) - \dfrac{b_4}{a_4}\right)e^{-a_4 k} + \dfrac{b_4}{a_4} = 101.0672e^{0.0911k} - 95.9495 \\ \hat{x}_u^{(0)}(k+3) = \hat{x}_u^{(1)}(k+3) - \hat{x}_u^{(0)}(k+2) \end{cases}$$

That is, $\hat{x}_u^{(0)}(k+3) = 101.0672e^{0.0911k} - 101.0672e^{0.0911k-0.0911} = 8.8003e^{0.0911k}$. So we obtain the lowest predicted values: $\hat{x}_u^{(0)}(7)$= 12.6694, $\hat{x}_u^{(0)}(8)$= 13.8777, and $\hat{x}_u^{(0)}(9)$=15.2014. From the highest and lowest predicted values, we obtain the basic prediction values:

$$\hat{x}^{(0)}(7) = \frac{1}{2}[\hat{x}_s^{(0)}(7) + \hat{x}_u^{(0)}(7)] = 16.8862$$

$$\hat{x}^{(0)}(8) = \frac{1}{2}[\hat{x}_s^{(0)}(8) + \hat{x}_u^{(0)}(8)] = 21.3928$$

$$\hat{x}^{(0)}(9) = \frac{1}{2}[\hat{x}_s^{(0)}(9) + \hat{x}_u^{(0)}(9)] = 27.4004$$

Based on the qualitative analysis on the estimated amount of ownership of bicycles of the city and the obvious improvement on the public transportation system, we conclude that the lowest predicted values are more reliable.

4.5.3 Disaster Predictions

4.5.3.1 Grey Disaster Predictions

The basic idea of *grey disaster predictions* is essentially the prediction of abnormal values. As for what kinds of values are considered abnormal, it is commonly determined based on individuals' experiences. The task of grey disaster predictions is to provide the time moments of the forthcoming abnormal values so that relevant parties can prepare for the worst ahead of time. In particular, let

$X = (x(1), x(2), \cdots, x(n))$ be a sequence of raw data. For a given upper abnormal (or catastrophe) value ξ, the subsequence of X

$$X_\xi = (x[q(1)], x[q(2)], \cdots, x[q(m)]) = \{x[q(i)] | x[q(i)] \geq \xi; i = 1,2,\cdots,m\}$$

is known as the *upper catastrophe sequence.* For a given lower abnormal (or catastrophe) value ζ, the subsequence

$$X_\zeta = (x[q(1)], x[q(2)], \cdots, x[q(l)]) = \{x[q(i)] | x[q(i)] \geq \zeta; i = 1,2,\cdots,l\}$$

is known as the *lower catastrophe sequence.* Together, these upper and lower catastrophe sequences are referred to as *catastrophe sequences.* Because the idea behind the discussion of catastrophe sequences is the same, in the following presentation, we will not distinguish upper and lower catastrophe sequences.

Now, assume $X_\xi = (x[q(1)], x[q(2)], \cdots, x[q(m)]) \subset X$ is a catastrophe sequence. Then, $Q^{(0)} = (q(1), q(2), \cdots, q(m))$ will be referred to as the *catastrophe date sequence.* The so-called disaster prediction is about finding patterns, if any, through the study of catastrophe date sequences to predict future dates of occurrences of catastrophes. In grey system theory, each disaster prediction is realized through establishing GM(1,1) models for relevant catastrophe date sequences. In particular, if the accumulation generation $Q^{(1)} = (q(1)^{(1)}, q(2)^{(1)}, \cdots, q(m)^{(1)})$ of the catastrophe date sequence $Q^{(0)}$ is employed to establish the GM(1,1) model $q(k) + az^{(1)}(k) = b$, where $Z^{(1)}$ is the adjacent neighbor mean generated sequence of $Q^{(1)}$, then this model is referred to as a catastrophe GM(1,1) model.

For the available sequence $X = (x(1), x(2), \cdots, x(n))$ of raw data, if n stands for the present and the last entry $q(m)(\leq n)$ in the corresponding catastrophe date sequence $Q^{(0)}$ represents when the last catastrophe occurred, then the predicted value $\hat{q}(m+1)$ means when the next catastrophe will happen and for any $k > 0$, $\hat{q}(m+k)$ stands for the predicted date for the *k*th catastrophe to occur in the future.

4.5.3.2 Seasonal Disaster Predictions

Let $\Omega = [a,b]$ be the overall time interval of a study. If $\omega_i = [a_i, b_i] \subset \Omega = [a,b], i = 1,2,\cdots,s$, satisfy $\Omega = \bigcup_{i=1}^{s} \omega_i$ and $\omega_i \cap \omega_j = \varnothing$ for any $i \neq j$, then $\omega_i (i = 1,2,\cdots,s)$ is referred to as a season (or a *time zone* or *time interval*) in Ω. Let $\omega_i \subset \Omega$ be a season. For an available sequence of raw data $X = (x(1), x(2), \cdots, x(n)) \subset \omega_i$ and a chosen value ξ of abnormality, the corresponding catastrophe sequence $X_\xi = (x[q(1)], x[q(2)], \cdots, x[q(m)])$ is referred to as a *seasonal catastrophe sequence*, and $Q^{(0)} = (q(1), q(2), \cdots, q(m))$ the date sequence of seasonal catastrophes.

Proposition 4.5. Given $\omega_i = [a_i, b_i] \subset \Omega$ with $a_i > 0$ and a sequence of raw data

$$X = (x(1), x(2), \cdots, x(n)) \subset \omega_i = [a_i, b_i]$$

let

$$y(k) = x(k) - a_i \ ; \ k = 1, 2, \cdots, n$$

then the distinguishing rate of the data in the sequence $Y = (y(1), y(2), \cdots, y(n))$ is greater than that of the original sequence X .

Proof. For details see (Liu and Lin, 2006, p. 295). QED.

In general, a *seasonal disaster prediction* is conducted by walking through the following steps: 1) collect the data $X = (x(1), x(2), \cdots, x(n))$; 2) determine the range or season $\omega_i = [a_i, b_i]$ of the available data; 3) Let $y(k) = x(k) - a_i$ so that the original data sequence is reduced to $Y = (y(1), y(2), \cdots, y(n))$ with a better rate of distinction; 4) choose the value ξ of abnormality and define the seasonal catastrophe sequence $X_\xi = (x[q(1)], x[q(2)], \cdots, x[q(m)])$ and the corresponding seasonal catastrophe date sequence $Q^{(0)} = (q(1), q(2), \cdots, q(m))$; 5) establish the catastrophe GM(1,1) model; and 6) check the simulation accuracy and make prediction.

4.5.4 Stock-Market-Like Predictions

When the available data sequence vibrates widely with large magnitudes, it is often difficult, if not impossible, to find an appropriate simulation model. In this case, one can consider making use of the pattern of fluctuation of the data to predict the future development of the wavy movement. This kind of prediction is known as a *stock-market-like prediction*. In particular, if $X = (x(1), x(2), \cdots, x(n))$ is the sequence of raw data, then $x_k = x(k) + (t - k)[x(k+1) - x(k)]$ is known as a *k-piece zigzagged line* of the sequence *X*, and

$\{x_k = x(k) + (t-k)[x(k+1) - x(k)] \| k = 1,2,\cdots,n-1\}$ the zigzagged line, still denoted by using X. Let $\sigma_{\max} = \max_{1 \le k \le n}\{x(k)\}$ and $\sigma_{\min} = \min_{1 \le k \le n}\{x(k)\}$. For any $\forall \xi \in [\sigma_{\min}, \sigma_{\max}]$, $X = \xi$ is known as the ξ *-contour* (line); and the solutions $(t_i, x(t_i))(i = 1,2,\cdots)$ of system of equations

$$\begin{cases} X = \{x(k) + (t-k)[x(k+1) - x(k)] \| k = 1,2,\cdots,n-1\} \\ X = \xi \end{cases}$$

the ξ*-contour points*. These points are the intersections of the zigzagged line X and the ξ-contour line.

Proposition 4.6. If on the ith segment of X is there a ξ-contour point, then the coordinates of this point is given by $\left(i + \frac{\xi - x(i)}{x(i+1) - x(i)}, \xi\right)$.

Let $X_\xi = (P_1, P_2, \cdots, P_m)$ be the sequence of ξ -contour points of X such that point P_i is located on the ith segment . Let

$$q(i) = t_i + \frac{\xi - x(t_i)}{x(t_i + 1) - x(t_i)},\ i = 1,2,\cdots,m$$

then $Q^{(0)} = (q(1), q(2), \cdots, q(m))$ is known as the ξ *-contour time moment sequence*. By establishing a GM(1,1) model using this ξ -contour moment sequence, one can produce the predicted values for future ξ -contour time moments: $\hat{q}(m+1), \hat{q}(m+2), \cdots, \hat{q}(m+k)$. The lines $X = \xi_i (i = 0,1,2,\cdots,s)$, where $\xi_0 = \sigma_{\min}$, $\xi_1 = \frac{1}{s}(\sigma_{\max} - \sigma_{\min}) + \sigma_{\min}$, … $\xi_i = \frac{i}{s}(\sigma_{\max} - \sigma_{\min}) + \sigma_{\min}$, …, $\xi_{s-1} = \frac{s-1}{s}(\sigma_{\max} - \sigma_{\min}) + \sigma_{\min}$, $\xi_s = \sigma_{\max}$, are known as *equal time distanced contours*. When taking contour lines, one needs to make sure that the corresponding contour moments satisfy the conditions for establishing valid GM(1,1) models.

Let $X = \xi_i (i = 1,2,\cdots,s)$ be s different contours, $Q_i^{(0)} = (q_i(1), q_i(2), \cdots, q_i(m_1)$, $i = 1,2,\cdots,s$, stand for the sequence of

ξ_i -contour time moments, and $\hat{q}_i(m_i+1), \hat{q}_i(m_i+2), \cdots, \hat{q}_i(m_i+k_i)$, $i=1,2,\cdots,s$, the GM(1,1) predicted ξ_i-contour time moments. If there are $i \neq j$ such that $\hat{q}_i(m_i+l_i) = \hat{q}_j(m_j+l_j)$, then these values are known as a pair of *invalid moments*.

Proposition 4.7. Let $\hat{q}_i(m_i+j)$, $j=1,2,\cdots,k_i$, $i=1,2,\cdots,s$, be the GM(1,1) predicted ξ_i-contour time moments. After deleting all invalid predictions, order the rest in terms of their magnitudes as follows: $\hat{q}(1) < \hat{q}(2) < \ldots < \hat{q}(n_s)$, where $n_s \le k_1 + k_2 + \cdots + k_s$. If $X = \xi_{\hat{q}(k)}$ is the contour line corresponding to $\hat{q}(k)$, then the predicted wavy curve of $X^{(0)}$ is given below:

$$X = \hat{X}^{(0)} = \{\xi_{\hat{q}(k)} + [t - \hat{q}(k)][\xi_{\hat{q}(k+1)} - \xi_{\hat{q}(k)}] \| k = 1,2,\cdots,n_s\}.$$

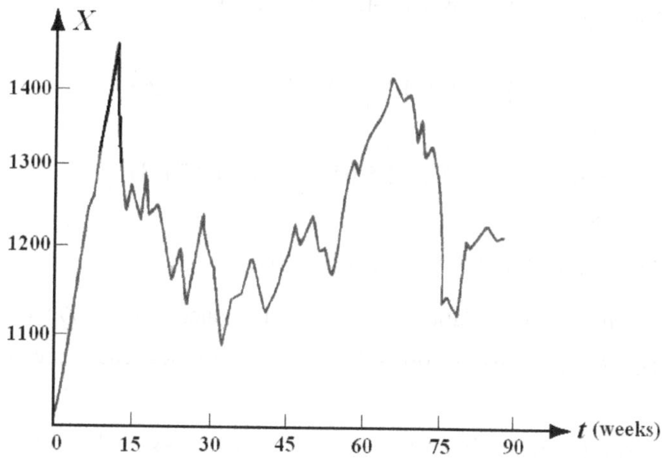

Fig. 4.2 Shanghai stock exchange index (Feb. 21, 1997, to Oct. 31, 1998)

Example 4.4. Let us look at a wavy curve prediction for the (synthetic) stock index of Shanghai stock exchange.

Using the stock index data of the stock index weekly closes of Shanghai stock exchange, the time series plot from February 21, 1997, to October 31, 1998, is shown in Figure 4.2. Let us take

$\xi_1 = 1140$, $\xi_2 = 1170$, $\xi_3 = 1200$, $\xi_4 = 1230$, $\xi_5 = 1260$,
$\xi_6 = 1290$, $\xi_7 = 1320$, $\xi_8 = 1350$, $\xi_9 = 1380$.

Then the corresponding ξ_i-contour time moment sequences are given below:

(1) For $\xi_1 = 1140$,

$$Q_1^{(0)} = \{q_1(k)\}_1^7 = (4.4, 31.7, 34.2, 41, 42.4, 76.8, 78.3)$$

(2) For $\xi_2 = 1170$,

$$Q_2^{(0)} = \{q_2(k)\}_2^{12} = (5.2, 19.8, 23, 25.6, 26.9, 31.2, 34.8, 39.5, 44.6, 76, 76.2, 79.2)$$

(3) For $\xi_3 = 1200$,

$$Q_3^{(0)} = \{q_3(k)\}_3^{11} = (5.9, 19.5, 24.8, 25.2, 26.5, 30.3, 46.2, 53.4, 55.4, 75.5, 79.7)$$

(4) For $\xi_4 = 1230$,

$$Q_4^{(0)} = \{q_4(k)\}_4^{10} = (6.5, 19.2, 28.3, 29.5, 49.7, 50.8, 56.2, 76.4, 82.9, 85)$$

(5) For $\xi_5 = 1260$,

$$Q_5^{(0)} = \{q_5(k)\}_5^7 = (7, 14.2, 16.5, 16.4, 18.8, 56.7, 75.2)$$

(6) For $\xi_6 = 1290$,

$$Q_6^{(0)} = \{q_6(k)\}_6^5 = (8.3, 13.4, 16.9, 56.2, 74.6)$$

(7) For $\xi_7 = 1320$,

$$Q_7^{(0)} = \{q_7(k)\}_7^6 = (8.8, 12.8, 60.2, 71.8, 72.7, 73.6)$$

(8) For $\xi_8 = 1350$,

$$Q_8^{(0)} = \{q_8(k)\}_8^6 = (9.6, 12.5, 61.8, 69.8, 70.9, 71.8)$$

(9) For $\xi_9 = 1380$,

$$Q_9^{(0)} = \{q_9(k)\}_9^4 = (10.8, 12.4, 64.1, 69)$$

Applying the accumulation generation once on $Q_i^{(0)} (i = 1,2,\cdots,9)$ produces $Q_i^{(1)} (i = 1,2,\cdots,9)$, whose GM(1,1) response sequences are respectively given by

$$\hat{q}_1^{(1)}(k+1) = 113.91e^{0.215k} - 109.51,\ \hat{q}_2^{(1)}(k+1) = 98.58e^{0.159k} - 93.83,$$
$$\hat{q}_3^{(1)}(k+1) = 102.08e^{0.166k} - 96.18,\ \hat{q}_4^{(1)}(k+1) = 151.66e^{0.160k} - 145.16,$$
$$\hat{q}_5^{(1)}(k+1) = 13e^{0.435k} - 6,\ \hat{q}_6^{(1)}(k+1) = 21.94e^{0.539k} - 13.64,$$
$$\hat{q}_7^{(1)}(k+1) = 185.08e^{0.192k} - 176.28,\ \hat{q}_8^{(1)}(k+1) = 193.19e^{0.186k} - 182.57,$$
$$\hat{q}_9^{(1)}(k+1) = 45.22e^{0.490k} - 35.39.$$

By letting $\hat{q}_i(k+1) = \hat{q}_i^{(1)}(k+1) - \hat{q}_i^{(1)}(k)$, we obtain the following ξ_i-contour prediction sequences, $i = 1,2,\cdots,9$,

$$\hat{Q}_1^{(0)} = (\hat{q}_1(12), \hat{q}_1(13)) = (99.8, 127.7)$$
$$\hat{Q}_2^{(0)} = (\hat{q}_2(13), \hat{q}_2(14), \hat{q}_2(15)) = (96.8, 116.7, 131.4)$$

$$\hat{Q}_3^{(0)} = (\hat{q}_3(12), \hat{q}_3(13), \hat{q}_3(14)) = (95.7, 114.2, 133.8)$$
$$\hat{Q}_4^{(0)} = (\hat{q}_4(11), \hat{q}_4(12), \hat{q}_4(13)) = (110.9, 134.2, 152.8)$$
$$\hat{Q}_5^{(0)} = (\hat{q}_5(8), \hat{q}_5(9)) = (94.2, 148.8)$$
$$\hat{Q}_6^{(0)} = (\hat{q}_6(6)) = (135.5)$$
$$\hat{Q}_7^{(0)} = (\hat{q}_7(7), \hat{q}_7(8), \hat{q}_7(9)) = (101.9, 123.4, 149.5)$$
$$\hat{Q}_8^{(0)} = (\hat{q}_8(7), \hat{q}_8(8), \hat{q}_8(9)) = (105, 119.8, 144.6)$$
$$\hat{Q}_9^{(0)} = (\hat{q}_9(5)) = (122.3)$$

Based on these predictions, we construct the predicted wavy curve for Shanghai stock exchange index for the time period from November 1998 to the end of 1999, Figure 4.3.

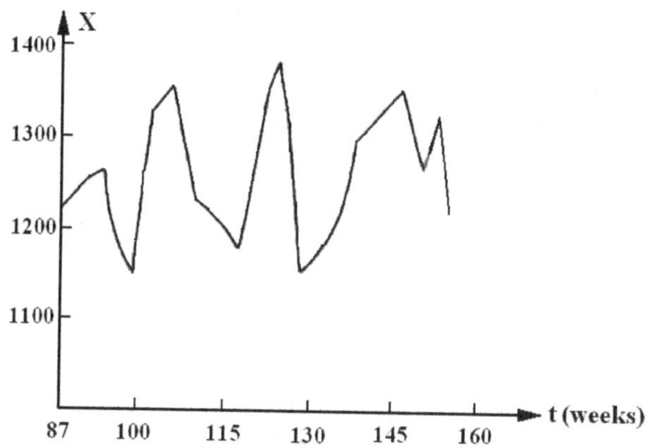

Fig. 4.3 The predicted wavy curve of Shanghai stock exchange index (Nov. 1998 to March 2000)

4.5.5 Systems Predictions

4.5.5.1 The Thought of Five-Step Modeling

When studying a system, generally one should first establish a mathematical model on which the overall functionality of the system, abilities of coordination, and incidence relations, causal relations, and dynamic relationships between different parts can be quantitatively investigated. This kind of study has to be guided by an early qualitative analysis, and close connection between the quantitative and qualitative studies. As for the development of the system's model, one generally goes through the following five steps: development of thoughts, analysis of relevant factors, quantification, dynamicalization, and optimization. This is the so-called five-step modeling.

Step 1: Develop thoughts and form concepts. Through an initial qualitative analysis, one clarifies his goal, goal, possible paths and specific procedures, and then verbally and precisely describes the desired outcomes. That is the initial language model of the problem.

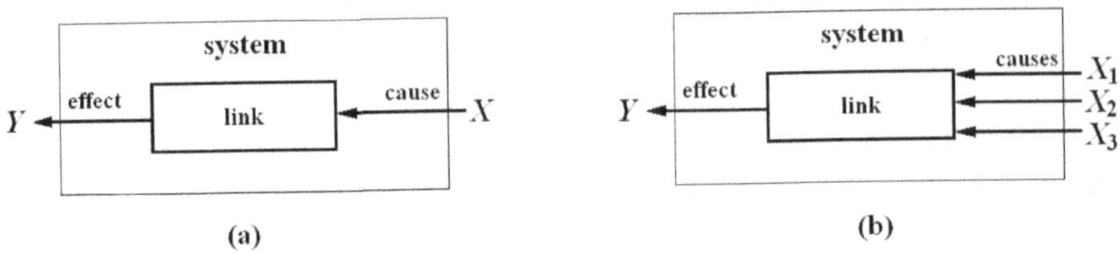

Fig. 4.4 Depicted causal relationships

Step 2: Examine all the factors involved in the language model and their mutual relationships in order to pinpoint out the causes and conclusions. Then, construct a line-drawing to depict the causal relationships, Figure 4.4. Each pair (or a group) of causes and effect form a link. A system might be made up of many of such links. At the same time, a quantity can be a cause of a link and also a consequence of another link. When several of these links are connected, one obtains a line drawing of many links that organically form the system of our concern, Figure 4.5.

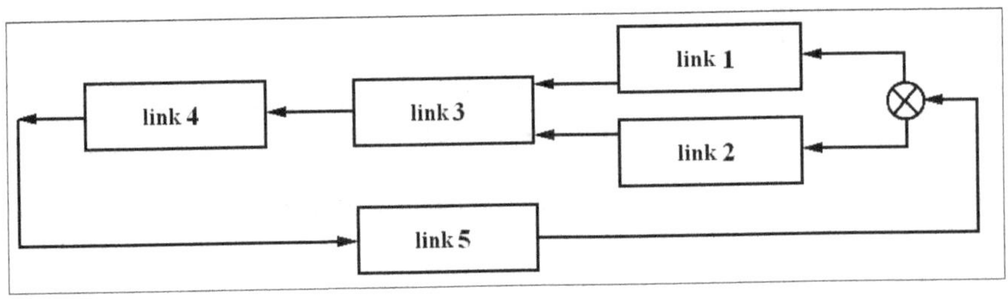

Fig. 4.5 Line drawing of an abstract system

Step 3: Quantitatively study each causality link and obtain an approximate quantitative relationship, which is a quantified model.

Step 4: For each link, collect additional input-output data, on which dynamic GM models are established. Such dynamic models are higher level quantitative models. They can further reveal the relationships between the input and output and their laws of transformation. They are the foundation of systems analysis and optimization.

Step 5: Systematically investigate the established dynamic models by adjusting their structures, mechanisms, and parameters in order to arrive at the purpose of optimizing the outcome and realizing the desired conclusions. Models obtained in this way are known as optimal models.

This procedure of five-step modeling is such a holistic process that at five different stages five different kinds of models are established: language models, network models, quantified models, dynamic models, and optimized models. In the entire process of modeling, the conclusions of the next level should be repeatedly fed back so that the modeling exercise itself becomes a feedback system making the model system as perfect as possible.

4.5.5.2 System of Prediction Models

For a system with many mutually related factors and many autonomous controlling variables, no single model can reflect adequately the development and change of the system. To effectively study such a system and to predict its future behaviors, one should consider establishing a system of models. In particular, let $X_i^{(0)}$, $i = 1,2,\ldots,m$, be sequences of raw data for the state variables of a system, and $U_j^{(0)}$, $j = 1,2,\ldots,s$, sequences of data of the control variables. Then the following

$$\frac{dx_i^{(1)}}{dt} = a_{21}x_1^{(1)} + a_{22}x_2^{(1)} + \cdots + a_{2m}x_m^{(1)} + b_{21}u_1^{(1)} + b_{22}u_2^{(1)} + \cdots + b_{2s}u_s^{(1)} ,$$
$i = 1,2,\ldots,m$

$$\frac{du_j^{(1)}}{dt} = c_2 u_2^{(1)} + d_2, \; j = 1,2,\ldots,s$$

is known as a *system of prediction models*. As a matter of fact, each system of prediction models consists of *m* GM(1,*m* + *s*) and *s* GM(1,1) models. If we write the previous system of prediction models using the terminology of matrices, we have

$$\begin{cases} \dot{X} = AX + BU \\ \dot{U} = CU + D \end{cases}$$

where $X = (x_1, x_2, \cdots, x_m)^T$, $U = (u_1, u_2, \cdots, u_s)^T$, $A = [a_{kl}]_{m\times m}$, $C = diag[c_j]_{s\times s}$, $B = [b_{pq}]_{m\times s}$, and $D = [d_j]_{s\times 1}$. *X* is known as the state vector, *U* the control vector, *A* the state matrix, *B* the control matrix, *C* the development matrix, and *D* the grey effect vector.

Proposition 4.8. For the previous system of prediction models, the time response sequences are given as follows:

$$\hat{x}_i^{(1)}(k+1)=\left\{x_i^{(1)}(0)+\frac{1}{a_{ii}}\left[\sum_{j\neq i}a_{ij}x_j^{(1)}(k+1)+\sum_{k=1}^{s}b_{ik}u_k^{(1)}(k+1)\right]\right\}\cdot e^{a_{ii}k},$$

$i=1,2,\ldots,m$

$$\hat{u}_j^{(1)}(k+1)=\left[u_j^{(1)}(0)+\frac{d_j}{c_j}\right]\cdot e^{c_jk}-\frac{d_j}{c_j},\ j=1,2,\ldots,s$$

where the response sequences of the state variables are only approximate.

Chapter 5
Discrete Grey Prediction Models

This chapter mainly studies the latest research results related to discrete grey models, optimization of these models, approximately nonhomogeneous exponential growths, and discrete grey models of multi-variables. In the process of establishing discrete grey models, discrete forms of equations are employed throughout the stages of parameter estimate, simulation, and prediction so that no attempt of substituting either continuous or discrete models by the other is made, leading to relatively high degrees of accuracy.

5.1 The Basics

5.1.1 Definitions on Discrete Grey Models

Definition 5.1. The following $x^{(1)}(k+1)=\beta_1 x^{(1)}(k)+\beta_2$ is referred to as a *discrete grey model* or a *discretization of the GM(1,1) model.*

Theorem 5.1. Let $X^{(0)}=\{x^{(0)}(1),x^{(0)}(2),\cdots,x^{(0)}(n)\}$ be a nonnegative sequence and its accumulation generation $X^{(1)}=\{x^{(1)}(1),x^{(1)}(2),\cdots,x^{(1)}(n)\}$. If $\hat{\beta}=(\beta_1,\beta_2)^T$ is the parametric sequence and

$$Y=\begin{bmatrix} x^{(1)}(2) \\ x^{(1)}(3) \\ \vdots \\ x^{(1)}(n) \end{bmatrix},\quad B=\begin{bmatrix} x^{(1)}(1) & 1 \\ x^{(1)}(2) & 1 \\ \vdots & \vdots \\ x^{(1)}(n-1) & 1 \end{bmatrix}$$

then the least squares estimates of the parameters of the discrete model $x^{(1)}(k+1)=\beta_1 x^{(1)}(k)+\beta_2$ satisfy $\hat{\beta}=(B^T B)^{-1}B^T Y$.

Proof. Substituting the data sequence into the grey differential equation $x^{(1)}(k+1)=\beta_1 x^{(1)}(k)+\beta_2$ produces

$$x^{(1)}(2)=\beta_1 x^{(1)}(1)+\beta_2$$
$$x^{(1)}(3)=\beta_1 x^{(1)}(2)+\beta_2$$
$$\cdots \quad \cdots \quad \cdots \quad \cdots$$
$$x^{(1)}(n)=\beta_1 x^{(1)}(n-1)+\beta_2$$

That is $Y=B\hat{\beta}$. For a pair of estimated values of β_1,β_2, using $\beta_1 x^{(1)}(k)+\beta_2$ to substitute $x^{(1)}(k+1)$, $k=1,2,\cdots,n-1$, on the left-hand side leads to the error sequence $\varepsilon=Y-B\hat{\beta}$. Assume that

$$S=\varepsilon^T\varepsilon=(Y-B\hat{\beta})^T(Y-B\hat{\beta})=\sum_{k=1}^{n-1}(x^{(1)}(k+1)-\beta_1 x^{(1)}(k)-\beta_2)^2$$

So, the values of β_1,β_2 making S the smallest possible satisfy

$$\begin{cases}\dfrac{\partial S}{\partial \beta_1}=-2\sum_{k=1}^{n-1}(x^{(1)}(k+1)-\beta_1 x^{(1)}(k)-\beta_2)\cdot x^{(1)}(k)=0\\[2ex] \dfrac{\partial S}{\partial \beta_2}=-2\sum_{k=1}^{n-1}(x^{(1)}(k+1)-\beta_1 x^{(1)}(k)-\beta_2)=0\end{cases}$$

Solving this system of equations produces

$$\begin{cases}\beta_1=\dfrac{\sum_{k=1}^{n-1}x^{(1)}(k+1)x^{(1)}(k)-\dfrac{1}{n-1}\sum_{k=1}^{n-1}x^{(1)}(k+1)\sum_{k=1}^{n-1}x^{(1)}(k)}{\sum_{k=1}^{n-1}(x^{(1)}(k))^2-\dfrac{1}{n-1}(\sum_{k=1}^{n-1}x^{(1)}(k))^2}\\[3ex] \beta_2=\dfrac{1}{n-1}[\sum_{k=1}^{n-1}x^{(1)}(k+1)-\beta_1\sum_{k=1}^{n-1}x^{(1)}(k)]\end{cases}$$

From $Y=B\hat{\beta}$, it follows that

$$B^TB\hat{\beta}=B^TY,\hat{\beta}=(B^TB)^{-1}B^TY$$

However, from

$$B^TB=\begin{bmatrix}x^{(1)}(1) & 1\\ x^{(1)}(2) & 1\\ \vdots & \vdots\\ x^{(1)}(n-1) & 1\end{bmatrix}^T\begin{bmatrix}x^{(1)}(1) & 1\\ x^{(1)}(2) & 1\\ \vdots & \vdots\\ x^{(1)}(n-1) & 1\end{bmatrix}=\begin{bmatrix}\sum_{k=1}^{n-1}(x^{(1)}(k))^2 & \sum_{k=1}^{n-1}x^{(1)}(k)\\ \sum_{k=1}^{n-1}x^{(1)}(k) & n-1\end{bmatrix}$$

$$(B^T B)^{-1} = \frac{1}{(n-1)\sum_{k=1}^{n-1}(x^{(1)}(k))^2 - [\sum_{k=1}^{n-1} x^{(1)}(k)]^2} \times \begin{bmatrix} n-1 & -\sum_{k=1}^{n-1} x^{(1)}(k) \\ -\sum_{k=1}^{n-1} x^{(1)}(k) & \sum_{k=1}^{n-1}(x^{(1)}(k))^2 \end{bmatrix}$$

$$B^T Y = \begin{bmatrix} x^{(1)}(1) & 1 \\ x^{(1)}(2) & 1 \\ \vdots & \vdots \\ x^{(1)}(n-1) & 1 \end{bmatrix}^T \begin{bmatrix} x^{(1)}(2) \\ x^{(1)}(3) \\ \vdots \\ x^{(1)}(n) \end{bmatrix} = \begin{bmatrix} \sum_{k=1}^{n-1} x^{(1)}(k) \cdot x^{(1)}(k+1) \\ \sum_{k=1}^{n-1} x^{(1)}(k+1) \end{bmatrix}$$

it follows that

$$\hat{\beta} = (B^T B)^{-1} B^T Y = \frac{1}{(n-1)\sum_{k=1}^{n-1}(x^{(1)}(k))^2 - [\sum_{k=1}^{n-1} x^{(1)}(k)]^2}$$

$$\times \begin{bmatrix} (n-1)\sum_{k=1}^{n-1} x^{(1)}(k)x^{(1)}(k+1) - \sum_{k=1}^{n-1} x^{(1)}(k)\sum_{k=1}^{n-1} x^{(1)}(k+1) \\ -\sum_{k=1}^{n-1} x^{(1)}(k)\sum_{k=1}^{n-1} x^{(1)}(k)x^{(1)}(k+1) + \sum_{k=1}^{n-1} x^{(1)}(k+1)\sum_{k=1}^{n-1}(x^{(1)}(k))^2 \end{bmatrix}$$

$$= \begin{bmatrix} \dfrac{\sum_{k=1}^{n-1} x^{(1)}(k+1)x^{(1)}(k) - \frac{1}{n-1}\sum_{k=1}^{n-1} x^{(1)}(k+1)\sum_{k=1}^{n-1} x^{(1)}(k)}{\sum_{k=1}^{n-1}(x^{(1)}(k))^2 - \frac{1}{n-1}(\sum_{k=1}^{n-1} x^{(1)}(k))^2} \\ \frac{1}{n-1}[\sum_{k=1}^{n-1} x^{(1)}(k+1) - \beta_1 \sum_{k=1}^{n-1} x^{(1)}(k)] \end{bmatrix} = \begin{bmatrix} \beta_1 \\ \beta_2 \end{bmatrix}. \quad \text{QED.}$$

Theorem 5.2. Let $B, Y, \hat{\beta}$ be the same as defined in Theorem 5.1, and $\hat{\beta} = [\beta_1, \beta_2] = (B^T B)^{-1} B^T Y$. Then the following hold true:

(1) If $x^{(1)}(1) = x^{(0)}(1)$, the recurrence relation is

$$\hat{x}^{(1)}(k+1) = \beta_1^k x^{(0)}(1) + \frac{1-\beta_1^k}{1-\beta_1} * \beta_2; k = 1,2,\cdots,n-1;$$

or

$$\hat{x}^{(1)}(k+1) = \beta_1^k (x^{(0)}(1) - \frac{\beta_2}{1-\beta_1}) + \frac{\beta_2}{1-\beta_1}; k = 1,2,\cdots,n-1$$

(2) The restored values are

$$\hat{x}^{(0)}(k+1)=\alpha^{(1)}\hat{x}^{(1)}(k+1)=\hat{x}^{(1)}(k+1)-\hat{x}^{(1)}(k);k=1,2,\cdots,n-1$$

Proof. (1) Substituting the obtained β_1,β_2 into the discrete form produces

$$\begin{aligned}\hat{x}^{(1)}(k+1)&=\beta_1\hat{x}^{(1)}(k)+\beta_2=\beta_1(\beta_1\hat{x}^{(1)}(k-1)+\beta_2)+\beta_2\\&=\cdots=\beta_1^k\hat{x}^{(1)}(1)+(\beta_1^{k-1}+\beta_1^{k-2}+\cdots+\beta_1+1)\cdot\beta_2\end{aligned}$$

Letting $x^{(1)}(1)=x^{(0)}(1)$ leads to

$$\hat{x}^{(1)}(k+1)=\beta_1^k x^{(0)}(1)+\frac{1-\beta_1^k}{1-\beta_1}\cdot\beta_2$$

(2) $\hat{x}^{(1)}(k+1)-\hat{x}^{(1)}(k)=\sum_{i=1}^{k+1}\hat{x}^{(0)}(i)-\sum_{i=1}^{k}\hat{x}^{(0)}(i)=\hat{x}^{(0)}(k+1)$. QED.

5.1.2 Relationship between Discrete Grey and GM(1,1) Models

Substituting $z^{(1)}(k+1)=\frac{1}{2}(x^{(1)}(k+1)+x^{(1)}(k))$ into the GM(1,1) model produces

$$x^{(1)}(k+1)=\frac{1-0.5a}{1+0.5a}x^{(1)}(k)+\frac{b}{1+0.5a},k=1,2,\cdots,n-1;$$

whose form is exactly the same as that of the discrete grey model. So, we can assume that the following relations hold true:

$$\frac{1-0.5a}{1+0.5a}=\beta_1,\frac{b}{1+0.5a}=\beta_2$$

from which, it follows that

$$a=\frac{2(1-\beta_1)}{1+\beta_1},b=\frac{2\beta_2}{1+\beta_1},\frac{b}{a}=\frac{\beta_2}{1-\beta_1}$$

So, corresponding to the solution of the GM(1,1) model, we also call the time response function

$$\hat{x}^{(1)}(k+1)=(x^{(0)}(1)-\frac{b}{a})e^{-ak}+\frac{b}{a},k=1,2,\cdots,n-1 \tag{5.1}$$

as the solution of the discrete grey model, and the relevant recurrence relation is

$$\hat{x}^{(1)}(k+1)=\beta_1^k x^{(0)}(1)+\frac{1-\beta_1^k}{1-\beta_1}\cdot\beta_2;k=1,2,\cdots,n-1 \tag{5.2}$$

In the following, we look at the relationship between equ. (5.1) and equ. (5.2). Separating the right-hand side of equ. (5.1) gives us

$$\hat{x}^{(1)}(k+1)=x^{(0)}(1)e^{-ak}+\frac{b}{a}(1-e^{-ak}),k=1,2,\cdots,n-1 \tag{5.3}$$

The Maclaurin series expansions of e^{-a} and $\beta_1=\dfrac{1-0.5a}{1+0.5a}$ are respectively

$$e^{-a}=1-a+\frac{a^2}{2!}-\frac{a^3}{3!}+\cdots+(-1)^n\frac{a^n}{n!}+o(a^n)$$

and

$$\begin{aligned}\beta_1=\frac{1-0.5a}{1+0.5a}&=1-a(1-\frac{a}{2}+\frac{a^2}{2^2}+\cdots+(-1)^n\frac{a^n}{2^n}+o(a^n))\\&=1-a+\frac{a^2}{2}-\frac{a^3}{2^2}+\cdots+(-1)^{n+1}\frac{a^{n+1}}{2^n}+o(a^{n+1})\end{aligned}$$

When the a value is relatively small, the effect of the high power terms can be ignored. If we consider to be accurate to a^3, then we have

$$e^{-a}=1-a+\frac{a^2}{2}-\frac{a^3}{6}$$

$$\beta_1=1-a+\frac{a^2}{2}-\frac{a^3}{4}$$

By letting $\Delta=e^{-a}-\beta_1$, we have

$$\Delta=e^{-a}-\beta_1=-\frac{a^3}{6}+\frac{a^3}{4}=\frac{a^3}{12}$$

Corresponding to different a values, the consequent $|\Delta|$ values of are listed in Table 5.1.

Table 5.1 Errors corresponding to different a values

a value	-0.1	-0.2	-0.3	-0.5	-0.8	-1
$\lvert\Delta\rvert$	0.000083	0.000667	0.00225	0.010417	0.04267	0.08333

Therefore, when a takes a relatively large value, $\beta_1 \approx e^{-a}$. By substituting e^{-a} with β_1, equ. (5.3) becomes

$$\hat{x}^{(1)}(k+1) = x^{(0)}(1)\cdot\beta_1^k + \frac{\beta_2}{1-\beta_1}\cdot(1-\beta_1^k) = x^{(0)}(1)\beta_1^k + \frac{1-\beta_1^k}{1-\beta_1}\cdot\beta_2, k = 1,2,\cdots,n-1$$

which in form is exactly same as that in equ. (5.2). Hence, the discrete grey model can be seen as a different representation of the GM(1,1) model. Especially when a takes relatively small values, they can be exchanged for each other.

5.1.3 Prediction Analysis of Completely Exponential Growths

In Section 4.3, the range of applicability of the GM(1,1) model was made clear based on the simulation results of total exponential growths. When the absolute value of a is relatively large, the GM(1,1) model experiences relatively large errors when employed to simulate exponential growths. On the contrary, the discrete grey model can accurately simulate exponential sequences.

Let $X^{(0)}$ be an equal ratio sequence $X^{(0)} = \{ac, ac^2, ac^3, \cdots, ac^n\}, c > 0$, that is,

$$x^{(0)}(k) = ac^k, k = 1,2,\cdots,n.$$

Generate

$$X^{(1)} = \left\{ac, a(c+c^2), a(c+c^2+c^3), \cdots, a\sum_{i=1}^{n} c^i\right\}$$

by using the accumulation operator on once $X^{(0)}$. So, we have

$$Y = \begin{bmatrix} a(c+c^2) \\ a(c+c^2+c^3) \\ \vdots \\ a\sum_{i=1}^{n} c^i \end{bmatrix}, B = \begin{bmatrix} ac & 1 \\ a(c+c^2) & 1 \\ \vdots & 1 \\ a\sum_{i=1}^{n-1} c^i & 1 \end{bmatrix}$$

$$\hat{\beta} = (B^T B)^{-1} B^T Y = \begin{bmatrix} c \\ ac \end{bmatrix}$$

$$\hat{x}^{(1)}(k+1) = x^{(0)}(1)\cdot\beta_1^k + \frac{\beta_2}{1-\beta_1}\cdot(1-\beta_1^k) = ac\cdot c^k + \frac{1-c^k}{1-c}\cdot ac = a\sum_{i=1}^{k+1} c^i$$

with the restored values

$$\hat{x}^{(0)}(k)=\hat{x}^{(1)}(k)-\hat{x}^{(1)}(k-1)=a\sum_{i=1}^{k}c^{i}-a\sum_{i=1}^{k-1}c^{i}=ac^{k}$$

Since $\hat{x}^{(0)}(k)$ and $x^{(0)}(k)$ are exactly the same, no bias is involved, while both a and c can take any values. So, as long as the original sequence of data represents an approximate exponential growth, the discrete grey model can be employed to simulate the data and make relevant predictions. From an analysis of why the discrete grey model and the GM(1,1) mode produce different prediction results, it can be seen that the reason the predictions out of the GM(1,1) model suffer from deviations is due to the existing difference between the term e^{-a} in the whitenization equation and the coefficient β_1 of the discrete GM model. When short-term predictions are concerned, or when the development coefficient ($-a$) is relatively small, the difference bears little effect on the entire prediction model. So, the accuracy of short-term predictions can be relatively high. However, when long-term predictions are concerned, or when the development coefficient ($-a$) is relatively large, the effect of the difference on the entire prediction model is many times magnified, leading to reduced prediction accuracy to such a degree that some of the results are not acceptable.

Example 5.1. For the given sequence of data

$$X^{(0)}=\{2.23, 8.29, 25.96, 84.88, 271.83\}$$

establish a discrete grey model.

Solution. Based on Theorem 5.1, we can obtain

$$\hat{\beta}=(B^{T}B)^{-1}B^{T}Y=\begin{bmatrix}3.2141\\3.3162\end{bmatrix}$$

Substituting this expression into equ. (5.2) leads to the simulated results of $\hat{x}^{(1)}(k)$ as shown in Table 5.2.

Table 5.2 Error checks

Ordinality	Actual data $x^{(0)}(k)$	Simulations $\hat{x}^{(0)}(k)$	Error $\varepsilon(k)=x^{(0)}(k)-\hat{x}^{(0)}(k)$	Relative error $\Delta_k=\frac{\lvert\varepsilon(k)\rvert}{x^{(0)}(k)}$
2	8.29	8.25	-0.04	0.44%
3	25.96	26.53	0.57	2.19%
4	84.88	85.26	0.38	0.45%
5	271.83	274.03	2.20	0.81%

From Table 5.2, it follows that the mean relative error is

$$\Delta = \frac{1}{4}\sum_{k=2}^{5}\Delta_k = 0.97\% \cdot$$

5.2 Generalization and Optimization of Discrete Grey Models

The discrete grey model, as studied in the previous section, uses $x^{(0)}(1)$ as the basis of iteration. However, in practical applications, this basis of iteration can be adjusted according to the specific circumstances involved in order to improve the degree of accuracy. For example, any one of the point or the end point in the given sequence can be employed as the basis of iteration. According to different choice of the initial value of iteration, we can establish three different kinds of discrete grey models.

5.2.1 *Three Forms of Discrete Grey Models*

Assume that the observed values of a system's behavioral characteristic are given in a sequence

$$X^{(0)} = \{x^{(0)}(1), x^{(0)}(2), \cdots, x^{(0)}(n)\}$$

Its accumulation generation is

$$X^{(1)} = \{x^{(1)}(1), x^{(1)}(2), \cdots, x^{(1)}(n)\}$$

Based on different choices of the iteration basis, that is the sequence data value that is assumed at the time of modeling to be the same as the simulated value, the consequent discrete grey models can take one of the following three forms:

$$(1)\ \begin{cases} \hat{x}^{(1)}(k+1) = \beta_1 \hat{x}^{(1)}(k) + \beta_2 \\ \hat{x}^{(1)}(1) = x^{(1)}(1) = x^{(0)}(1) \end{cases}$$

where $\hat{x}^{(1)}(k)$ stands for a simulated value of the original sequence value with β_1 and β_2 to be determined and $\hat{x}^{(1)}(1)$ the basis of iteration. This form of the discrete grey model is referred to that *with fixed starting points*, written in short as SDGM.

$$(2)\ \begin{cases} \hat{x}^{(1)}(k+1) = \beta_1 \hat{x}^{(1)}(k) + \beta_2 \\ \hat{x}^{(1)}(m) = x^{(1)}(m) = \sum_{i=1}^{m} x^{(0)}(i), 1 < \mathrm{m} < \mathrm{m} \end{cases}$$

where $\hat{x}^{(1)}(k)$, $k = 1, 2, \ldots, n$, stand for simulated values the original sequence with β_1 and β_2 to be determined and $\hat{x}^{(1)}(m)$ the basis of iteration. This kind of the discrete grey model is referred to as that *with a middle-point fixed*, denoted in short as MDGM.

$$(3)\ \begin{cases} \hat{x}^{(1)}(k+1) = \beta_1 \hat{x}^{(1)}(k) + \beta_2 \\ \hat{x}^{(1)}(n) = x^{(1)}(n) = \sum_{i=1}^{n} x^{(0)}(i) \end{cases}$$

where $\hat{x}^{(1)}(k)$, $k = 1, 2, \ldots, n$, stand for simulated values of the original sequence with β_1 and β_2 to be determined and $\hat{x}^{(1)}(n)$ the basis of iteration. This kind of the discrete grey model is referred to as that *with the end point fixed*, denoted in short as EDGM.

5.2.2 Impacts of Initial Values on Iterations

From the analysis of the previous subsection, it can be seen that for a given sequence of data, different choices of the initial value of iteration lead to different results, where a small vibration in the initial value might create a drastically different simulation sequence. In the following, we will use the SDGM model to do our analysis. As for the MDGM and EDGM models, similar results can be derived.

In the SDGM model, let us take $\hat{x}^{(1)}(1) = x^{(1)}(1) = x^{(0)}(1)$ as the basis of iteration. In this case, the curve in the Cartesian coordinate plane of the simulated values has to pass through the point $(1, x^{(0)}(1))$. However, according to the principle of least squares estimate, the optimal simulation curve does not to go through the basis point $(1, x^{(0)}(1))$ of iteration. So, we do not have a needed theoretical reason for choosing $\hat{x}^{(1)}(1) = x^{(1)}(1) = x^{(0)}(1)$ as out initial condition. Intuitively, we can analyze the relevant graphs, as shown in Figure 5.1.

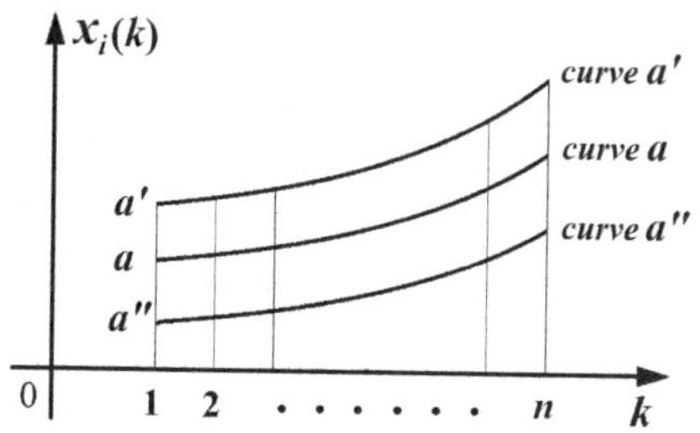

Fig. 5.1 Analysis of the starting point fixed model

We assume that curve a is the one that satisfies the principle of least squares estimate. Then, curve a should also pass through the basis point a, that is $(1, x^{(0)}(1))$. However, in reality, $\hat{x}^{(1)}(1)$ might be greater than or smaller than $x^{(0)}(1)$. That is, when the actual data value $x^{(0)}(1) = a'$, we obtain the simulated curve a', while the actual data value $x^{(0)}(1) = a''$, the obtained simulation curve is curve a''. In both of these cases, the simulated curve deviates from curve a. Theoretically, we can also prove this result. In particular, we know the recurrence relation is

$$\hat{x}^{(1)}(k+1) = \beta_1^k \hat{x}^{(1)}(1) + \frac{1-\beta_1^k}{1-\beta_1} * \beta_2; k = 1, 2, \cdots, n-1$$

So, the restored values are

$$\begin{aligned}\hat{x}^{(0)}(k+1) &= \hat{x}^{(1)}(k+1) - \hat{x}^{(1)}(k) \\ &= (\beta_1^k \hat{x}^{(1)}(1) + \frac{1-\beta_1^k}{1-\beta_1} * \beta_2) - (\beta_1^{(k-1)} \hat{x}^{(1)}(1) + \frac{1-\beta_1^{(k-1)}}{1-\beta_1} * \beta_2) \\ &= (\beta_1^k - \beta_1^{(k-1)}) * \hat{x}^{(1)}(1) + \frac{\beta_1^{(k-1)} - \beta_1^k}{1-\beta_1} * \beta_2\end{aligned}$$

Because both β_1 and β_2 are uniquely determined by using the method of least squares estimate, we know both $(\beta_1^k - \beta_1^{(k-1)})$ and $\frac{\beta_1^{(k-1)} - \beta_1^k}{1-\beta_1} * \beta_2$ are uniquely determined. So, when $x^{(1)}(1)$ is not located on the simulation curve obtained by using the least squares estimate, that is when $\hat{x}^{(1)}(1) > x^{(1)}(1)$ or $\hat{x}^{(1)}(1) < x^{(1)}(1)$, the computed $\hat{x}^{(0)}(k+1)$ should also be correspondingly too large or too small.

For the MDGM and EDGM models, similar conclusions can be obtained. More specifically, from Figures 5.2 and 5.3, it can be seen respectively that the iteration basis value $\hat{x}^{(1)}(m)$ of the MDGM deviates from $x^{(1)}(m)$, and that when the iteration basis value $\hat{x}^{(1)}(n)$ of the EDGM deviates from $x^{(1)}(n)$, the other simulated values $\hat{x}^{(0)}(k+1)$ will also be affected. These observations can also be theoretically shown. Therefore, we know that the simulations of the entire sequence are affected by the choice of the iteration basis $\hat{x}^{(1)}(1)$, $\hat{x}^{(1)}(m)$, or $\hat{x}^{(1)}(n)$. So, for the given circumstances, various iteration bases need to be tested before the the final decision is made.

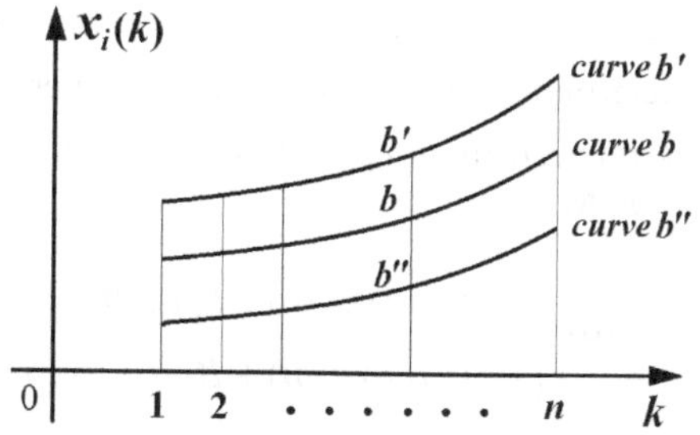

Fig. 5.2 Analysis of the middle-point fixed model

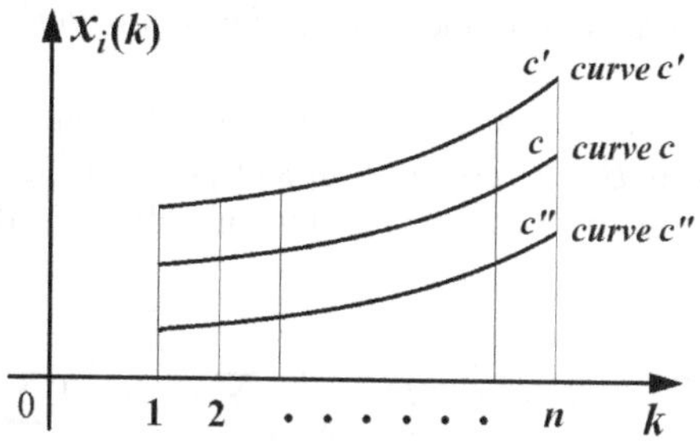

Fig. 5.3 Analysis of the end-point fixed model

5.2.3 *Optimization of Discrete Grey Models*

In order to eliminate the effect of the chosen initial value for iteration on the resultant simulation as much as possible, let us consider adding a small adjustment to the initial value, hoping that through using this adjustment the deviation caused by the specific choice of the initial value can be cancelled. When employing this idea, the original three forms of the discrete grey model respectively become:

$$(4)\ \begin{cases} \hat{x}^{(1)}(k+1) = \beta_1 \hat{x}^{(1)}(k) + \beta_2 \\ \hat{x}^{(1)}(1) = x^{(1)}(1) + \beta_3 \end{cases}$$

where $\hat{x}^{(1)}(k)$ stands of a simulated value of the data in the original sequence with β_1, β_2 and β_3 to be determined and $\hat{x}^{(1)}(1)$ the basis of iteration. This form of the discrete grey model is referred to the *optimized model with fixed starting point*, written in short as OSDGM.

$$(5)\ \begin{cases} \hat{x}^{(1)}(k+1) = \beta_1 \hat{x}^{(1)}(k) + \beta_2 \\ \hat{x}^{(1)}(m) = x^{(1)}(m) + \beta_3 \end{cases}$$

where $\hat{x}^{(1)}(k)$ stands of a simulated value of the data in the original sequence with β_1, β_2 and β_3 to be determined and $\hat{x}^{(1)}(m)$ the basis of iteration. This form of the discrete grey model is referred to an *optimized model with a middle point fixed*, written in short as OMDGM.

$$(6)\ \begin{cases} \hat{x}^{(1)}(k+1) = \beta_1 \hat{x}^{(1)}(k) + \beta_2 \\ \hat{x}^{(1)}(n) = x^{(1)}(n) + \beta_3 \end{cases}$$

where $\hat{x}^{(1)}(k)$ stands of a simulated value of the data in the original sequence with β_1, β_2 and β_3 to be determined and $\hat{x}^{(1)}(n)$ the basis of iteration. This form of the discrete grey model is referred to the *optimized model with fixed end point*, written in short as OEDGM.

For these optimized discrete grey models, among the three unknown parameters β_1, β_2, and β_3, we compute β_1 and β_2 by using the method of least squares estimate, as in the general situation of solving a discrete grey model, and obtain

$$\begin{cases} \beta_1 = \dfrac{\sum_{k=1}^{n-1} x^{(1)}(k+1)x^{(1)}(k) - \dfrac{1}{n-1}\sum_{k=1}^{n-1} x^{(1)}(k+1)\sum_{k=1}^{n-1} x^{(1)}(k)}{\sum_{k=1}^{n-1}(x^{(1)}(k))^2 - \dfrac{1}{n-1}(\sum_{k=1}^{n-1} x^{(1)}(k))^2} \\ \beta_2 = \dfrac{1}{n-1}[\sum_{k=1}^{n-1} x^{(1)}(k+1) - \beta_1 \sum_{k=1}^{n-1} x^{(1)}(k)] \end{cases}$$

Let us now look at how to compute the third parameter β_3 using the OSDGM model as our example. For the OMDGM and OEDGM models, similar conclusions can be drawn. For the purpose of solving for this third parameter β_3, we still employ the method of least squares estimate by establishing an optimization model without any constraint. By soling this problem for $\hat{x}^{(1)}(k)$ so that the sum of squared errors between $\hat{x}^{(1)}(k)$ and $x^{(1)}(k)$ is the minimum. That is, we solve the following optimization problem:

$$\min_{\beta_3} \sum_{k=1}^{n} [\hat{x}^{(1)}(k) - x^{(1)}(k)]^2$$

and substitute the obtained β_1 and β_2 into $\hat{x}^{(1)}(k+1) = \beta_1^k \hat{x}^{(1)}(1) + \dfrac{1-\beta_1^k}{1-\beta_1} * \beta_2$.

By letting

$$S=\sum_{k=1}^{n}[\hat{x}^{(1)}(k)-\hat{x}^{(1)}(k)]^2=\beta_3^2+\sum_{k=1}^{n-1}[(\beta_1^k\hat{x}^{(1)}(1)+\frac{1-\beta_1^k}{1-\beta_1}*\beta_2)-x^{(1)}(k+1)]^2$$
$$=\beta_3^2+\sum_{k=1}^{n-1}[\beta_1^k x^{(1)}(1)+\beta_1^k*\beta_3+\frac{1-\beta_1^k}{1-\beta_1}*\beta_2-x^{(1)}(k+1)]^2$$

we have

$$\frac{dS}{d\beta_3}=2\beta_3+2\sum_{k=1}^{n-1}[\beta_1^k x^{(1)}(1)+\beta_1^k*\beta_3+\frac{1-\beta_1^k}{1-\beta_1}*\beta_2-x^{(1)}(k+1)]*\beta_1^k=0$$

and consequently obtain

$$\beta_3=\frac{\sum_{k=1}^{n-1}[x^{(1)}(k+1)-\beta_1^k x^{(1)}(1)-\frac{1-\beta_1^k}{1-\beta_1}*\beta_2]*\beta_1^k}{1+\sum_{k=1}^{n-1}(\beta_1^k)^2}$$

That is, we have

$$\begin{cases}\beta_1=\dfrac{\sum_{k=1}^{n-1}x^{(1)}(k+1)x^{(1)}(k)-\dfrac{1}{n-1}\sum_{k=1}^{n-1}x^{(1)}(k+1)\sum_{k=1}^{n-1}x^{(1)}(k)}{\sum_{k=1}^{n-1}(x^{(1)}(k))^2-\dfrac{1}{n-1}(\sum_{k=1}^{n-1}x^{(1)}(k))^2}\\ \beta_2=\dfrac{1}{n-1}[\sum_{k=1}^{n-1}x^{(1)}(k+1)-\beta_1\sum_{k=1}^{n-1}x^{(1)}(k)]\\ \beta_3=\dfrac{\sum_{k=1}^{n-1}[x^{(1)}(k+1)-\beta_1^k x^{(1)}(1)-\dfrac{1-\beta_1^k}{1-\beta_1}*\beta_2]*\beta_1^k}{1+\sum_{k=1}^{n-1}(\beta_1^k)^2}\end{cases}\tag{5.4}$$

Evidently, after optimization, all the OSDGM, OMDGM, and OEDGM models can produce out of a given sequence of data the same simulation curve as the optimal curve. Due to this reason, in practical applications of simulation and prediction, any one of these models can be employed.

5.2.4 Recurrence Functions for Optimizing Discrete Grey Models

Based on the parameters β_1, β_2, and β_3, we can establish the formulas for directly computing the sequences of simulation and prediction. These formulas are also referred to as the recurrence relations for optimizing discrete grey models.

Theorem 5.3. For the OSDGM model, if the parameters are given as in equ. (5.4), then the recurrence relation is given by

$$\hat{x}^{(1)}(k+1)=\beta_1^k x^{(0)}(1)+\beta_1^k*\beta_3+\frac{1-\beta_1^k}{1-\beta_1}*\beta_2;k=1,2,\cdots,n-1;$$

or

$$\hat{x}^{(1)}(k+1)=\beta_1^k(x^{(0)}(1)+\beta_3-\frac{\beta_2}{1-\beta_1})+\frac{\beta_2}{1-\beta_1};k=1,2,\cdots,n-1$$

and the restored valued given by

$$\hat{x}^{(0)}(k+1)=\alpha^{(1)}\hat{x}^{(1)}(k+1)=\hat{x}^{(1)}(k+1)-\hat{x}^{(1)}(k);k=1,2,\cdots,n-1;$$

Proof. (1) Substituting the obtained β_1,β_2 values into the discrete form gives

$$\begin{aligned}\hat{x}^{(1)}(k+1)&=\beta_1\hat{x}^{(1)}(k)+\beta_2=\beta_1(\beta_1\hat{x}^{(1)}(k-1)+\beta_2)+\beta_2\\&=\cdots=\beta_1^k\hat{x}^{(1)}(1)+(\beta_1^{k-1}+\beta_1^{k-2}+\cdots+\beta_1+1)\cdot\beta_2\end{aligned}$$

Taking $\hat{x}^{(1)}(1)=x^{(1)}(1)+\beta_3=x^{(0)}(1)+\beta_3$ produces

$$\begin{aligned}\hat{x}^{(1)}(k+1)&=\beta_1^k\hat{x}^{(1)}(1)+\frac{1-\beta_1^k}{1-\beta_1}*\beta_2\\&=\beta_1^k(x^{(1)}(1)+\beta_3)+\frac{1-\beta_1^k}{1-\beta_1}*\beta_2\\&=\beta_1^k x^{(0)}(1)+\beta_1^k*\beta_3+\frac{1-\beta_1^k}{1-\beta_1}*\beta_2\end{aligned}$$

(2) $\hat{x}^{(1)}(k+1)-\hat{x}^{(1)}(k)=\sum_{i=1}^{k+1}\hat{x}^{(0)}(i)-\sum_{i=1}^{k}\hat{x}^{(0)}(i)=\hat{x}^{(0)}(k+1)$. QED.

Theorem 5.4. For the OMDGM model, if the parameters are given as in equ. (5.4), then the recurrence function is given below:

$$\hat{x}^{(1)}(k)=\beta_1^{(k-m)}(x^{(1)}(m)+\beta_3-\frac{\beta_2}{1-\beta_1})+\frac{\beta_2}{1-\beta_1};k=1,2,\cdots;\quad 1\le m\le n$$

and the restored values

$$\hat{x}^{(0)}(k+1)=\alpha^{(1)}\hat{x}^{(1)}(k+1)=\hat{x}^{(1)}(k+1)-\hat{x}^{(1)}(k);k=1,2,\cdots$$

Proof. The details are similar to those in the argument of Theorem 5.3 and omitted. QED.

Theorem 5.5. For the OEDGM model, if the parameters are given as in equ. (5.4), then the recurrence relation is given as follows:

$$\hat{x}^{(1)}(k) = \beta_1^{(k-n)}(x^{(1)}(n) + \beta_3 - \frac{\beta_2}{1-\beta_1}) + \frac{\beta_2}{1-\beta_1}; k = 1, 2, \cdots;$$

and the restored values

$$\hat{x}^{(0)}(k+1) = \alpha^{(1)}\hat{x}^{(1)}(k+1) = \hat{x}^{(1)}(k+1) - \hat{x}^{(1)}(k); k = 1, 2, \cdots$$

Proof. Omitted. QED.

5.3 Approximately Nonhomogeneous Exponential Growth

No matter whether it is the GM(1,1) model, the discrete grey model, or the optimized discrete grey model, the very foundation of establishing a model is the assumption that the original sequence $X^{(0)} = \{x^{(0)}(1), x^{(0)}(2), \cdots, x^{(0)}(n)\}$ of data satisfies the *law of homogeneous exponential growth*. In symbols, the sequence $X^{(0)}$ satisfies that condition that $x^{(0)}(k) \approx ac^k$, $k = 1, 2, \cdots$ However, in reality, the number of such sequences of data are very limited. In this section, we will mainly look at the case of non-homogeneous exponential growths. That is, the original sequence $X^{(0)}$ roughly satisfies $x^{(0)}(k) \approx ac^k + b$, $k = 1,2,3,...$

Definition 5.2. For a given sequence $X^{(0)} = \{x^{(0)}(1), x^{(0)}(2), \cdots, x^{(0)}(n)\}$ of raw data, let its accumulation generation be $X^{(1)} = \{x^{(1)}(1), x^{(1)}(2), \cdots, x^{(1)}(n)\}$. Then the following

$$\begin{cases} \hat{x}^{(1)}(k+1) = \beta_1 \hat{x}^{(1)}(k) + \beta_2 * k + \beta_3 \\ \hat{x}^{(1)}(1) = x^{(1)}(1) + \beta_4 \end{cases}$$

is referred to a *non-homogeneous discrete grey model* (NDGM), where $\hat{x}^{(1)}(k)$ stands for the simulated value of the corresponding given sequence value,

β_1, β_2, β_3, and β_4 are parameters to be determined, and $\hat{x}^{(1)}(1)$ the basis value of iteration.

Similar to the optimized discrete grey models, the parameters β_1, β_2, and β_3 can be determined by using the method of least squares estimate, while can be solved for using an optimization model without any constraint. In particular, if $\hat{\beta} = (\beta_1, \beta_2, \beta_3)^T$ is a sequence of parameters and

$$Y = \begin{bmatrix} x^{(1)}(2) \\ x^{(1)}(3) \\ \vdots \\ x^{(1)}(n) \end{bmatrix}, \quad B = \begin{bmatrix} x^{(1)}(1) & 1 & 1 \\ x^{(1)}(2) & 2 & 1 \\ \vdots & \vdots & \vdots \\ x^{(1)}(n-1) & n-1 & 1 \end{bmatrix}$$

then according to the principle of least squares estimate, the parametric sequence should satisfy

$$\hat{\beta} = (\beta_1, \beta_2, \beta_3)^T = (B^T B)^{-1} B^T Y$$

from which we can obtain the more detailed expressions for the parameters as follows:

$$\begin{aligned}
P_1 &= (\beta_1, \beta_2, \beta_3)^T \\
&= \frac{1}{(n-1)\sum_{k=1}^{n-1} k^2 \sum_{k=1}^{n-1} (x^{(1)}(k))^2 + 2\sum_{k=1}^{n-1} k \sum_{k=1}^{n-1} k x^{(1)}(k) \sum_{k=1}^{n-1} x^{(1)}(k) - (\sum_{k=1}^{n-1} k)^2 \sum_{k=1}^{n-1} (x^{(1)}(k))^2 - (n-1)(\sum_{k=1}^{n-1} k x^{(1)}(k))^2 - \sum_{k=1}^{n-1} k^2 (\sum_{k=1}^{n-1} x^{(1)}(k))^2} \\
&\times \begin{bmatrix}
(n-1)\sum_{k=1}^{n-1} k^2 \sum_{k=1}^{n-1} x^{(1)}(k) x^{(1)}(k+1) + \sum_{k=1}^{n-1} k \sum_{k=1}^{n-1} x^{(1)}(k) \sum_{k=1}^{n-1} k x^{(1)}(k+1) + \sum_{k=1}^{n-1} k \sum_{k=1}^{n-1} k x^{(1)}(k) \sum_{k=1}^{n-1} x^{(1)}(k+1) \\
\quad - (\sum_{k=1}^{n-1} k)^2 \sum_{k=1}^{n-1} x^{(1)}(k) x^{(1)}(k+1) - (n-1)\sum_{k=1}^{n-1} k x^{(1)}(k) \sum_{k=1}^{n-1} k x^{(1)}(k+1) - \sum_{k=1}^{n-1} k^2 \sum_{k=1}^{n-1} x^{(1)}(k) \sum_{k=1}^{n-1} x^{(1)}(k+1) \\
\sum_{k=1}^{n-1} k \sum_{k=1}^{n-1} x^{(1)}(k) \sum_{k=1}^{n-1} x^{(1)}(k) x^{(1)}(k+1) + (n-1)\sum_{k=1}^{n-1} (x^{(1)}(k))^2 \sum_{k=1}^{n-1} k x^{(1)}(k+1) + \sum_{k=1}^{n-1} k x^{(1)}(k) \sum_{k=1}^{n-1} x^{(1)}(k) \sum_{k=1}^{n-1} x^{(1)}(k+1) \\
\quad - (n-1)\sum_{k=1}^{n-1} k x^{(1)}(k) \sum_{k=1}^{n-1} x^{(1)}(k) x^{(1)}(k+1) - \sum_{k=1}^{n-1} k x^{(1)}(k+1) (\sum_{k=1}^{n-1} x^{(1)}(k))^2 - \sum_{k=1}^{n-1} k \sum_{k=1}^{n-1} (x^{(1)}(k))^2 \sum_{k=1}^{n-1} x^{(1)}(k+1) \\
\sum_{k=1}^{n-1} k \sum_{k=1}^{n-1} k x^{(1)}(k) \sum_{k=1}^{n-1} x^{(1)}(k) x^{(1)}(k+1) + \sum_{k=1}^{n-1} k x^{(1)}(k) \sum_{k=1}^{n-1} x^{(1)}(k) \sum_{k=1}^{n-1} k x^{(1)}(k+1) + \sum_{k=1}^{n-1} k^2 \sum_{k=1}^{n-1} (x^{(1)}(k))^2 \sum_{k=1}^{n-1} x^{(1)}(k+1) \\
\quad - \sum_{k=1}^{n-1} k^2 \sum_{k=1}^{n-1} x^{(1)}(k) \sum_{k=1}^{n-1} x^{(1)}(k) x^{(1)}(k+1) - \sum_{k=1}^{n-1} k \sum_{k=1}^{n-1} (x^{(1)}(k))^2 \sum_{k=1}^{n-1} k x^{(1)}(k+1) - \sum_{k=1}^{n-1} x^{(1)}(k+1) (\sum_{k=1}^{n-1} k x^{(1)}(k))^2
\end{bmatrix}
\end{aligned}$$

For solving for the parameter β_4, we establish and solve an optimization model without any constraint so that the sun of squared errors between $\hat{x}^{(1)}(k)$ and $x^{(1)}(k)$ is the minimum. That is, solving

$$\min_{\beta_4} \sum_{k=1}^{n} [\hat{x}^{(1)}(k) - x^{(1)}(k)]^2$$

for β_4 value as follows

$$\beta_4 = \frac{\sum_{k=1}^{n-1}[x^{(1)}(k+1) - \beta_1^k x^{(1)}(1) - \beta_2 \sum_{j=1}^{k} j\beta_1^{k-j} - \frac{1-\beta_1^k}{1-\beta_1} * \beta_3] * \beta_1^k}{1 + \sum_{k=1}^{n-1}(\beta_1^k)^2}$$

Similar to the situation with the SDGM model, in the following, we use a completely non-homogeneous exponential sequence of data to test the effects of simulation and prediction of the NDGM model. In particular, assume that the initial development sequence of data is

$$X^{(0)} = \left\{ac+b, ac^2+b, ac^3+b, \cdots, ac^n+b\right\}, c>0$$

then the future development tendency is $x^{(0)}(k) = ac^k + b, k = n+1, n+2, \cdots\cdots$ Applying the accumulation generation on $X^{(0)}$ once produces

$$X^{(1)} = \left\{ac+b, a(c+c^2)+2b, a(c+c^2+c^3)+3b, \cdots, a\sum_{i=1}^{n} c^i + nb\right\}$$

$$Y = \begin{bmatrix} a(c+c^2)+2b \\ a(c+c^2+c^3)+3b \\ \vdots \\ a\sum_{i=1}^{n} c^i + nb \end{bmatrix}, \quad B = \begin{bmatrix} ac+b & 1 & 1 \\ a(c+c^2)+2b & 2 & 1 \\ \vdots & \vdots & \vdots \\ a\sum_{i=1}^{n-1} c^i + (n-1)b & n-1 & 1 \end{bmatrix}$$

from which we obtain

$$\hat{\beta} = (\beta_1, \beta_2, \beta_3)^T = (B^T B)^{-1} B^T Y = \begin{bmatrix} c \\ b(1-c) \\ ac+b \end{bmatrix}, \quad \beta_4 = 0$$

$$\begin{aligned} \hat{x}^{(1)}(k+1) &= \beta_1^k (x^{(1)}(1) + \beta_4) + \beta_2 \sum_{j=1}^{k} j\beta_1^{k-j} + \frac{1-\beta_1^k}{1-\beta_1} * \beta_3 \\ &= c^k(ac+b) + b(1-c)\sum_{j=1}^{k} jc^{k-j} + \frac{1-c^k}{1-c} * (ac+b) \\ &= a\sum_{i=1}^{k+1} c^i + (k+1)b \end{aligned}$$

with the restored values:

$$\hat{x}^{(0)}(k)=\hat{x}^{(1)}(k)-\hat{x}^{(1)}(k-1)=a\sum_{i=1}^{k}c^i+kb-a\sum_{i=1}^{k-1}c^i-(k-1)b=ac^k+b$$

where $\hat{x}^{(0)}(k)$ and $x^{(0)}(k)$ are identical to each other. This fact implies that there is no bias involved in the simulation and prediction with the arbitrarily chosen a , b , and c . So, it can be seen that as long as the original sequence of data follows the *law of approximate non-homogeneous exponential growth*, we can employ the NDGM model to conduct our simulation and make prediction.

5.4 Discrete Grey Models of Multi-variables

The purpose of this section is to establish the discrete grey model of multi-variables and to compare this model with the high order GM(r,h) model of multi-variables.

Definition 5.3. Let the first order accumulation generated sequence of the original sequence $X_i^{(0)}$ be $X_i^{(1)}$, $i=1,2,...,h$. Then, the following

$$x_1^{(1)}(k)+\alpha_1x_1^{(1)}(k-1)+\alpha_2x_1^{(1)}(k-2)+\cdots+\alpha_rx_1^{(1)}(k-r)=\sum_{j=1}^{h-1}\beta_jx_{j+1}^{(1)}(k)+\beta_h$$

is referred to as the *discrete grey model of the nth order in h variables*, denoted as DGM(r,h), with $\alpha_1,\alpha_2,\cdots,\alpha_r;\beta_1,\beta_2,\cdots,\beta_h$ being the parameters.

Proposition 5.1. For $i=1,2,\cdots,n$, $\alpha^{(r)}x_1^{(1)}(k)=\sum_{i=0}^{r}C_r^i(-1)^ix_1^{(1)}(k-i)$.

Proof. By mathematical induction, when $r=1,2$, we have

$$\alpha^{(1)}x_1^{(1)}(k)=x_1^{(1)}(k)-x_1^{(1)}(k-1)=\sum_{i=0}^{1}C_1^i(-1)^ix^{(1)}(k-i)$$

$$\begin{aligned}\alpha^{(2)}x_1^{(1)}(k)&=\alpha^{(1)}x_1^{(1)}(k)-\alpha^{(1)}x_1^{(1)}(k-1)\\&=[x_1^{(1)}(k)-x_1^{(1)}(k-1)]-[x_1^{(1)}(k-1)-x_1^{(1)}(k-2)]\\&=x_1^{(1)}(k)-2x_1^{(1)}(k-1)+x_1^{(1)}(k-2)\\&=\sum_{i=0}^{2}C_2^i(-1)^ix^{(1)}(k-i)\end{aligned}$$

Assume that the proposition holds true when $r=j$. That is, we have

$$\alpha^{(j)}x_1^{(1)}(k)=\sum_{i=0}^{j}C_j^i(-1)^ix_1^{(1)}(k-i),$$

then when $r = j+1$, we have

$$\begin{aligned}
\alpha^{(j+1)} x_1^{(1)}(k) &= \alpha^{(j)} x_1^{(1)}(k) - \alpha^{(j)} x_1^{(1)}(k-1) \\
&= \sum_{i=0}^{j} C_j^i (-1)^i x_1^{(1)}(k-i) - \sum_{i=0}^{j} C_j^i (-1)^i x_1^{(1)}(k-1-i) \\
&= C_j^0 x_1^{(1)}(k) + [C_j^1 (-1)^1 x_1^{(1)}(k-1) - C_j^0 (-1)^0 x_1^{(1)}(k-1)] \\
&\quad + [C_j^2 (-1)^2 x_1^{(1)}(k-2) - C_j^1 (-1)^1 x_1^{(1)}(k-2)] \\
&\quad + \cdots + [C_j^j (-1)^j x_1^{(1)}(k-j) - C_j^{j-1} (-1)^{j-1} x_1^{(1)}(k-j)] - C_j^j (-1)^j x_1^{(1)}(k-1-j)
\end{aligned}$$

because

$$\begin{aligned}
C_j^p + C_j^{p-1} &= \frac{j!}{p!(j-p)!} + \frac{j!}{(p-1)!(j-p+1)!} \\
&= \frac{j!}{(p-1)!(j-p)!}\left(\frac{1}{p} + \frac{1}{j-p+1}\right) \\
&= \frac{(j+1)!}{p!(j+1-p)!} \\
&= C_{j+1}^p
\end{aligned}$$

Therefore, we have

$$\begin{aligned}
\alpha^{(j+1)} x_1^{(1)}(k) &= C_{j+1}^0 x_1^{(1)}(k) + C_{j+1}^1 (-1)^1 x_1^{(1)}(k-1) + C_{j+1}^2 (-1)^2 x_1^{(1)}(k-2) \\
&\quad + \cdots + C_{j+1}^{j+1} (-1)^{j+1} x_1^{(1)}(k-j-1) \\
&= \sum_{i=0}^{j+1} C_j^i (-1)^i x_1^{(1)}(k-i)
\end{aligned}$$

So, the result is proved. QED.

Proposition 5.2. Both GM(r,h) and DGM(r,h) models are equivalent.

Proof. Substituting the results of Proposition 5.1 into the GM(r,h) model produces

$$\begin{aligned}
&\sum_{i=0}^{n} C_r^i (-1)^i x_1^{(1)}(k-i) + a_1 \sum_{i=0}^{n-1} C_{r-1}^i (-1)^i x_1^{(1)}(k-i) + a_2 \sum_{i=0}^{n-2} C_{r-2}^i (-1)^i x_1^{(1)}(k-i) \\
&\qquad + \cdots + a_{r-1} \sum_{i=0}^{1} C_1^i (-1)^i x_1^{(1)}(k-i) + a_r z_1^{(1)}(k) = \sum_{j=1}^{h-1} b_j x_{j+1}^{(1)}(k) + b_h
\end{aligned}$$

By combining like-terms on the left-hand side, we obtain

$$
\begin{aligned}
&(C_r^0 + a_1C_{r-1}^0 + a_2C_{r-2}^0 \cdots + a_{r-1}C_1^0)(-1)^0 x_1^{(1)}(k) \\
&+(C_r^1 + a_1C_{r-1}^1 + a_2C_{r-2}^1 \cdots + a_{r-1}C_1^1)(-1)^1 x_1^{(1)}(k-1) \\
&+(C_r^2 + a_1C_{r-1}^2 + a_2C_{r-2}^2 \cdots + a_{r-2}C_2^2)(-1)^2 x_1^{(1)}(k-2) \\
&+\cdots+(C_r^{r-1} + a_1C_{r-1}^{r-1})(-1)^{r-1} x_1^{(1)}(k-r+1) + C_r^r(-1)^r x_1^{(1)}(k-n) \\
&+a_r(0.5x_1^{(1)}(k) + 0.5x_1^{(1)}(k-1)) \\
&=\sum_{j=1}^{h-1} b_j x_{j+1}^{(1)}(k) + b_h
\end{aligned}
$$

Denote $\alpha_0 = (C_r^0 + a_1C_{r-1}^0 + a_2C_{r-2}^0 \cdots + a_{r-1}C_1^0)(-1)^0 + 0.5a_r$, divide the equation by α_0, and let

$$\alpha_1 = \frac{1}{\alpha_0}[(C_n^1 + a_1C_{r-1}^1 + a_2C_{r-2}^1 \cdots + a_{r-1}C_1^1)(-1)^1 + 0.5a_r]\frac{r!}{i!(r-i)!}$$

$$\alpha_p = \frac{1}{\alpha_0}(C_r^p + a_1C_{r-1}^p + a_2C_{r-2}^p \cdots + a_{r-1}C_p^p)(-1)^p, \quad p=2,3,\cdots,r;$$

$$\beta_j = \frac{b_j}{\alpha_0}, \quad j=1,2,\cdots,h$$

then the previous equation becomes

$$x_1^{(1)}(k) + \alpha_1 x_1^{(1)}(k-1) + \alpha_2 x_1^{(1)}(k-2) + \cdots + \alpha_r x_1^{(1)}(k-r) = \sum_{j=1}^{h-1} \beta_j x_{j+1}^{(1)}(k) + \beta_h$$

whose form is identical to that of the DGM(r,h). QED.

In particular, when $r = 1$ and $h = 1$, the GM(r,h) model reduces into the GM(1,1) model, while the DGM(r,h) model simplifies into the general discrete grey model.

Chapter 6
Combined Grey Models

Along with the disciplinary development of systems science and systems engineering, methods and modeling techniques established for systems evaluation, prediction, decision-making, and optimization are enriched constantly. Generally, each method and every model have their strengths and weaknesses so in practical applications, several different methods and modeling techniques are combined to form hybrid methods and/or techniques in order to successfully deal with the problems in hand. Here, the meaning of combination and mixture is to absorb the strengths and advantages of different methods so that they complement each other and the weaknesses of individual methods and modeling techniques are improved. That explains why in general such combined or mixed systems are more superior to each of the individual compoenent methods. Additionally, the avaiablbility of many different methods and modeling techniques also provides us with different ways to deal with information and systems. Therefore, how to combine and mix different methods and techniques has become a research direction with wide-ranging applicability in areas of data mining and knowledge discovery.

Grey systems theory and methods complement strongly with many of the traditional technologies and soft computing techniques. In this chapter, we will mainly explore various combinations, mixtures and applications of grey systems models and those models developed in econometrics, production functions, artificial neural networks, linear regression, Markov models, rough sets, etc.

6.1 Grey Econometrics Models

6.1.1 Determination of Variables Using Principles of Grey Incidence

In analyzing systems, due to the complications of mutually crossing influences of the endogenous variables, at the very start of modeling, the first problem that needs to be addressed is how to select the variables that will be part of the eventual model. To revolve this problem, the researcher needs not only rely on his qualitative analysis of the system, but also have sufficiently adequate tools for conducting quantitative analysis. Grey incidences provide an effective method for this class of problems.

Let y be an endogenous variable of the system of our concern (for systems with many endogenous variables, these variables can be studied individually), and

$x_1, x_2, \cdots, x_n$ be pre-images of influencing factors that are correlated either positively or negatively to y. First study the degree of incidence ε_i between y and x_i, $i = 1,2,\cdots,n$. For a chosen lower threshold value ε_0, when $\varepsilon_i < \varepsilon_0$, remove the variable x_i out of consideration. By doing so, some of the system's endogenous variables with weak degrees of incicence with y can be removed from further consideration. Assume that the remaining illustrative variables of y are $x_{i_1}, x_{i_2}, \cdots, x_{i_m}$. Nest, consider the degrees $\varepsilon_{i_j i_k}$ $(i_j, i_k = i_1, i_2, \cdots, i_m)$ of incidence between these remaining variables. For a chosen threshold value ε_0', when $\varepsilon_{i_j i_k} \geq \varepsilon_0'$, the variables x_{i_j} and x_{i_k} are seen as the same kind so that the remaining variables are divided into several subsets. Now, choose one representative from each of these subsets to enter into the eventual model. By going through this prossess, the resultant econometric model established can be greatly simplified without losing the needed power of explanation. At the same time, to a certain degree the difficult problem of collinearity of the variables can be avoided.

6.1.2 Grey Econometrics Models

In econometrics, there are many different kinds of models, such as linear regression models in one or multiple variables, nonlinear models, systems of equations, etc. When estimiating the parameters of these models, one often faces with some phenomena that are difficult to explain. For instance, the coefficients of the major illustrative variables are nearly zero; the signs of some estimated values of the parameters do not agree with the reality or contradict the theoretical economic analysis; small vibrations in some individual observations cause drastic changes in most of other estimated parametric values; etc. Among the main reasons underlying these difficulties are

(1) During the time period the observations are done, the internal structure of the system went through major changes;
(2) There is a problem of collinearity between the illustrative variables; and
(3) There are randomness and noise in the observed data.

For the first two scenarios, there is a need to repeat the investigation of the model structure or a need to recheck the illustrative variables. For the third scenario, one can consider establishing models using the GM(1,1) simulated values of the original observations to eliminate the effect of the randomness or noise existing in the available data. The *combined grey econometric model*, obtained this way, can more accurately reflect the relationship between the system's variables. At the same time, the prediction results made on the endogenous variables of the grey econometric model system based on the GM(1,1) predicted values of the illustrative variables possess more solid scientific foundation. Besides, by comparing the results of grey predictions of the endogenous variables with those obtained out of econometric models, one can further improve the reliability of the predictions.

The steps for establishing and applying grey econometric models are as follows:

Step 1: Design the theoretical model. Study the economic activity of interest closely. According to the purpose of the investigation, select the variables that will potentially enter the model. Discover the relationships between these variables based on theories of economic behaviors and experience and/or by analyzing the sampled data. Develop the mathematical expressions, which are the theoretical model, that describe the relationships between these variables. This stage is the most important and most difficult period of time of the entire modeling process; and the following works need to be done:

(1) Study the relevant theories of economics

Theoretical models summarize the foundamental characteristics and laws of development of the objective matters. They are abstract pictures of the reality. Therefore, in the stage of model design, one first needs to conduct a qualitative analysis using economic theories. With different theories, various models can be established. For instance, according to the theory of equilibriumof labor markets, the rate y of wage increase are related to the unemployment rate x_1 and that x_2 of inflation, that is, $y = f(x_1, x_2)$. The greater the unemployement rate increases, the smaller the rate of wage increase due to the fact that the supply of labor is clearly greater than the demand. This is the famous Alban W. Phillips curve, which has been widely accepted and applied in the economics models of the Western countries. However, this model may not necessarily hold true in the socialist market economy of China. As a second example, according to Keynes's theory of consumption, it is believed that on the average, when income grows, people tend to consume; however, the degree of increase in consumption is not as much as that in income. Assume that y stands for consumption, x the income. Then, a mathematical expression for the relationship between these variables is

$$y = f(x) = b_0 + b_1 x + \varepsilon$$

where the parameter $b_1 = dy / dx$ stands for the marginal consumption tendency, and ε a random noise, representing the randomness inherently existing in the consumption. According to Keynes, $0 < b_1 < 1$. However, Simon Kuznets does not agree with Keynes's opinion of a declining marginal consumption tendency. His work indicates that there is a stable proportion of increase between consumption and income. That is, the previous model is only a product produced out of Keynes's theory.

(2) Variables and the form of the eventual model

The established model should reflect the objective economic activity. However, it is impossible for such a reflection to include all details. That is the reason why we need reasonable assumptions. Employing the method of this section to select the major variables to be included in the model using grey incidences will help to eliminate minor relationships and factors. It focuses on the dominant connections, while simplifying the eventual model, making it convenient to handle and apply.

The specific works of this stage of model design include: (i) Determine which variables to include, which ones are dependent variables, and which ones are independent. Here, each inpedendent variable is also known as illustrative variable. (ii) Determine the number of parameters to be included in the model and their (positive or negative) signs. (iii) Determine the mathematical form of the model expression. Is it linear or nonlinear?

(3) Collection and organization of statistical data

After having decided on which variables to consider, one needs to collect all the relevant data. That is the foundation of establishing models. Generally speaking, all the collected raw data need to be statistically categorized and organized so that they become the empirical evidence of the characteristics of the problem of concern and are systematically usable for the purpose of modeling. The basic types of statistical data, as discussed in Chapter 3, include behavioral sequences, time series, index sequences, horizontal sequences, etc.

Step 2: Establish the GM(1,1) model and obtain its simulated values. In order to eliminate the random effect or error noise existing in the observational values of individual variables of the model, establish the GM(1,1) models for the inidividually observed sequences and then apply the simulated values of these GM(1,1) models as the base sequences on which to construct the eventual model.

Step 3. Estimiate the parameters. After having designed the econometric model, the next task is to estimate the parameters, which are the constant coefficients of the quantitative relationship of the model between the chosen variables. They connect the individual variables within the model. More specifically, these parameters explain how independent variables affect the dependent variable. Before using observed data to make estimations, these parameters are unknown. After the form of the model is established, on the basis of the GM(1,1) model simuated sequences solve for the estimated values of the parameters using an appropriate method, such as that of least squares estimate. As soon as the parameters are clearly specified, the relationships between the variables in the model are known definitely so that the model in turn is determined.

The estimated values of the parameters provide realistic and empirical contents for the theories of economics and also verification of these theories. For instance, in the previously mentioned consumption model, if the estimated value of the parameter b_1 is $\hat{b}_1 = 0.8$, it not only classifies the realistic content of the marginal consumption tendency, but also at the same time provides a piece of evidence for the assumption of Keynes's theory of consumption that this parameter is between 0 and 1.

Step 4: Test the model. After the parameters are estimated, the abstract model becomes specific and determined. However, to determine whether or not the model agrees with the objective reality, and whether or not it can explain realistic economic processes, it still has to go through tests. The tests consist of two asepcts, the test of economic meanings and statistical tests. The test of economic meanings is to check whether or not the individual estimated values of the parameters agree with the economic theories and the relevant experiences. The statistical tests are about checking the reliability of the estimate, the effectiveness of the data sequence simulation, the correctness of varous econometric assumptions, and the overall structure of the model and its prediction ability using the principles of statistical reasoning. Only after the model passes through these tests, it can be practically applied in practice. If the model does not pass the tests, then the model needs to be modified and improved.

Step 5: Apply the established model. Grey econometric models have been mainly employed to analyze economic structures, evaluate policies and decisions, simulate economic systems, and predict economic development. Each process of application is also a process of verifying the model and its underlying theory. If the prediction contains small errors, it means that the model is of high accuracy and quality with strong ability to explain the reality and the fact that its underlying theory agrees with the reality. Otherwise, the model and the economic theory on which the model was initially developed need to be modified.

Combined grey econometric models not only can be employed to the situations of kwnon system structures, but also is especially appropriate for the situations of system structures that need further study and exploration.

6.2 Combined Grey Linear Regression Models

Combined grey linear regression models can improve the weakness of the original linear regression models that no exponential growth is considered and the weakness of the GM(1,1) model that no enough linear factors are involved. So, such combined models are suited for studying sequences with both linear tendencies and exponential growth tendencies. For such a sequence, the modeling prossess can be described as follows.

Let $X^{(0)} = \{x^{(0)}(1), x^{(0)}(2), ..., x^{(0)}(n)\}$ be a sequence of raw data. Its first order accumulation generation sequence is $X^{(1)} = \{x^{(1)}(1), x^{(1)}(2), ..., x^{(1)}(n)\}$. From the GM(1,1) model, we can obtain

$$\hat{X}^{(1)}(t+1) = (X^{(0)}(1) - b/a)\exp(-at) + b/a \tag{6.1}$$

whose form can be written as below:

$$\hat{X}^{(1)}(t+1) = C_1 \exp(vt) + C_2 \tag{6.2}$$

Now, use the sum of the linear regression equation $Y = aX + b$ and the exponential equation $Y = a * \exp(X)$ to simulate this accumulated sequence $X^{(1)}(t)$ producing a new sequence

$$\hat{X}^{(1)}(t) = C_1 \exp(vt) + C_2 t + C_3 \tag{6.3}$$

where v, C_1, C_2, C_3 are parameters that need to be estimated.

To determine the values of these parameters, assume the following parametric sequence

$$\begin{aligned} Z(t) &= \hat{X}^{(1)}(t+1) - \hat{X}^{(1)}(t) \\ &= C_1 \exp[v(t+1) + C_2(t+1) + C_3 - C_1 \exp(vt) - C_2 t - C_3 \\ &= C_1 \exp(vt)[\exp(v) - 1] + C_2,\ (t = 1,2,...,n-1) \end{aligned} \tag{6.4}$$

Let

$$\begin{aligned} Y_m(t) &= Z(t+m) - Z(t) \\ &= C_1 \exp[v(t+m)][\exp(v-1)] + C_2 - C_1 \exp(vt)[\exp(v) - 1] - C_2 \\ &= C_1 \exp(vt)[\exp(v_m) - 1][\exp(v) - 1] \end{aligned} \tag{6.5}$$

Similarly, we have

$$Y_m(t+1) = C_1 \exp[v(t+1)][\exp(v_m) - 1][\exp(v) - 1] \tag{6.6}$$

The ratio of the previous two equation is

$$Y_m(t+1) / Y_m(t) = \exp(v) \tag{6.7}$$

From this equation, we solve for v and obtain:

$$v = \ln[Y_m(t+1) / Y_m(t) \tag{6.8}$$

By replacing the values of $\hat{X}^{(1)}$ in equ. (6.4) by those of $X^{(1)}$, from equ. (6.8) we can obtain an approximate solution $\tilde{V}$ for v. For different values of m, we obtain different values of $\tilde{V}$. We use the mean $\hat{V}$ of these values $\tilde{V}$ to be the estimated value of v.

For equ. (6.4) with $Z(t) = X^{(1)}(t+1) - X^{(1)}(t)$, $t = 1,2,...,n-1$, if $m = 1$, we have

$$Y_1(t) = Z(t+1) - Z(t),\ t = 1,2,...,n-2$$

$$\tilde{V}_1(t) = \ln[Y_1(t+1) / Y_1(t)],\ t = 1,2,...,n-3$$

For $m = 2$, we have

$$Y_2(t) = Z(t+2) - Z(t),\ t = 1,2,...,n-3$$

$$\tilde{V}_2(t) = \ln[Y_2(t+1) / Y_2(t)],\ t = 1,2,...,n-4$$

....

For $m = n-3$, we have

$$Y_{n-3}(t) = Z(t+n-3) - Z(t)\,,\ t = 1,2$$

$$\tilde{V}_{n2}(t) = \ln[\,Y_2(t+1)\,/\,Y_2(t)]\,,\ t = 1$$

Here, we computed $(n-3)+(n-4)+...+2+1=(n-2)(n-3)/2$ many $\tilde{V}$ values. Now, we take the mean of these $\tilde{V}$ values as the estimate of v. That is,

$$\hat{V} = \frac{\sum_{m=1}^{n-3}\sum_{t=1}^{n-2-m}\tilde{V}_m(t)}{(n-2)(n-3)/2} \tag{6.9}$$

By letting $L(t) = \exp(\hat{V}\,t)$, equ. (6.3) can be written as follows:

$$\hat{X}^{(1)}(t) = C_1 L(t) + C_2 t + C_3 \tag{6.10}$$

By making use of the method of least squares estimate, we can obtain the values of C_1, C_2, C_3.

Let

$$X^{(1)} = \begin{bmatrix} x^{(1)}(1) \\ x^{(2)}(2) \\ \vdots \\ x^{(1)}(n) \end{bmatrix},\ C = \begin{bmatrix} C_1 \\ C_2 \\ C_3 \end{bmatrix},\ A = \begin{bmatrix} L(1) & 1 & 1 \\ L(2) & 2 & 1 \\ \vdots & \vdots & \vdots \\ L(n) & n & 1 \end{bmatrix}$$

then we have

$$X^{(1)} = AC \tag{6.11}$$

Therefore, we have

$$C = (A^T A)^{-1} A^T X^{(1)} \tag{6.12}$$

Now, the predicted values of the accumulation generated sequence are given by

$$\hat{X}^{(1)}(t) = C_1 \exp(\hat{V}\,t) + C_2 t + C_3 \tag{6.13}$$

From equ. (6.13), it can be seen that if $C_1 = 0$, then the first order accumulation generation sequence stands for a linear regression model. If $C_2 = 0$, then the accumulation generation stands for a GM(1,1) model. This new model improves the differenties that in the original linear regression model no exponential growth is included and that in the GM(1,1) model no linear factors are considered.

By applying the inverse accumulation generation on equ. (6.13), we obtain the simulated and predicted values $\hat{X}^{(0)}$ of the oringal sequence.

Example 6.1. At a certain observation station of ore and rock movement, the sequence of recorded subsides of a specific location from February 1995 to April 1996 is given in Table 6.1. Try to make predictions for the sinking dynamics of this specific location.

Table 6.1 The original sequence of recorded subsides

Time	9502	9504	9506	9508	9510	9512	9602	9604
Amount of subside	12	22	31	43	51	57	75	83

Solutioon. Due to the small amount of available data, grey system models are more appropriate for this task of prediction. However, grey systems models employ exponential functions to simulate accumulation generated sequences. So, they in general are only suitable for modeling situations of exponential development; and it is difficult for them to describe linear tendency of change. Because of this reason, in this case study, we will apply a grey linear, exponential regression model to predict the subsides of the specified location.

The original sequence of data is

$$X^{(0)} = (12,22,31,43,51,57,75,83)\,.$$

Its first order accumulation sequence is

$$X^{(1)} = (12,34,65,108,159,216,291,374)\,.$$

For different m values, from equs. (6.6) and (6.7), we obtain the estimated value $\hat{V} = 0.02058096$ for v. And from equ. (6.10), we obtain the estimated value of C:

$$C = (A^T A)^{-1} A^T X^{(1)} = (21750.995, -439.9523, -21751.078)$$

So, the combined model of the first order accumulation generation sequence is

$$\hat{X}^{(1)}(t) = 21750.995\exp(0.02058096t) - 439.9523t - 21751.078$$

Out of this model, we obtain the simulated and predicted values for each of the time moments as listed in Table 6.2.

Table 6.2 Simulated and predicted values and their errors

Time	9502	9504	9506	9508	9510	9512	9602	9604	9606	9608
$x^{(0)}$	12	22	31	43	51	57	75	83		
$\hat{x}^{(0)}$	12.34	21.75	31.35	41.15	51.15	61.36	71.79	82.43	93.29	104.38
Error (%)	-2.85	1.15	-1.12	4.31	-0.30	-7.66	4.28	0.69		

6.3 Grey Cobb-Douglas Model

In this section, we study the Cobb-Douglas or production function model. In particular, Let K be the capital input, L the labor input, and Y the production output. Then,

$$Y = A_0 e^{\gamma t} K^{\alpha} L^{\beta}$$

is known as the C-D production function model, where α stands for the capital elasticity, γ the labor elasticity, and the parameter for the progress of technology. The log-linear form of this production function model is given below:

$$\ln Y = \ln A_0 + \gamma t + \alpha \ln K + \beta \ln L$$

For given time series data of the production output Y , capital input K , and labor input L ,

$$Y = (y(1), y(2), \cdots, y(n)),\ K = (k(1), k(2), \cdots, k(n)), \text{ and } L = (l(1), l(2), \cdots, l(n)),$$

one can employ the method of multivariate least squares estimate to approximate the parameters $\ln A_0$, γ, α, and β.

When Y , K , and L represent the time series of a specific depertment, district, or business, it is ofen the case that due to severe fluctuations existing in the data the estimated parameters contain errors leading to obviously incorrect results. For instance, the estimated coefficient γ for progress of technology is too small or becomes a negative number; the estimated values for the elasticities α and β go beyond their reasonable ranges. Under such circumstances, if one considers using the GM(1,1) simulated data of Y , K , and L as the original data for his least squares estimates, then to a certain degree he can eliminate some of the random fluctuations, produce more reasonable estimated parameter values, and obtain his model that can more accurately reflect the relationship between the production output and labor and capital inputs and the progress of technology.

Assume that

$$\hat{Y} = (\hat{y}(1), \hat{y}(2), \cdots, \hat{y}(n)),$$
$$\hat{K} = (\hat{k}(1), \hat{k}(2), \cdots, \hat{k}(n)), \text{ and}$$
$$\hat{L} = (\hat{l}(1), \hat{l}(2), \cdots, \hat{l}(n))$$

are respectively the GM(1,1) simulated sequences of Y , K , and L . Then $\hat{Y} = A_0 e^{\gamma t} \hat{K}^{\alpha} \hat{L}^{\beta}$ is known as the grey production function model.

In the grey production function model, although no grey parameters appear explicitly, it stands for an expression that combines the idea of grey systems modeling into the C-D production function model. That is, this model possesses a very deep intension of the greyness. It embodies the non-uniquness principle of solutions and the absoluteness principle of greyness. That is why in practical applications, this model has produced satisfactory results. To this end, please consult with (Liu and Lin, 2006, p. 256 – 258) for how applications are carried out.

6.4 Grey Artificial Neural Network Models

6.4.1 BP Artificial Neural Model and Computational Schemes

Each *articifical neural network* is made up of a large amount of elementary *information processors*, known as *neurons* or *nodes*. The model with *multi-layered nodes* or the scheme, known as error back propagation, represents the currently well developed and widely employed artificial neural network system and computational method. It translates the input-output problem of an available sample into a nonlinear optimization problem. It is a powerful tool that can be employed to uncover the laws and patterns hidden in large amounts of data. Using the method of artificial neural networks to simulate data sequences has several latent advantages: First of all, it has the ability to model multi kinds of functions, including nonlinear functions, piecewise defined functions, etc. Secondly, unlike the traditional methods of distinguishing data sequences, which, to work properly, have to have presumed types of functional relationships between the data sequences, artificial reural networks can establish the needed functional relationship by using the attributes and intension naturally existing in the provided data variables without presuming the kinds of distributions the parameters satisfy. Thirdly, this method possesses the advantage of making use of the available information very efficiently, while avoiding the problem that the real meanings and pictures of the data are lost due to

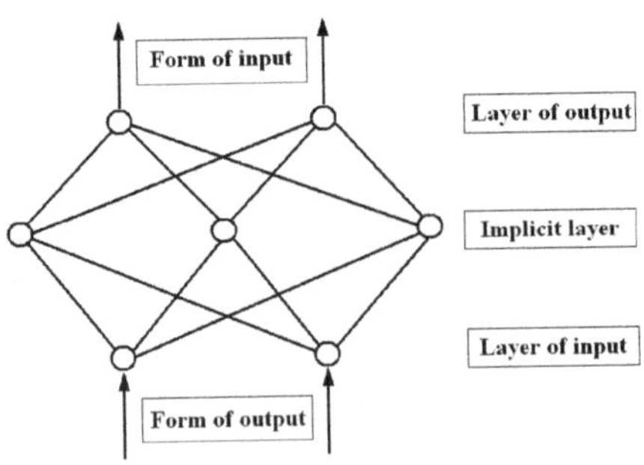

Fig. 6.1 A back propagation neural network

various combinations, such as additions of positive and negative values, of the data mining methods. That is, the method of artificial neural networks is especially useful for improving the GM(1,1) model.

Figure 6.1 shows a back propagation network with three layers. The network consists of an input layer, an implict (or latent) layer, and an output layer. An entire process of learning consists of forward and backward propagations. The particular scheme of learning is given below:

(1) Apply random numbers to initialize W_{ij} (the connection weight between nodes i and j of different layers) and θ_j (the threshold value of node j);
(2) Feed in the preprocessed training samples $\{X_{PL}\}$ and $\{Y_{PK}\}$;
(3) Compute the output of of the nodes of each layer, $O_{pj} = f\sum_i (W_{ij} I_{pi} - \theta_j)$ for the pth sample point, where I_{pi} stands for the output of node i and the input of node j;
(4) Compute the information error of each layer. For the input layer, $\delta_{pk} = O_{pk}(y_{pk} - O_{pk})(1 - O_{pk})$; for the latent layer, $O_{pi} = O_{pi}(1 - O_{pi})\sum_i \delta_{pi} W_{ij}$
(5) For the backward propagation, the modifiers of the weights are $W_{ij}(t+1) = \alpha\delta_{pi} O_{pi} + W_{ij}(t)$, and the modifiers of the thresholds $\theta_j(t+1) = \theta_j(t) + \beta\delta_{pi}$, where α stands for the learning factor and β the momentum factor for accelerated convergence.
(6) Calculate the error: $E_p = (\sum_p \sum_k)(O_{pk} - Y_{pk})^2 / 2$

6.4.2 Principle and Method for Grey BP Neural Network Modeling

Assume that a time series $\{x^{(0)}(i)\}$, $i = 1,2,\cdots,n$, is given. We then obtain the restored values $\hat{x}^{(0)}(t)$, $i = 1,2,\cdots,n$, using the outputs of the GM(1,1) model

$$\frac{dx^{(1)}}{dt} = ax^{(1)} + b.$$

Definition 6.1. The difference between the restored value $x^{(0)}(L)$ and the GM(1,1) model simulated value $\hat{x}^{(0)}(L)$ at time moment L is referred to as the *error of moment L*, denoted $e^{(0)}(L)$. In symbols, $e^{(0)}(L) = x^{(0)}(L) - \hat{x}^{(0)}(L)$.

1. Establish the back propagation network model for the error sequence $\{e^{(0)}(L)\}$.

Assume that $e^{(0)}(L)$ stands for the error sequence, $L=1,2,\cdots,n$. If the order of prediction is S, meaning that we use the information of $e^{(0)}(i-1)$, $e^{(0)}(i-2)$, …, $e^{(0)}(i-S)$ to predict the value at the ith moment, we will treat $e^{(0)}(i-1)$, $e^{(0)}(i-2)$, …, $e^{(0)}(i-S)$ as the input sample points of the back propagation network training, while using the value of $e^{(0)}(i)$ as the expected prediction of the back propagation network training. By using the back propagation computational scheme outline earlier, train this network through enough amount of cases of error sequences so that output values (along with empirical test values) are produced corresponding to different input vectors. The resultant weights and thresholds represent the correct internal representations through the self-learning and adaptation of the network. A well trained back propagation network model can be an effective tool for error sequence prediction.

(2) Determine new prediction values of $\{e^{(0)}(L)\}$.

Assume that out of a well trained back propagation network the predicted error sequence of $\{e^{(0)}(L)\}$ is $\left\{\hat{e}^{(0)}(L)\right\}$. We now construct a new predicted sequence $\hat{X}^{(0)}(i,1)$ on the basis of the previous prediction as follows:

$$\hat{X}^{(0)}(i,1)=\hat{X}^{(0)}(i)+\hat{e}^{(0)}(1)$$

which is the predicted sequence of the grey artificial neural network model.

Example 6.2. Given the actual yearly investments in environment protection over a period of time of a certain location, and the GM(1,1) simulations and relevant errors in Table 6.3, establish an artificial neural network model for the error sequence.

Table 6.3 The GM(1,1) simulations and errors

Year	Investment $x^{(0)}(L)$	GM(1,1) simulation $\hat{x}^{(0)}(L)$	GM(1,1) error $e^{(0)}(L)$
1985	110.20	110.20	0
1986	146.34	164.39	-19.05
1987	185.36	187.65	-2.29
1988	221.14	214.22	6.92
1989	255.16	244.54	10.52
1990	289.18	279.17	9.01
1991	320.54	319.69	1.85
1992	352.79	363.81	-11.02

Solution. Based on and using the GM(1,1) error sequence data given in Table 6.3, we apply the previously outline method to establish a back propagation network model. Our projected back propagation network will have three characteristic parameters, one latent layer, within which there are 6 nodes, and one input layer within which there is one node. Let the learning rate be 0.6, the convergence rate 0.001, and the variance limited within the range of 0.01. Let us conduct the training and testing of the network on a computer. Then, Table 6.4 lists the simulation results of the combined back propagation network model.

Table 6.4 Simulation results of the grey artificial neural network model

Year	Actual value $x^{(0)}(L)$	Simulated value $\hat{x}^{(0)}(i,1)$	Relative errors (%)
1988	221.14	221.12	0.01
1989	255.16	255.29	0.05
1990	289.18	289.11	0.02
1991	320.54	320.79	0.08
1992	352.79	352.70	0.03

6.5 Grey Markov Model

6.5.1 Grey Moving Probability Markov Model

Assume that $\{X_n, n \in T\}$ is a stochastic process. If for any whole number $n \in T$ and any states $i_0, i_1, \cdots, i_{n+1} \in I$, the following conditional probability satisfies

$$P(X_{n+1} = i_{n+1} \mid X_0 = i_0, X_1 = i_1, \cdots, X_n = i_n) = P(X_{n+1} = i_{n+1} \mid X_n = i_n) \quad (6.14)$$

then $\{X_n, n \in T\}$ is known as a *Markov chain*. Equ. (6.14) is seen as without any post-effect. It means that the future state of the system at $t = n+1$ is only related to the current state at $t = n$ without any influence from any other earlier state $t \le n-1$.

For any $n \in T$ and states $i, j \in I$, the following

$$p_{ij}(n) = P(X_{n+1} = j \mid X_n = i) \quad (6.15)$$

is known as the *moving probability* of the Markov chain. If the moving probability $p_{ij}(n)$ in this equation does not have anyting to do with the index n, then $\{X_n,\ n \in T\}$ is known as a *homogeneous Markov chain*. For such a Markov chain, the moving probability $p_{ij}(n)$ is often denoted as p_{ij}. Because our discussion will be mainly on homogeneous Markov chains, the word "homogeneous" will be omitted. When all the moving probabilities $p_{ij}(n)$ placed in a matrix, such as $P=[p_{ij}]$, this matrix is referred to as the *moving probability matrix* of the system's state.

Proposition 6.1. The entries of the moving probability matrix P satisfy

1) $p_{ij} \geq 0,\ i, j \in I$; and (6.16)

2) $\sum_{j\in I} p_{ij} = 1,\ i \in I$. (6.17)

The probability $p_{ij}^{(n)} = P(X_{m+n} = j \mid X_m = i), i, j \in I, n \geq 1$ is known as the *n-th step moving probability* of the given Markov chain, and $P^{(n)} = [p_{ij}^{(n)}]$ the *n-th step moving probability matrix.*

Proposition 6.2. The *n*-th step moving probability matrix $P^{(n)}$ satisfies

1) $p_{ij}^{(n)} \geq 0,\ i, j \in I$; (6.18)

2) $\sum_{j\in I} p_{ij}^{(n)} = 1,\ i \in I$; and (6.19)

3) $P^{(n)} = P^n$. (6.20)

Any Markov chain with grey moving probability(-ies) is known as a *grey Markov chain*. When studying practical problems, due to lack of sufficient information, it is often difficult to determine the exact values of the moving probabilities. In this case, based on the available information it might be possible to determine the grey ranges $p_{ij}(\otimes)$ of these uncertain probabilitites. When the moving probability matrix is grey, the entries of its whitenization $\tilde{P}(\otimes) = [\tilde{P}_{ij}(\otimes)]$ are generally required to satisfy the following properties:

1) $\tilde{P}_{ij}(\otimes) \geq 0,\ i, j \in I$; and

2) $\sum_{j\in I} \tilde{P}_{ij}(\otimes) = 1,\ i \in I$ ·

Proposition 6.3. Assume the the initial distribution of a finite-state grey Markov chain is $P^T(0) = (p_1, p_2, \cdots, p_n)$ and the moving probability matrix $P(\otimes) = [P_{ij}(\otimes)]$. Then, the system's distribution of the next *s*th state is

$$P^T(s) = P^T(0) P^s(\otimes) \tag{6.21}$$

That is, when the system's initial distribution and the moving probability matrix are known, one can predict the system's distribution for any future state.

6.5.2 Grey State Markov Model

Let $X^{(0)}(k)(k=1,2,\cdots,p)$ be a sequence of raw data, $\hat{Y}(k)$ the GM(1,1) simulated or predicted value at the time moment k based on the original data. For an instable stochastic sequence $X^{(0)}$ that saistifes the condition of Markov chains, if we divide it into n states and each of the state $\otimes_i$ is expressed by

$$\otimes_i = [\tilde{\otimes}_{1i}, \tilde{\otimes}_{2i}],\ \tilde{\otimes}_i \in \otimes_i,$$

with $\tilde{\otimes}_{1i} = \hat{y}(k) + A_i$, $\tilde{\otimes}_{2i} = \hat{y}(k) + B_i$, $(i=1,2,\cdots,n)$, then because $\hat{Y}(k)$ is a function of time k , the grey elements $\tilde{\otimes}_{1i}$ and $\tilde{\otimes}_{2i}$ also change with time.

If $M_{ij}(m)$ is the size of the sample of data representing the development from state $\otimes_i$ to state $\otimes_j$ through m steps, and M_i the size of the sample of data for staying at the state $\otimes_i$, then

$$P_{ij}(m) = \frac{M_{ij}(m)}{M_i},\ (i=1,2,\cdots,n) \tag{6.22}$$

is known as the *state moving probability*. In practical applications, one generally considers the matrix P of one-step moving probabilities. Assume that the object of prediction is located at state $\otimes_k$, then consider the *k*th row of P. If

$$\max_j p_{kj} = p_{kl}$$

then it can be thought of that at the next time moment, the system will most likely transform from state $\otimes_k$ to state $\otimes_l$. If there are two or more entries in the *k*th row of P that are equal or roughly equal, then the direction of change in the system's state is difficult to determine. In this case, one needs to look at the two-step or *n*-step matrix $P^{(2)}$ or $P^{(n)}$ of moving probabilities, where $n \geq 3$.

6.6 Combined Grey-Rough Models

Both grey systems theory and rough set theory are two mathematical tools developed to deal with uncertain and incomplete information. They to a certain degree complement each other. They both apply the idea of lowering the preciseness of expression of the available data to gain the extra generality of the expression. In particular, grey systems theory employs the method of grey sequence generations to reduce the accuracy of data expressions, while rough set theory makes use of the idea of data scattering, so that they make it possible to uncover patterns hidden in the data by ignoring the unnecessary details. Neither grey systems theory nor rought set theory requires any prior knowledge, such as probability distribution, degree of membership, etc. What rough set theory investigates includes rough, non-intersecting classes and concepts of roughness with emphasis placed on the indistinguishability of objects. What grey systems theory focuses on is the grey sets with clear extension and unclear intension with its emphasis placed on uncertainties caused by sufficient information. Thus, if rough set theory and grey systems methodology are mixed, their individual weaknesses both in theory and application can be mightily improved so that greater theoretical strength and practical applicability can be achieved.

6.6.1 Rough Membership, Grey Membership and Grey Numbers

Rough set theory can be seen as an expansion of the classical set theory. When facing the concept of sets, it makes use of rough membership functions to define rough sets, where each membership function is explained and understood as those of conditional probabilities.

The concepts of rough approximation sets and rough membership functions of the rough set theory are closely related to that of greyness of grey numbers. When either $\mu_X(x)=0$ or $\mu_X(x)=1$, the object is assured either to belong to the set *X* or not to belong to *X*. In such cases, the classification is definite and clear; the involved greyness is the smalles. If $0<\mu_X(x)<1$, then the object x belongs to the set *X* with the degree of confidence $\mu_X(x)$. In this case, the object x projects a kind of grey state of transition between being definitely in the set *X* and being definitely not in *X*. When $\mu_X(x)=0.5$, the probability for the object x either to belong to the set *X* or not to belong to *X* is all 50%. For this situation, the degree of uncertinaty is the highest. That is the degree of greyness is the highest. When the rough membership function $\mu_X(x)$ is near 1 or 0, the uncertainty for the object x to belong to or not to belong to the set *X* is decreased, and the corresponding degree of greyness should also decrease. The closer to 0.5 the rough membership is, the greater the uncertainty for the object x to belong to or not to belong to the set *X*, and the cooresponding degree of greyness is also greater. We categorize all rough membership functions into two groups: upper and lower rough membership functions, where a rough membership function is *upper* if its values come from the interval [0.5,1], denoted $\overline{\mu}_X(x)$; the corresponding grey membership function is also referred to as upper and denoted by $\overline{g}_X(x)$. A *lower rough membership function* is one that takes values from the interval [0,0.5], denoted $\underline{\mu}_X(x)$. The corresponding grey membership function is referred to as a lower grey membership function, denoted $\underline{g}_X(x)$.

Evidently, upper, lower and general rough membership functions satisfy the following properties:

(1) $\overline{\mu}_X(x) = 1 - \underline{\mu}_X(x)$;

(2) $\mu_{X \cup Y}(x) = \mu_X(x) + \mu_Y(x) - \mu_{X \cap Y}(x)$; and

(3) $\max(0, \mu_X(x) + \mu_Y(x) - 1) \le \mu_{X \cap Y}(x) \le \min(1, \mu_X(x) + \mu_Y(x))$.

Based on the discussion above, we introduce the following definition of grey membership functions using the concept of rough membership functions.

Definition 6.2. Assume that x is an object with its field of discourse U. That is, $x \in U$. Let X be a subset of U. Then mappings from U to the closed interval [0,1]:

$$\overline{\mu}_X : U \to [0.5,1],\ \mu \mapsto \overline{g}_X(x) \in [0,1], \text{ and}$$

$$\underline{\mu}_X : U \to [0,0.5],\ \mu \mapsto \underline{g}_X(x) \in [0,1]$$

are respectively referred to as an *upper* and a *lower grey membership functions* of X, where $\overline{\mu}_X \ge \underline{\mu}_X$, and $\overline{g}_X(x)$ and $\underline{g}_X(x)$ are respectively referred to as an upper and lower grey membership functions of the object x with respect to X.

The concept of grey membership functions defined based on rough membership functions is depicted in Figure 6.2.

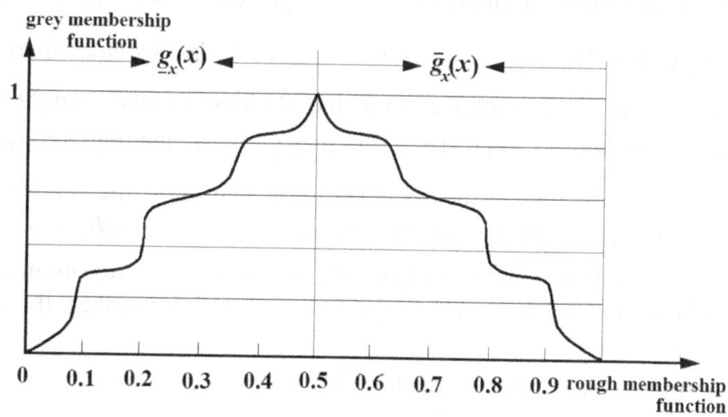

Fig. 6.2 A conceptual depiction of grey membership function

Definition 6.3. Assume that $x \in U$, $X \subseteq U$, the *grey number scale* of the uncertainty for x to belong to X is g_c , the grey number scale of the upper grey membership function $\overline{g}_X(x)$ is $\overline{g}_c$, the grey number scale of the lower grey membership function $\underline{g}_X(x)$ is $\underline{g}_c$. Then the greyness scales $\overline{g}_c$ and $\underline{g}_c$ of the upper grey number and the lower grey number of the greyness scales g_c of different grey numbers are respectively given as follows:

The greyness of white numbers ($g_c = 0$): If $\mu_X(x) = 0$, then $\underline{g}_c = 0$; if $\mu_X(x) = 1$, then $\overline{g}_c = 0$.

For the first class grey numbers ($g_c = 1$): If $\mu_X(x) \in (0,0.1]$, then $\underline{g}_c = 1$; if $\mu_X(x) \in [0.9,1)$, then $\overline{g}_c = 1$.

For the second class grey numbers ($g_c = 2$) : If $\mu_X(x) \in (0.1,0.2]$, then $\underline{g}_c = 2$; if $\mu_X(x) \in [0.8,0.9)$, then $\overline{g}_c = 2$.

For the third class gey numbers ($g_c = 3$) : If $\mu_X(x) \in (0.2,0.3]$, then $\underline{g}_c = 3$; if $\mu_X(x) \in [0.7,0.8)$, then $\overline{g}_c = 3$.

For the fourth clas grey numbers ($g_c = 4$) : if $\mu_X(x) \in (0.3,0.4]$, then $\underline{g}_c = 4$; if $\mu_X(x) \in [0.6,0.7)$, then $\overline{g}_c = 4$.

For the fifth class grey numbers ($g_c = 5$) : if $\mu_X(x) \in (0.4,0.5)$, then $\underline{g}_c = 5$; if $\mu_X(x) \in (0.5,0.6)$, then $\overline{g}_c = 5$.

The greyness of black numbers ($g_c > 5$) : if $\mu_X(x) = 0.5$, then $\underline{g}_c = \overline{g}_c > 5$.

When $\mu_X(x) \in [0,1]$, $\underline{g}_X(x) = 0$ and $\overline{g}_X(x) = 1$. In this case, there is no uncertain information. So it is referred to as the *greyness of white numbers*. That is, $g_c = \underline{g}_c = \overline{g}_c = 0$. When $\mu_X(x) = 0.5$, $\underline{g}_X(x) = \overline{g}_X(x) = 1$, the degree of uncertainty for the object x to belong to the set X or not to belong to the set X reaches the maximum, which is referred to as the *greyness of black numbers* $g_c > 5$. From Definition 6.2, it follows that the higher the greyness of a grey number, the less clear the information is; the lower the greyness of a grey number, the clearer the information is.

From Definition 6.2, it can be readily obtained that $\overline{\mu}_X(x) = 1 - \underline{\mu}_X(x)$. If we use the greyness of the upper grey number to represent the degree of uncertainty for the object x to belong to the set X, and the greyness of the lower grey number to illustrate the degree of uncertainty for the object x not to belong to the set X, then these two degrees of uncertainty are supplementary to each other.

According to Definition 6.3, the scale of the greyness of a grey number is determined the grey interval the maximum rough membership value of the information granularity could possibly belong to. So, the whitenizations of grey numbers of different degrees of greyness are defined as the maximum possible rough membership value of the grey numbers of corresponding scales. For example, if the possible maximum rough membership value of a certain conditional subset, computed out of the available decision-making table, is $\mu_X(x) = 0.75$, because $0.75 \in [0.7,0.8)$, then $\mu_X(x) = 0.75$ stands for the whitenized value of such a grey number whose upper greyness is $\overline{g}_c = 3$.

6.6.2 Grey Rough Approximation

Definition 6.3. Assume that $S = (U, A, V, f)$, $A = C \cup D$, $X \subseteq U$, $P \subseteq C$, and given the greyness scale $g_c \leq 5$ of a grey number. Then

$$\underline{apr}_P^{g_c}(X) = \bigcup \left\{ \frac{|I_P(x) \cap X|}{|I_P(x)|} \leq \overline{g}_c \right\} \tag{6.23}$$

and

$$\overline{apr}_P^{g_c}(X) = \bigcup \left\{ \frac{|I_P(x) \cap X|}{|I_P(x)|} > \underline{g}_c \right\} \tag{6.24}$$

are respectively referred to as the g_c*-lower approximation* and g_c*-upper approximation* of X with respect to I_P, where the upper rough membership function corresponding to the upper scale $\overline{g}_c$ of grey-number greyness satisfies $\overline{\mu}_X(x) \in (0.5,1]$, and the lower rough membership function corresponding to the lower scale $\underline{g}_c$ of grey-number greyness satisfies $\underline{\mu}_X(x) \in [0,0.5)$,

The g_c-lower approximation of the set $X \subseteq U$ under the grey-number greyness scale g_c equals the union of all the equivalence classes of U that belong to X with grey-number greyness scales less than or equal to the upper grey-number greyness scale $\overline{g}_c$. The g_c-upper approximation is equal to the intersection of all the equivalence classes of U that belong to X with grey-number greyness scales greater than the lower grey-number greyness scale $\underline{g}_c$.

Definition 6.5. The quality of g_c*-classification* is

$$\gamma_P^{g_c}(P,D)=\frac{|\bigcup\left\{\frac{|X\cap I_P(x)|}{|I_P(x)|}\leq \overline{g}_c\right\}|}{|U|} \tag{6.25}$$

The *classification quality* $\gamma_P^{g_c}(P,D)$ measures the percentage of the knowledge in the field of discourse that can be clearly classified for a given grey-number greyness scale $g_c \leq 5$, in the totality of current knowledge.

For a given grey-number greyness scale $g_c \leq 5$, let approximate reduction $red_P^{g_c}(C,D)$ stand for the set of the attributes with the minimum condition that still produces clear classification without containing any extra attributes.

In rough set theory, the classification of the elements located along the boundary regions is not clear. Whether or not an element in such a region can be clearly classified is determined most commonly by the pre-fixed greyness scale. The concept of grey rough approximation so defined is analogous to that of variable precision rough approximation. When the interval grey numbers in which the upper greyness scale $\overline{g}_c$ and the lower greyness scale $\underline{g}_c$ of the grey-number greyness g_c of the grey rough approximation respectively belongs to take their corresponding whitenized values, the grey rough approximation is consequently transformed into the rough approximation under the meaning of variable precision rough sets. Evidently, variable precision rough approximation can be seen as a special case of grey rough approximation. When compared to models of variable precision rough sets, of the sets of variable precision whether or not elements in a relatively rough set X can be correctly classified is mostly determined by the pre-fixed maximum critical confidence threshold parameter β, where classification can be done if smaller than or equal to the upper bound of the β value and indistinguishbility appears when this upper bound is surpassed. However, the parameter of the maximum critical confidence threshold β in general is difficult to determine before hand, especially for large data sets. In other words, the parameter of maximum critical confidence threshold β generally stands for a grey number. So, the concept of interval grey numbers provides a practical quantitative tool by appointing the upper and lower endpoints. For the cases where we cannot obtain much information about the degree of accuracy of the actual data, this method of representation becomes extremely useful.

Proposition 6.4. Given greyness scale $g_c \le 5$, the following hold true:

(1) $\overline{apr}_P^{g_c}(X \cup Y) \supseteq \overline{apr}_P^{g_c}(X) \cup \overline{apr}_P^{g_c}(Y)$;

(2) $\underline{apr}_P^{g_c}(X \cap Y) \subseteq \underline{apr}_P^{g_c}(X) \cap \underline{apr}_P^{g_c}(Y)$;

(3) $\underline{apr}_P^{g_c}(X \cup Y) \supseteq \underline{apr}_P^{g_c}(X) \cup \underline{apr}_P^{g_c}(Y)$; and

(4) $\overline{apr}_P^{g_c}(X \cap Y) \subseteq \overline{apr}_P^{g_c}(X) \cap \overline{apr}_P^{g_c}(Y)$.

Proof. (1) For any $X \subseteq U$ and $Y \subseteq U$, and given greyness scale g_c, we have

$$\frac{|I_P(x) \cap (X \cup Y)|}{|I_P(x)|} \ge \frac{|I_P(x) \cap X|}{|I_P(x)|}$$

and

$$\frac{|I_P(x) \cap (X \cup Y)|}{|I_P(x)|} \ge \frac{|I_P(x) \cap Y|}{|I_P(x)|}$$

Therefore, $\overline{apr}_P^{g_c}(X \cup Y) \supseteq \overline{apr}_P^{g_c}(X) \cup \overline{apr}_P^{g_c}(Y)$.

(2) For any X, $Y \subseteq U$, and given greyness scale $g_c \le 5$, we have

$$\frac{|I_P(x) \cap (X \cap Y)|}{|I_P(x)|} \le \frac{|I_P(x) \cap X|}{|I_P(x)|}$$

and

$$\frac{|I_P(x) \cap (X \cap Y)|}{|I_P(x)|} \le \frac{|I_P(x) \cap Y|}{|I_P(x)|}$$

Therefore, $\underline{apr}_P^{g_c}(X \cap Y) \subseteq \underline{apr}_P^{g_c}(X) \cap \underline{apr}_P^{g_c}(Y)$.

Similarly, we can prove (3) and (4). QED.

Proposition 6.5. $\underline{apr}_P^{g_c}(X) \subseteq \overline{apr}_P^{g_c}(X)$.

Proof. Let $x \in \underline{apr}_P^{g_c}(X)$. Because the equivalence relation I_P is reflective, we have $x \in I_P(x)$. From Definition 6.3, it follows that $g_c \le 5$ and that the interval grey number to which the rough membership value corresponding to the upper grey-number greyness scale belongs is greater than the interval grey number to which the rough membership value corresponding to the lower grey-number greyness scale belongs. Hence, we have $x \in \overline{apr}_P^{g_c}(X)$ and consequently $\underline{apr}_P^{g_c}(X) \subseteq \overline{apr}_P^{g_c}(X)$. QED.

6.6.3 Combined Grey Clustering and Rough Set Model

When employing the expansion dominant rough set model to make probobalistic decision making, one needs to have a given multi-criteria decision-making table. However, in many practical applications involving uncertain multi-criteria decision-making, instead of being able to obtain his multi-criteria decision making table, the researcher has to rely on the available sets of data to generate his multi-criteria information table. For instance, we can easily collect the financial data of a publically traded company, such as per share income, per share net asset, net profit, reliability, opetrating profit, etc. Based on the collected financial data, we can establish a multi-criteria information table. Because the class to which company's style of decision-making belongs, such as high risk, moderate risk, and low risk, is unknown ahead of time, it is difficult, if not impossible, to generate the relevant multi-criteria decision-making table. So, the dominant rough set models and expanded dominant rough set models cannot be directly employed to conduct decision-making analysis of these problems. On the other hand, the method of grey clustering of the grey systems theory generally groups objects into different preference categories by considering preference information of attributes and preference behaviors of the decision-makers. In particular, the method of grey fixed weight clustering provides an effective way to transform a multi-criteria information table, which is made of preferred attributes of various dimensions, into a multi-criteria decision-making table. For instance, based on the collected financial data of companies, the distributions of the preferred attributes' values of the criteria, and the preferred behaviors of the decision-makers, we can establish whitenizations weight functions. On this basis, we can group the companies into different risk classes, such as high risk class, moderate risk class, and low risk class, etc.

Considering these strengths of the methods of dominant rough sets and grey fixed weight clustering, we can construct a hybrid method combining the grey fixed weight clustering and dominant rough sets, where grey fixed weight clustering can be seen as a tool of processing used before the method of dominant rough sets is employed. The purpose of doing so is to generalize the dominant rough sets to such a method that can be employed to conduct decision-making analysis based on multi-criteria information tables and to extract the most precise expression of knowledge from the multi-criteria information table.

By following the following steps, one establishes his needed model combining grey fixed weight clustering and dominant rough sets:

(1) Develop a system of knowledge expressions using the values of preferred conditional attributes (criteria);
(2) Determine the ordered decision-making evaluation grey classes g according to the specific circumstances;
(3) Establish the whitenization weight function for the field of each criterion. Let the whitenization nweight function of the kth subclass of the jth criteria be $f_i^k(\cdot)(j=1,2,\cdots,m;k=1,2,\cdots,g)$.
(4) determine the clustering weight η_j, $j = 1, 2, \ldots, m$, for each criterion;
(5) Based on the observed value x_{ij}, $i = 1, 2, \ldots, n$, $j = 1, 2, \ldots, m$, of object i with respect to criterion j, compute the coefficients $\sigma_i^k = \sum_{j=1}^{m} f_j^k(x_{ij})\eta_j$ of the grey fixed weight clustering, $i = 1, 2, \ldots, n$, $k = 1, 2, \ldots, g$.
(6) Obtain the clustering coefficient vector:

$$\sigma_i = (\sigma_i^1, \sigma_i^2, \ldots, \sigma_i^g) = (\sum_{j=1}^{m} f_j^1(x_{ij})\eta_j, \sum_{j=1}^{m} f_j^2(x_{ij})\eta_j, \ldots \sum_{j=1}^{m} f_j^g(x_{ij})\eta_j)$$

(7) Generate the clustering coefficient matrix

$$\Sigma = (\sigma_i^k) = \begin{bmatrix} \sigma_1^1 & \sigma_1^2 & \cdots & \sigma_1^g \\ \sigma_2^1 & \sigma_2^2 & \cdots & \sigma_2^g \\ \vdots & \vdots & \vdots & \vdots \\ \sigma_n^1 & \sigma_n^2 & \cdots & \sigma_n^g \end{bmatrix}$$

(8) Based on the clustering coefficient matrix Σ, determine the classes individual objects belong to. If $\max_{1\le k\le g}\{\sigma_i^k\} = \sigma_i^{k^*}$, then the object i belongs to grey class k^*.
(9) Establish the decision-making table using preferred condtional attributes and preferred decision-making grey classes.
(10) Employ the method of dominant rough sets to conduct decision-making analysis.

Example 6.3. Let us look at how to choose regional key technologies using the hybrid model combining the methods of grey fixed weight clustering and dominant rough sets.

For a specific geographic area, the evaluation criteria system and the relevant evaluation values for its key technologies are given in Table 6.5.

Table 6.5 The criteria system for evaluating regional key technologies

Code	Meaning of criterion	Criterion weight	Evaluation values
a_1	Time lag of technology	0.1	A: > 10 years; B: 5 – 10 years; C: 3 – 5 years; D: < 3 years
a_2	Time length technological a bottleneck existed	0.09	A: > 10 years; B: 5 – 10 years; C: 3 – 5 years; D: < 3 years
a_3	Ability to create own knowledge right	0.14	A: complete own right; B: partial right; C: no right at all
a_4	Coverage of technology	0.09	A: widely applicable; B: applied in profession; C: special tecnique
a_5	Promotion and lead of technological fields	0.11	A: strong; B: relatively strong; C: general; D: weak
a_6	Time needed for technology transfer	0.07	A: within 1 year; B: 1 – 3 years; C: 4 – 5 years; D: > 5 years
a_7	Input/ouput ratio	0.13	A: high; B: relatively high; C: normal; D: low
a_8	Effect on environment protection	0.12	A: strong; B: relatively strong; C: normal; D: weak

Based on the evaluations of the relevant experts on 11 candidates of key technologies, we generate the knowledge expression system as shown in Table 6.6.

Table 6.6 The knowledge system on the regional key technologies

U	a_1	a_2	a_3	a_4	a_5	a_6	a_7	a_8
n_1	B	B	C	B	D	C	B	A
n_2	D	D	B	B	C	B	C	D
n_3	D	D	B	B	C	A	D	A
n_4	B	C	C	B	B	B	A	C
n_5	B	B	B	B	C	B	B	C
n_6	D	D	B	B	B	B	B	C
n_7	D	D	B	C	D	A	C	C
n_8	C	B	B	C	C	C	B	C
n_9	B	B	B	B	A	B	A	B
n_{10}	C	B	B	B	B	B	B	B
n_{11}	B	B	B	B	C	B	C	B

In the following, we will conduct a decision-making analysis for this region's candidates of key technologies.

For the evaluation criteria of the region's key technologies, the preference orders are all the same as A > B > C > D. Quantify the set of criteria evaluations by letting the set be $V = (A, B, C, D) = (7,5,3,1)$.

According to the practical needs, we divide each criterion into three grey classes of decision-making: the class of weak need for key technologies (coded with 1), the class of general need for key technologies (coded with 2), and the class of strong need for key technologies (coded with 3). Let us take the whitenization weight

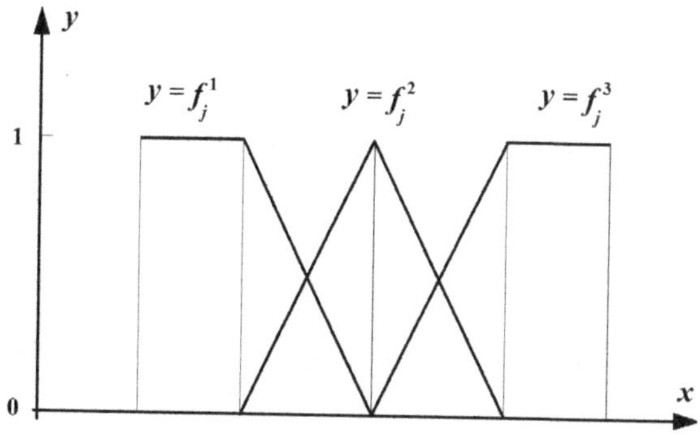

Fig. 6.3 Whitenization weight functions of the three grey classes

Table 6.7 Construction of evaluation classes for regional key technologies

Criterion name	Weak need class (1)	Moderate need class (2)	Strong need class (3)
Time lag of technology	$f(x)=\begin{cases}1 & 1\le x<2\\ 3-x & 2\le x<3\\ 0 & \text{otherwise}\end{cases}$	$f(x)=\begin{cases}x-2 & 2\le x<3\\ 4-x & 3\le x<4\\ 0 & \text{otherwise}\end{cases}$	$f(x)=\begin{cases}0.5x-1.5 & 3\le x<5\\ 1 & 5\le x\le 7\\ 0 & \text{otherwise}\end{cases}$
Time length technological a bottleneck existed	$f(x)=\begin{cases}1 & 1\le x<2\\ 3-x & 2\le x<3\\ 0 & \text{otherwise}\end{cases}$	$f(x)=\begin{cases}x-2 & 2\le x<3\\ 4-x & 3\le x<4\\ 0 & \text{otherwise}\end{cases}$	$f(x)=\begin{cases}0.5x-1.5 & 3\le x<5\\ 1 & 5\le x\le 7\\ 0 & \text{otherwise}\end{cases}$
Ability to create own knowledge right	$f(x)=\begin{cases}1 & 3\le x<4\\ 5-x & 4\le x<5\\ 0 & \text{otherwise}\end{cases}$	$f(x)=\begin{cases}x-4 & 4\le x<5\\ 6-x & 5\le x<6\\ 0 & \text{otherwise}\end{cases}$	$f(x)=\begin{cases}x-5 & 5\le x<6\\ 1 & 6\le x\le 7\\ 0 & \text{otherwise}\end{cases}$
Coverage of technology	$f(x)=\begin{cases}1 & 3\le x<4\\ 5-x & 4\le x<5\\ 0 & \text{otherwise}\end{cases}$	$f(x)=\begin{cases}x-4 & 4\le x<5\\ 6-x & 5\le x<6\\ 0 & \text{otherwise}\end{cases}$	$f(x)=\begin{cases}x-5 & 5\le x<6\\ 1 & 6\le x\le 7\\ 0 & \text{otherwise}\end{cases}$
Promotion and lead of technological fields	$f(x)=\begin{cases}1 & 1\le x<2\\ 2-0.5x & 2\le x<4\\ 0 & \text{otherwise}\end{cases}$	$f(x)=\begin{cases}0.5x-1 & 2\le x<4\\ 3-0.5x & 4\le x<6\\ 0 & \text{otherwise}\end{cases}$	$f(x)=\begin{cases}0.5x-2 & 4\le x<6\\ 1 & 6\le x\le 7\\ 0 & \text{otherwise}\end{cases}$
Time needed for technology transfer	$f(x)=\begin{cases}1 & 1\le x<2\\ 2-0.5x & 2\le x<4\\ 0 & \text{otherwise}\end{cases}$	$f(x)=\begin{cases}0.5x-1 & 2\le x<4\\ 3-0.5x & 4\le x<6\\ 0 & \text{otherwise}\end{cases}$	$f(x)=\begin{cases}0.5x-2 & 4\le x<6\\ 1 & 6\le x\le 7\\ 0 & \text{otherwise}\end{cases}$
Input/ouput ratio	$f(x)=\begin{cases}1 & 1\le x<3\\ 4-x & 3\le x<4\\ 0 & \text{otherwise}\end{cases}$	$f(x)=\begin{cases}x-3 & 3\le x<4\\ 3-0.5x & 4\le x<6\\ 0 & \text{otherwise}\end{cases}$	$f(x)=\begin{cases}0.5x-2 & 4\le x<6\\ 1 & 6\le x\le 7\\ 0 & \text{otherwise}\end{cases}$
Effect on environment protection	$f(x)=\begin{cases}1 & 1\le x<3\\ 2.5-0.5x & 3\le x<5\\ 0 & \text{otherwise}\end{cases}$	$f(x)=\begin{cases}0.5x-1.5 & 3\le x<5\\ 3.5-0.5x & 5\le x<7\\ 0 & \text{otherwise}\end{cases}$	$f(x)=\begin{cases}0.5x-25 & 5\le x\le 7\\ 0 & \text{otherwise}\end{cases}$

function of the class of weak need as the measurement of the low bound, that of the class of general need as the merderate measurement, and that of the class of strong need as the measurement of the upper bound. For details, see Figure 6.3.

Based on the goals of decision-making and specific distributions of the experts' exvaluation values, we introduce the whitenization functions for each grey classes as shown in Table 6.7.

From the formula $\sigma_i^k = \sum_{j=1}^{m} f_j^k(x_{ij})\eta_j$, we can compute the clustering coefficient for each grey class of each key technology, based on which we establish the evaluation decision-making Table 6.8 for choosing key technologies for the rgion.

Table 6.8 Evaluation decision-making table for regional key technologies

U	a_1	a_2	a_3	a_4	a_5	a_6	a_7	a_8	Cl
n_1	B	B	C	B	D	C	B	A	3
n_2	D	D	B	B	C	B	C	D	1
n_3	D	D	B	B	C	A	D	A	1
n_4	B	C	C	B	B	B	A	C	3
n_5	B	B	B	B	C	B	B	C	2
n_6	D	D	B	B	B	B	B	C	1
n_7	D	D	B	C	D	A	C	C	1
n_8	C	B	B	C	C	C	B	C	2
n_9	B	B	B	B	A	B	A	B	3
n_{10}	C	B	B	B	B	B	B	B	2
n_{11}	B	B	B	B	C	B	C	B	2

Because the values of all the conditional attributes have the preference order A > B > C > D, these attributes contain preference information. Based on the decision-making attributes, the comprehensive evaluation can divided into three preference ordered classes: $Cl_1 = \{1\}$, $Cl_2 = \{2\}$, $Cl_3 = \{3\}$. Based on this result, we divide the field of discourse and obtain the following unions of the decision-making classes:

$Cl_1^{\leq} = Cl_1$, with the comprehensive evaluation 1 (the need for key technologies is weak);

$Cl_2^{\leq} = Cl_1 \cup Cl_2$, with the comprehensive evaluation ≤ 2 (the need for key technologies is at most moderate);

$Cl_2^{\geq} = Cl_2 \cup Cl_3$, with the comprehensive evaluation ≥ 2 (the need for key technologies is at least moderate);

$Cl_3^{\leq} = Cl_1 \cup Cl_2 \cup Cl_3$, with the comprehensive evaluation ≤ 3 (the need for key technologies is at most strong); and

$Cl_3^{\geq} = Cl_3$, with the comprehensive evaluation 3 (the need for key technologies is strong).

A reduction found by using the method of dominant rough sets is $\{a_2, a_7\}$. The sets $D_{\geq}$ and $D_{\leq}$ of the least amounts of preference rules generated from this reduction are respectively given in Tables 6.9 and 6.10.

Table 6.9 Set $D_{\geq}$ of preference rules

Rule	Confidence	Support number
If the length of time for technology bottleneck to exist $\geq$ C and input/output ratio = A, then the urgency for needing key technologies = 3 (strong)	100%	2
If the length of time for technology bottleneck to exist $\geq$ B and input/output ratio $\geq$ C, then the urgency for needing key technologies $\geq$ 2 (moderate)	100%	5
If the length of time for technology bottleneck to exist = D, then the urgency for needing key technologies = 1 (weak)	100%	4

Table 6.10 Set $D_{\leq}$ of preference rules

Rule	Confidence	Support number
If the length of time for technology bottleneck to exist $\leq$ C and input/output ratio = A, then the need for key technologies = 3 (strong)	100%	2
If the length of time for technology bottleneck to exist $\leq$ B and input/output ratio $\leq$ C, then the need for key technologies $\leq$ 2 (moderate)	100%	1
If the length of time for technology bottleneck to exist = D, then the need for key technologies = 1 (weak)	100%	4

Based on the set $D_{\geq}$ of preference decision-making rules generated by employing our hybrid model that combines grey fixed weight clustering and dominant rough sets, all the 11 key technologies considered are correctly classified. That is, the quality of classification is 100%. Based on the set $D_{\leq}$ of preference decision-making rules, a total of 7 key technologies are classified correctly so that the quality of classification is 63.6%.

Chapter 7

Grey Models for Decision Making

Deciding on what actions to take based on the actual circumstances and pre-determined goals is known as *decision making*. The essential meaning of decision-making is to make a decision or to choose a course of actions. Decision-making not only plays an important part of various kinds of management activities, but also appears throughout every person's daily life. The understanding of the concept of decision-making can be divided into two categories: general and specific. In the general category, each decision-making stands for the entire process of activities, including posting questions, collecting the data, establishing the goal, making, analyzing, and evaluating the plan of action, , implementing the plan, looking back, and modifying the plan. In the specific category, decision-making only represents the step of choosing a specific plan of action out of the entire decision-making process. Also, some scholars understand decision-making as choosing and picking a plan of action under uncertain conditions. In this case, the choice can be most likely influenced by the decision maker's prior experience, attitude, and willingness to take a certain amount of risk. Grey decision-making is about making decision using such decision models that involve grey elements or combine the general decision model and grey systems models. Its focus of study is on the problem of choosing a specific plan.

In the rest of this chapter, we will call the problem waiting to be resolved, the event needing to be handled, a current state of a system's behavior, etc., an *event*. Events are where we begin our investigation. In particular, events, *countermeasures*, *objectives*, and *effects* are known as the four key elements of decision-making. The totality of all events within the range of a research is known as the set of events of the study, denoted $A = \{a_1, a_2, ... a_n\}$, where $a_i, i = 1,2,3,\cdots,n$, stands for the *i*th event. The totality of all possible countermeasures is known as the set of countermeasures, denoted $B = \{b_1, b_2, \cdots, b_m\}$ with b_j, $j = 1,2,...m$, be the *j*th countermeasure. The Cartesian product $A \times B = \{(a_i, b_j) | a_i \in A, b_j \in B\}$ of the event set *A* and the countermeasure set *B* is known as the *situation set*, written $S = A \times B$, where each ordered pair $s_{ij} = (a_i, b_j)$, for any $a_i \in A, b_j \in B$, is known as a *situation*.

For a given situation $s_{ij} \in S$, evaluating the effects under a set pre-determined objectives and deciding on the basis of the evaluation on what to take and what to let go is the decision-making we will study in this chapter. In the following sections, we will study several different kinds of grey decision-making methods.

7.1 Different Approaches for Grey Decisions

7.1.1 Grey Target Decisions

Let $S = \{s_{ij} = (a_i, b_j) | a_i \in A, b_j \in B\}$ be the situation set, $u_{ij}^{(k)}$ the *effect value of situation* s_{ij} with respect to objective k, R the set of all real numbers. Then $u_{ij}^{(k)}$: $S \mapsto R$, defined by $s_{ij} \mapsto u_{ij}^{(k)}$, is known as the *effect mapping* of S with respect to object *k*. If $u_{ij}^{(k)} = u_{ih}^{(k)}$, then we say that the countermeasures b_j and b_h of event a_i are equivalent with respect to objective *k*, written $b_j \cong b_h$; and the set

$$B_i^{(k)} = \{b | b \in B, b \cong b_h\}$$

is known as the effect *equivalence class* of countermeasure b_h of event a_i with respect to objective *k*.

If *k* is such an objective that the greater the effect value is the better, and $u_{ij}^{(k)} > u_{ih}^{(k)}$, then we say that the countermeasure b_j is *superior* to b_h in terms of event a_i with respect to objective *k*, written as $b_j \succ b_h$. The set $B_{ih}^{(k)} = \{b | b \in B, b \succ b_h\}$ is known as the superior set of the countermeasure b_h of event a_i with respect to objective *k*. Similarly, the concepts of superior classes of countermeasures for the situations that the closer to a fixed moderate value the effect value is the better, and that the smaller the effect value is the better.

If $u_{ij}^{(k)} = u_{ih}^{(k)}$, then the events a_i and a_j are said to be *equivalent* in terms of the countermeasure b_h with respect to objective *k*, written $a_i \cong a_j$. And the set

$$A_{jh}^{(k)} = \{a | a \in A, \quad a \cong a_i\}$$

is known as the *effect equivalence class* of events of the countermeasure b_h with respect to objective *k*. And, if *k* is such an objective that the greater the effect value is the better, and $u_{ih}^{(k)} > u_{jh}^{(k)}$, then we say that event a_i is *superior* to event a_j in terms of countermeasure b_h with respect to objective *k*, denoted $a_i \succ a_j$. The set

$$A_{jh}^{(k)} = \{a | a \in A, a \succ a_j\}$$

is known as the superior class of the event a_j in terms of countermeasure b_h with respect to objective *k*. Similarly, the concepts of superior classes can be defined for the situations that the closer to a fixed moderate value the effect value is the better and that the smaller the effect value is the better.

If $u_{ij}^{(k)} = u_{hl}^{(k)}$, then the situation s_{ij} is *equivalent* to situation s_{hl} under the objective *k*, denoted $s_{ij} \cong s_{hl}$. The set

$$S^{(k)} = \{s | s \in S, s \cong s_{hl}\}$$

is known as the *effect equivalence class* of the situation s_{hl} under objective *k*. If *k* is such an objective that the greater the effect value is the better, and $u_{ij}^{(k)} > u_{hl}^{(k)}$, then the situation s_{ij} is said to be *superior* to situation s_{hl} under objective *k*, denoted $s_{ij} \succ s_{hl}$. The set

$$S_{hl}^{(k)} = \{s | s \in S, s \succ s_{hl}\}$$

is known as the *effect superior class* of the situation s_{hl} under objective *k*. Similarly, the concepts of superior classes for situation effects can be defined for the scenarios that the closer to a fixed moderate value the effect value of a situation is the better and that the smaller the effect value of situation is the better.

Proposition 7.1. Assume that $S = \{s_{ij} = (a_i, b_j) | a_i \in A, b_j \in B\} \neq \varnothing$ and $U^{(k)} = \{u_{ij}^{(k)} | a_i \in A, b_j \in B\}$ is the set of effects under objective *k*, and $\{S^{(k)}\}$ the set of effect equivalence classes of situations under objective *k*. Then the mapping $u^{(k)} : \{S^{(k)}\} \to U^{(k)}$, defined by $S^{(k)} \mapsto u_{ij}^{(k)}$, is bijective.

Let $d_1^{(k)}$ and $d_2^{(k)}$ be the upper and lower threshold values of the situational effects of under objective *k*. Then $S^1 = \{r | d_1^{(k)} \le r \le d_2^{(k)}\}$ is known as the one-dimensional *grey target* of the objective *k*, $u_{ij}^{(k)} \in [d_1^{(k)}, d_2^{(k)}]$ a satisfactory effect under objective *k*, the corresponding s_{ij} a desirable situation with respect to objective *k*, and b_j a desirable countermeasure of the event a_i with respect to objective *k*.

Proposition 7.2. Assume that $u_{ij}^{(k)}$ stands for the effect value of situation s_{ij} with respect objective *k*. If $u_{ij}^{(k)} \in S^1$, that is, s_{ij} is a desirable situation with respect to objective *k*. Then for any $s \in S_{ij}^{(k)}$, s is also a desirable situation. That is, when s_{ij} is desirable, all situations in its effect superior class are desirable.

What has been discussed above is about the case of a single objective. Similarly, the *grey targets* of decision-making with multi-objectives can be addressed.

Assume that $d_1^{(1)}$ and $d_2^{(1)}$ are the threshold values of situational effects of objective 1, $d_1^{(2)}$ and $d_2^{(2)}$ the threshold values of situational effects of objective 2. Then

$$S^2 = \{(r^{(1)}, r^{(2)}) | d_1^{(1)} \le r^{(1)} \le d_2^{(1)}, d_1^{(2)} \le r^{(2)} \le d_2^{(2)}\}$$

is known as a grey target of two-dimensional decision-making. If the effect vector of situation s_{ij} satisfies $u_{ij} = \{u_{ij}^{(1)}, u_{ij}^{(2)}\} \in S^2$, then s_{ij} is seen as a *desirable situation* with respect to objectives 1 and 2, and b_j a *desirable countermeasure* for event a_i with respect to objectives 1 and 2.

Generally, assume that $d_1^{(1)}, d_2^{(1)}$; $d_1^{(2)}, d_2^{(2)}; \cdots; d_1^{(s)}, d_2^{(s)}$ are respectively the threshold values of situational effects under objectives 1, 2, …, *s*. Then the following region of the s–dimensional Euclidean space

$$S^s = \{(r^{(1)}, r^{(2)}, \cdots, r^{(s)}) | d_1^{(1)} \le r^{(1)} \le d_2^{(1)}, d_1^{(2)} \le r^{(2)} \le d_2^{(2)}, \cdots, d_1^{(s)} \le r^{(s)} \le d_2^{(s)}\}$$

is known as a *grey target of an s–dimensional decision-making*. If the effect vector of situation s_{ij} satisfies

$$u_{ij} = (u_{ij}^{(1)}, u_{ij}^{(2)}, \cdots, u_{ij}^{(s)}) \in S^s$$

where $u_{ij}^{(k)}$ stands for the effect value of the situation s_{ij} with respect to objective *k*, $k = 1, 2, \cdots, s$, then s_{ij} is known as a *desirable situation* with respect to the objectives 1, 2, …, *s*, and b_j a *desirable countermeasure* of the event a_i with respect to the objectives 1, 2, …, *s*.

Intuitively, the grey targets of a decision-making essentially represent the location of satisfactory effects in terms of relative optimization. In many practical circumstances, it is impossible to obtain the absolute optimization so that people are happy if they can achieve a satisfactory outcome. Of course, based on the need, one can gradually shrink the grey targets of his decision-making to a single point in order to obtain the ultimate optimal effect, where the corresponding situation is the most desirable, and the corresponding countermeasure the optimal countermeasure.

The following

$$R^s = \left\{(r^{(1)}, r^{(2)}, \cdots, r^{(s)}) \middle| (r^{(1)} - r_0^{(1)})^2 + (r^{(2)} - r_0^{(2)})^2 + \cdots + (r^{(s)} - r_0^{(s)})^2 \le R^2 \right\}$$

is known as an s-dimensional spherical grey target centered at $r_0 = (r_0^{(1)}, r_0^{(2)}, \cdots, r_0^{(s)})$ with radius R. The vector $r_0 = (r_0^{(1)}, r_0^{(2)}, \cdots, r_0^{(s)})$ is seen as the *optimum effect vector*.

For $r_1 = (r_1^{(1)}, r_1^{(2)}, \cdots, r_1^{(s)}) \in R$,

$$|r_1 - r_0| = \left[(r_1^{(1)} - r_0^{(1)})^2 + (r_1^{(2)} - r_0^{(2)})^2 + \cdots + (r_1^{(s)} - r_0^{(s)})^2\right]^{1/2}$$

is known as the *bull's-eye distance* of the vector r_1. The values of this distance reflect the superiority of the corresponding situational effect vectors.

Let s_{ij} and s_{hl} be two different situations, and $u_{ij} = (u_{ij}^{(1)}, u_{ij}^{(2)}, \cdots, u_{ij}^{(s)})$ and $u_{hl} = (u_{hl}^{(1)}, u_{hl}^{(2)}, \cdots, u_{hl}^{(s)})$ respectively their effect vectors. If

$$|u_{ij} - r_0| \ge |u_{hl} - r_0| \tag{7.1}$$

then the situation s_{hl} is said to be *superior* to s_{ij}, denoted $s_{hl} \succ s_{ij}$. When the equal sign in equ. (7.1) holds true, the situations s_{ij} and s_{hl} are said to be *equivalent*, written $s_{hl} \cong s_{ij}$.

If for $i = 1, 2, \cdots, n$ and $j = 1, 2, \cdots, m$, $u_{ij} \ne r_0$ always holds true, then the optimum situation does not exist, and the event does not have any optimum countermeasure. If the optimum situation does not exists, however, there are such h and l such that for any $i = 1, 2, \cdots, n$ and $j = 1, 2, \cdots, m$, $|u_{hl} - r_0| \le |u_{ij} - r_0|$ holds true, that is, for any $s_{ij} \in S$, $s_{hl} \succ s_{ij}$ holds, then s_{hl} is known as a *quasi-optimum situation*, a_h a *quasi-optimum event*, and a *quasi-optimum countermeasure*.

Theorem 7.1. Let $S = \{s_{ij} = (a_i, b_j) | a_i \in A, b_j \in B\}$ be the set of situations, $R^s = \left\{(r^{(1)}, r^{(2)}, \cdots, r^{(s)}) \middle| (r^{(1)} - r_0^{(1)})^2 + (r^{(2)} - r_0^{(2)})^2 + \cdots + (r^{(s)} - r_0^{(s)})^2 \le R^2 \right\}$

an s-dimensional spherical grey target. The S becomes an ordered set with "superiority" as its order relation $\prec$.

Theorem 7.2. There must be quasi-optimum situations in the situation set $(S,\succ)$. Proof. This is a restatement of Zorn's Lemma in set theory. QED.

Example 7.1. Consider the event a_1 of reconstructing an old building. There are three possibilities: b_1 = renovate the building completely; b_2 = tear down the building and reconstruct another one; and b_3 = simply maintain what the building is by fixing up minor details. Let us make a grey target decision making using three objectives: cost, functionality, and construction speed.

Solution. Let us denote the cost as objective 1, the functionality as objective 2, and the construction speed as objective 3. Then, we have the following three situations:

$$s_{11} = (a_1, b_1) = \text{(reconstruction, renovation)},$$
$$s_{12} = (a_{1,} b_2) = \text{(reconstruction, new building), and}$$
$$s_{13} = (a_{1'} \ \ b_3) = \text{(reconstruction, maintenance)}.$$

Evidently, different situations with respect to different objectives have different effects; and the standards for measuring the effects are also accordingly different. For instance, for the cost, it will be the lesser the better; for the functionality, it will be the higher the better; and for the speed, it will be the faster the better. Let us divide the effects of the situations into three classes: good, okay, and poor.

The effect vectors of the situations are respectively defined as follows:

$$u_{11} = \left(u_{11}^{(1)}, u_{11}^{(1)}, u_{11}^{(3)}\right) = (2,2,2),$$
$$u_{12} = \left(u_{12}^{(1)}, u_{12}^{(2)}, u_{12}^{(3)}\right) = (3,1,3), \text{ and}$$
$$u_{13} = \left(u_{13}^{(1)}, u_{13}^{(2)}, u_{13}^{(3)}\right) = (1,3,1).$$

Let the bull's eye be located at $r_0 = (1,1,1)$ and compute the bull's-eye distances

$$\left|u_{11} - r_0\right| = \left[\left(u_{11}^{(1)} - r_0^{(1)}\right)^2 + \left(u_{11}^{(2)} - r_0^{(2)}\right)^2 + \left(u_{11}^{(3)} - r_0^{(3)}\right)^2\right]^{1/2}$$
$$= \left[(2-1)^2 + (2-1)^2 + (2-1)^2\right]^{1/2} = 1.73$$
$$\left|u_{12} - r_0\right| = \left[\left(u_{12}^{(1)} - r_0^{(1)}\right)^2 + \left(u_{12}^{(2)} - r_0^{(2)}\right)^2 + \left(u_{12}^{(2)} - r_0^{(3)}\right)^2\right]^{1/2}$$
$$= \left[(3-1)^2 + (1-1)^2 + (3-1)^2\right]^{1/2} = 2.83$$
$$\left|u_{13} - r_0\right| = \left[\left(u_{13}^{(1)} - r_0^{(1)}\right)^2 + \left(u_{13}^{(2)} - r_0^{(2)}\right)^2 + \left(u_{13}^{(3)} - r_0^{(3)}\right)^2\right]^{1/2}$$
$$= \left[(1-1)^2 + (3-1)^2 + (1-1)^2\right]^{1/2} = 2$$

where $|u_{11} - r_0|$ is the smallest. So, the effect vector $u_{11} = (2,2,2)$ of the situation S_{11} enters the grey target. Hence, taking renovation is a satisfactory decision.

7.1.2 Grey Incidence Decisions

The bull's-eye distance between a situational effect vector and the center of the target measures the superiority of the situation in comparison with other situations. At the same time, the degree of incidence between the effect vector of a situation and the optimum effect vector can be seen as another way to evaluate the superiority of the situation. In particular, let $S = \{s_{ij} = (a_i, b_j) | a_i \in A, b_j \in B\}$ be the set of situations, and $u_{i_0 j_0} = \{u_{i_0 j_0}^{(1)}, u_{i_0 j_0}^{(2)}, \cdots, u_{i_0 j_0}^{(s)}\}$ the optimum effect vector. If the situation corresponding to $u_{i_0 j_0}$ satisfies $u_{i_0 j_0} \notin S$, then $u_{i_0 j_0}$ is known as an *imagined optimum effect vector*, and $s_{i_0 j_0}$ the *imagined optimum situation*.

Proposition 7.3. Let *S* be the same as above and the effect vector of situation s_{ij} is $u_{ij} = \{u_{ij}^{(1)}, u_{ij}^{(2)}, \cdots, u_{ij}^{(s)}\}$, for $i = 1,2,\cdots,n$, $j = 1,2,\cdots,m$.

(1) When *k* is an objective such that the greater its effect value is the better, let $u_{i_0 j_0}^{(k)} = \max_{1 \le i \le n, 1 \le j \le m} \{u_{ij}^{(k)}\}$;

(2) When *k* is an objective such that the closer to a fixed moderate value u_0 its effect value is the better, let $u_{i_0 j_0}^{(k)} = u_0$; and

(3) When *k* is an objective such that the smaller its effect value is the better, let $u_{i_0 j_0}^{(k)} = \min_{1 \le i \le n, 1 \le j \le m} \{u_{ij}^{(k)}\}$,

then $u_{i_0 j_0} = \{u_{i_0 j_0}^{(1)}, u_{i_0 j_0}^{(2)}, \cdots, u_{i_0 j_0}^{(s)}\}$ is the imagined optimum effect vector.

Proposition 7.4. Assume the same as in Proposition 7.3 and let $u_{i_0 j_0} = \left\{u_{i_0 j_0}^{(1)}, u_{i_0 j_0}^{(2)}, \cdots, u_{i_0 j_0}^{(s)}\right\}$ be the imagined optimum effect vector, ε_{ij} the absolute degree of grey incidence between u_{ij} and $u_{i_0 j_0}$, for $i = 1,2,\cdots,n$,

$j=1,2,\cdots,m$. If for any $i\in\{1,2,\cdots,n\}$ and $j\in\{1,2,\cdots,m\}$ satisfying $i\neq i_1$ and $j\neq j_1$, $\varepsilon_{i_1j_1}\geq\varepsilon_{ij}$ always holds true, then $u_{i_1j_1}$ is a quasi-optimum effect vector and $s_{i_1j_1}$ a quasi-optimum situation.

Grey incidence decisions can be made by following the following steps:

Step 1: Determine the set of events $A=\{a_1,a_2,\cdots,a_n\}$ and the set of countermeasures $B=\{b_1,b_2,\cdots,b_m\}$. And then construct the set of situations $S=\{s_{ij}=(a_i,b_j)|a_i\in A,b_j\in B\}$.

Step 2: Choose the objectives 1, 2, …, *s*, for the decision-making.

Step 3: Compute the effect values $u_{ij}^{(k)}$ of the individual situations s_{ij}, $i=1,2,\cdots,n$, $j=1,2,\cdots,m$, with respect to objective k: $u_{ij}=(u_{11}^{(k)},u_{12}^{(k)},\cdots u_{1m}^{(k)};u_{21}^{(k)},u_{22}^{(k)},\cdots u_{2m}^{(k)};\cdots;u_{n1}^{(k)},u_{n2}^{(k)},\cdots u_{nm}^{(k)})$; $k=1,2,\cdots,s$.

Step 4: Compute the average image of the situation effect sequence $u^{(k)}$ with respect to objective *k*, which is still written the same as

$$u_{ij}=(u_{11}^{(k)},u_{12}^{(k)},\cdots u_{1m}^{(k)};u_{21}^{(k)},u_{22}^{(k)},\cdots u_{2m}^{(k)};\cdots;u_{n1}^{(k)},u_{n2}^{(k)},\cdots u_{nm}^{(k)});\ k=1,2,\cdots,s$$

Step 5: Based on the results of Step 4, write out the effect vector $u_{ij}=\left\{u_{ij}^{(1)},u_{ij}^{(2)},\cdots,u_{ij}^{(s)}\right\}$ of the situation s_{ij}, for $i=1,2,\cdots,n$, $j=1,2,\cdots,m$.

Step 6: Compute the imagined optimum effect vector $u_{i_0j_0}=\left\{u_{i_0j_0}^{(1)},u_{i_0j_0}^{(2)},\cdots,u_{i_0j_0}^{(s)}\right\}$.

Step 7: Calculate the absolute degree ε_{ij} between u_{ij} and $u_{i_0j_0}$, $i=1,2,\cdots,n$, $j=1,2,\cdots,m$.

Step 8: From $\max\limits_{1\leq i\leq n,1\leq j\leq m}\{\varepsilon_{ij}\}=\varepsilon_{i_1j_1}$, the quasi-optimum effect vector $u_{i_1j_1}$ and the quasi-optimum situation $s_{i_1j_1}$ are obtained.

Example 7.2. Let us look at the grey incidence decision-making regarding the evaluation of looms.

Solution. Let us denote the event of evaluating loom models by a_1. Then the event set is $A=\{a_1\}$. There are three models of the looms under considerations:

Model 1: purchase projectile loom, which is treated as countermeasure b_1 ; Model 2: select air jet loom, which is treated as countermeasure b_2 ; Model 3: choose rapier loom, which is treated as countermeasure b_3 . So, the set of countermeasure is $B=\{b_1,b_2,b_3\}$, and the set of situations is $S=\{s_{ij}=(a_i,b_j)|a_i\in A,b_j\in B\}=\{s_{11},s_{12},s_{13}\}$.

Now, let us determine the objectives. According to the functionalities of looms, eleven (11) objectives are chosen. The weft-insertion rate (m/min) of the looms is objective 1. The efficiency of the looms is objective 2. The total investment (in ten thousand US$) on the looms is objective 3. The total energy cost (W/a) is objective 4. The total area (m^2) of the land to be occupied by the looms is objective 5. The total manpower (person) is objective 6. The quantity of weft yarn waste (cm/weft) is objective 7. The cost of replacement parts (ten thousand Yuan/a) is objective 8. Noise (dB) is objective 9. The quality of the produced fabric is objective 10. And, the adaptability of the type of loom is objective 11.

Under the assumptions that the afore-mentioned three models of looms produce the same kind of grey fabric meeting a same set of requirements, and that these looms will produce the same amount of annual output, let us conduct the associated computations for the said models of looms. Our quantitative calculations lead to the relevant values of the objectives with some of which determined from the literature and field investigations. See Table 7.1.

Table 7.1 Objective values of the looms of concern

Model	projectile loom	air jet loom	rapier loom
weft-insertion rate (m/min)	1000	1200	800
Efficiency (%)	92	90	92
Total investment (10 K US$)	880	336	612
Total energy consumption (W/a)	374	924	816
Total land needed (m^2)	1760	1092	2124
Total manpower (person)	18	22	24
Quantity of weft yarn waste (cm/weft)	5	6	10
Cost of parts (10K ¥/a)	37	35	75
Noise (dB)	85	91	91
Quality	Best	Good	Fine
Adaptability	good	Better	Best

In the following, we compute situational effect sequences $U^k(k=1,2,\cdots,11)$ with respect to the objectives.

For objective 1, we have $U^{(1)}=(u_{11}^{(1)},u_{12}^{(1)},u_{13}^{(1)})=(1000,1200,800)$.

For objective 2, we have $U^{(2)}=(u_{11}^{(2)},u_{12}^{(2)},u_{13}^{(2)})=(92,90,92)$.

For objective 3, we have $U^{(3)}=(u_{11}^{(3)},u_{12}^{(3)},u_{13}^{(3)})=(880,336,612)$.

For objective 4, we have $U^{(4)} = (u_{11}^{(4)}, u_{12}^{(4)}, u_{13}^{(4)}) = (374,924,816)$.

For objective 5, we have $U^{(5)} = (u_{11}^{(5)}, u_{12}^{(5)}, u_{13}^{(5)}) = (1760,1092,2124)$.

For objective 6, we have $U^{(6)} = (u_{11}^{(6)}, u_{12}^{(6)}, u_{13}^{(6)}) = (18,22,24)$.

For objective 7, we have $U^{(7)} = (u_{11}^{(7)}, u_{12}^{(7)}, u_{13}^{(7)}) = (5,6,10)$.

For objective 8, we have $U^{(8)} = (u_{11}^{(8)}, u_{12}^{(8)}, u_{13}^{(8)}) = (37,35,75)$.

For objective 9, we have $U^{(9)} = (u_{11}^{(9)}, u_{12}^{(9)}, u_{13}^{(9)}) = (85,91,91)$.

For objective 10, we have $U^{(10)} = (u_{11}^{(10)}, u_{12}^{(10)}, u_{13}^{(10)}) = (\text{best, good, fine})$.

For objective 11, we have $U^{(11)} = (u_{11}^{(11)}, u_{12}^{(11)}, u_{13}^{(11)}) = (\text{good, better, best})$.

Quantify the last two qualitative objectives as follows:

$$U^{(10)} = (u_{11}^{(10)}, u_{12}^{(10)}, u_{13}^{(10)}) = (9,8,7)$$

$$U^{(11)} = (u_{11}^{(11)}, u_{12}^{(11)}, u_{13}^{(11)}) = (8,7,9)$$

We now compute the average images of the situational effect sequences for each of the objectives:

$U^{(1)} = (1,1.2,0.8)$; $U^{(2)} = (1.01,0.98,1.01)$; $U^{(3)} = (1.44,0.55,1.01)$;

$U^{(4)} = (0.53,1.31,1.16)$; $U^{(5)} = (1.06,0.66,1.28)$; $U^{(6)} = (0.84,1.03,1.13)$;

$U^{(7)} = (0.71,0.86,1.43)$; $U^{(8)} = (0.76,0.71,1.53)$; $U^{(9)} = (0.96,1.02,1.02)$;

$U^{(10)} = (1.13,1,0.87)$; and $U^{(11)} = (1,0.87,1.13)$,

and the effect vectors U_{ij} of the situations s_{ij}, $i=1$, $j=1,2,3$:

$$U_{11} = (u_{11}^{(1)}, u_{11}^{(2)}, \cdots, u_{11}^{(11)}) = (1,1.01,1.44,0.53,1.06,0.84,0.71,0.76,0.96,1.13,1),$$

$$U_{12} = (u_{12}^{(1)}, u_{12}^{(2)}, \cdots, u_{12}^{(11)}) = (1.2,0.98,0.55,1.31,0.66,1.03,0.86,0.71,1.02,1,0.87),$$

and

$$U_{13} = (u_{13}^{(1)}, u_{13}^{(2)}, \cdots, u_{13}^{(11)}) = (0.8,1.01,1.01,1.16,1.28,1.13,1.43,1.53,1.02,0.87,1.13)$$

According to the principle of constituting the optimum reference sequences, from the average images of the situational effect sequences of the objectives, it follows that

For objective 1, the greater the effect value is the better, so $U_{i_0 j_0}^{(1)} = \max\{u_{ij}^{(1)}\} = u_{12}^{(1)} = 1.2$;

For objective 1, the higher the effect value is the better, so $U_{i_0 j_0}^{(2)} = \max\{u_{ij}^{(2)}\} = u_{11}^{(2)} = 1.01$;

For objective 3, the smaller effect value is the better, so $U_{i_0j_0}^{(3)} = \min\{u_{ij}^{(3)}\} = u_{12}^{(3)} = 0.55$;

For objective 4, the smaller effect value is the better, so $U_{i_0j_0}^{(4)} = \min\{u_{ij}^{(4)}\} = u_{11}^{(4)} = 0.53$;

For objective 5, the smaller effect value is the better, so $U_{i_0j_0}^{(5)} = \min\{u_{ij}^{(5)}\} = u_{11}^{(5)} = 0.66$;

For objective 6, the smaller effect value is the better, so $U_{i_0j_0}^{(6)} = \min\{u_{ij}^{(6)}\} = u_{11}^{(6)} = 0.84$;

For objective 7, the smaller effect value is the better, so $U_{i_0j_0}^{(7)} = \min\{u_{ij}^{(7)}\} = u_{11}^{(7)} = 0.71$;

For objective 8, the smaller effect value is the better, so $U_{i_0j_0}^{(8)} = \min\{u_{ij}^{(8)}\} = u_{12}^{(8)} = 0.71$;

For objective 9, the smaller effect value is the better, so $U_{i_0j_0}^{(9)} = \min\{u_{ij}^{(9)}\} = u_{11}^{(9)} = 0.96$;

For objective 10, the higher effect value is the better, so $U_{i_0j_0}^{(10)} = \max\{u_{ij}^{(10)}\} = u_{11}^{(10)} = 1.13$; and

For objective 11, the higher effect value is the better, so $U_{i_0j_0}^{(11)} = \max\{u_{ij}^{(11)}\} = u_{13}^{(11)} = 1.13$.

That is, we obtain the following optimum reference sequence:

$$U_{i_0j_0}^{(11)} = (u_{i_0j_0}^{(1)}, u_{i_0j_0}^{(2)}, \cdots, u_{i_0j_0}^{(11)}) = (1.2,1.01,0.55,0.53,0.66,0.84,0.71,0.71,0.96,1.13,1.13)$$

From u_{ij} and $u_{i_0j_0}$, we compute the absolute degrees of grey incidence:

$$\varepsilon_{11} = 0.628, \varepsilon_{12} = 0.891, \varepsilon_{13} = 0.532$$

From the definition of grey incidence decision-making, it follows that because $\max\{\varepsilon_{ij}\} = \varepsilon_{12} = 0.891$, U_{12} is the quasi-optimum vector and s_{12} the quasi-optimum situation. That is to say, in terms of producing the general grey fabric, the air jet loom is most ideal choice among all the available models of the looms.

7.1.3 Grey Development Decisions

Grey development decision-making is done based on the development tendency or the future behaviors of the situation of concern. It does not necessarily place specific emphasis on the current effect of the situation. Instead it focuses more on the change of the situational effect over time. This method of decision-making can be and has been employed for long-term planning and the decision-makings of large scale engineering projects and urban plannings. It looks at problems from the angle of development while attempting to make feasible arrangements and avoid repetitious constructions so that great savings of capital and manpower can be materialized.

Assume that $A = \{a_1, a_2, \cdots, a_n\}$ is a set of events, $B = \{b_1, b_2, \cdots, b_m\}$ a set of countermeasures, and $S = \{s_{ij} = (a_i, b_j) | a_i \in A, b_j \in B\}$ the set of situations. Then,

$$u_{ij}^{(k)} = (u_{ij}^{(k)}(1), u_{ij}^{(k)}(2), \cdots, u_{ij}^{(k)}(h))$$

is known as the *situational effect time series* of the situation s_{ij} with respect to objective k. As of now, what we have discussed earlier are static situations with a fixed time moment. Because we now involve the concept of time, as time moves, constantly changing situational effects are considered.

Proposition 7.5. Let the situational effect time series of the situation s_{ij} with respect to objective k be

$$u_{ij}^{(k)} = (u_{ij}^{(k)}(1), u_{ij}^{(k)}(2), \cdots, u_{ij}^{(k)}(h))$$

$\hat{a}_{ij}^{(k)} = [a_{ij}^{(k)}, b_{ij}^{(k)}]^T$ the least squares estimate of the parameters of the GM(1,1) model of $u_{ij}^{(k)}$. Then the inverse accumulation restoration of the GM(1,1) time response of $u_{ij}^{(k)}$ is given by

$$\hat{u}_{ij}^{(k)}(l+1) = [1 - \exp(a_{ij}^{(k)})] \cdot \left[u_{ij}^{(k)}(1) - \frac{b_{ij}^{(k)}}{a_{ij}^{(k)}} \right] \exp(-a_{ij}^{(k)} \cdot l)$$

Assume that the restored sequence through inverse accumulation of the GM(1,1) time response of the situational effect time series of the situation s_{ij} with respect to objective k is

$$\hat{u}_{ij}^{(k)}(l+1) = [1 - \exp(a_{ij}^{(k)})] \cdot \left[u_{ij}^{(k)}(1) - \frac{b_{ij}^{(k)}}{a_{ij}^{(k)}} \right] \exp(-a_{ij}^{(k)} \cdot l)$$

When objective k satisfies that the greater the effect value is the better, if

(1) $\max_{1 \le i \le n, 1 \le j \le m} \{-a_{ij}^{(k)}\} = -a_{i_0 j_0}^{(k)}$, then $s_{i_0 j_0}$ is known as the *optimum situation of development coefficients* with respect to objective k;

(2) $\max_{1\le i\le n,1\le j\le m}\{\hat{u}_{ij}^{(k)}(h+l)\}=\hat{u}_{i_0j_0}^{(k)}(h+l)$, then $s_{i_0j_0}$ is known as the *optimum situation of predictions* with respect to objective k.

Similarly, the concepts of optimum situations of development coefficients and predictions can be defined for the cases of objectives satisfying that the smaller the effect value is the better, and that the closer to a moderate value the effect value is the better, respectively. In particular, for objectives satisfying that the smaller the effect value is the better, one only needs to replace "max" in the items (1) and (2) above by "min"; if k is an objective satisfying that the closer to a fixed moderate value the effect value is the better, one can first take the means of the development coefficients and predicted values; then define the optimum situation based on the distances of the development coefficients and predicted values to their means.

If, additionally, k is such an objective that the closer to a moderate value the effect value is the better, then

(1) When $\min_{1\le i\le n,1\le j\le m}\left\{\frac{1}{m+n}\sum_{j=1}^{m}\sum_{i=1}^{n}a_{ij}^{(k)}-a_{ij}^{(k)}\right\}=\frac{1}{m+n}\sum_{j=1}^{m}\sum_{i=1}^{n}a_{ij}^{(k)}-a_{i_0j_0}^{(k)}$, $s_{i_0j_0}$ is known as the *optimum situation of development coefficients* with respect to objective k;

(2) When

$$\min_{1\le i\le n,1\le j\le m}\left\{\hat{u}_{ij}^{(k)}(h+l)-\frac{1}{m+n}\sum_{j=1}^{m}\sum_{i=1}^{n}\hat{u}_{ij}^{(k)}(h+l)\right\}=\hat{u}_{i_0j_0}^{(k)}(h+l)-\frac{1}{m+n}\sum_{j=1}^{m}\sum_{i=1}^{n}\hat{u}_{ij}^{(k)}(h+l)$$

$s_{i_0j_0}$ is known as the *optimum situation of predictions* with respect to objective k.

In practical applications, one may face the scenarios of either that both the optimum situations of development coefficients and predictions are the same or that they are different. Even so, the following theorem tells us that eventually these optimum situations would converge to one.

Theorem 7.3. Assume that k is such an objective that the greater its effect value is the better, $s_{i_0j_0}$ the optimum situation of development coefficients, that is, $-a_{i_0j_0}^{(k)}=\max_{1\le i\le n,1\le j\le m}\{-a_{ij}^{(k)}\}$, and $\hat{u}_{i_0j_0}^{(k)}(h+l+1)$ the predicted value for the situation effect of $s_{i_0j_0}$. Then there must be $l_0>0$ such that

$$\hat{u}_{i_0j_0}^{(k)}(h+l_0+1)=\max_{1\le i\le n,1\le j\le m}\{\hat{u}_{ij}^{(k)}(h+l_0+1)\}$$

That is, in a sufficiently distant future, $s_{i_0j_0}$ will also be the optimum situation of predictions.

Proof. See (Liu and Lin, 2006, p. 340 – 341) for details. QED.

Similar results hold true for those objectives satisfying either that the smaller the effect value is the better or that the closer to a fixed moderate value the effect value is the better.

At this junction, careful readers might have noticed that Theorem 7.3 does not state the case that there are some increasing and decreasing sequences among situational effect time series at the same time. As a matter of fact, for objectives satisfying that the greater the effect value is the better, there is no need to consider decreasing situational effect time series. For objective satisfying that the smaller the effect value is the better, all increasing situational effect time series are deleted in advance in all discussions. As for objectives satisfying that the closer to a moderate value the effect value is the better, one can consider only either increasing or decreasing situational effect time series depending on the circumstances involved.

7.1.4 Grey Cluster Decisions

Grey cluster decision is useful for synthetic evaluations about some objects with respect to several different criteria so that decisions can be made about whether or not an object meets the given standards for inclusion or exclusion. This method has been often employed for the classification decision-making regarding objects or people. For instance, school students can be classified based on their individual capabilities to receive information, to comprehend what is provided, and to grow consequently so that different teaching methods can be applied and that different students can be enrolled in different programs. As a second example, based on different sets of criteria, comprehensive evaluations can be done for general employees, technicians, and administrators respectively so that decisions can be made regarding who is qualified for his/her job, who is ready for a promotion, etc.

Assume that there are *n* objects to make decisions on, *m* criteria, *s* different grey classes, the quantified evaluation value of object *i* with respect to criterion *j* is x_{ij}, $f_j^k(*)$ the whitenization weight function of the *k*th grey class with respect to the *j*th criterion, w_j the synthetic decision-making weight of criterion *j* such that $\sum_{j=1}^{m} w_j = 1$, $i = 1,2,\ldots, n, j = 1, 2, \ldots, m, k = 1,2,\cdots, s$, then

$$\sigma_i^k = \sum_{j=1}^{m} f_j^k(x_{ij}) w_j$$

is known as the *decision coefficient* for the object *i* to belong to grey class *k*; $\sigma_i = (\sigma_i^1, \sigma_i^2, \cdots, \sigma_i^s)$ is known as the *decision coefficient vector* of object *i*,

$i=1,2,\cdots,n$; and $\sum=(\sigma_i^k)_{n\times s}$ the decision coefficient matrix. If $\max\limits_{1\le k\le s}\{\sigma_i^k\}=\sigma_i^{k^*}$, then the decision is that the object *i* belongs to grey class k^*.

In practical applications, it is quite often that many objects belong to the same decision grey class at the same time, while there is a constraint on how many objects are allowed in the grey class. When this occurs, we will need to order the objects in order to decide on which ones to take and which ones to delete from the grey class. Because decision coefficient vectors are generally not constructed against a same background or dimension, it is often difficult to actually compare these vectors. To this end, let us first normalize all the decision coefficient vectors by letting

$$\delta_i^k=\frac{\sigma_i^k}{\sum\limits_{k=1}^{s}\sigma_i^k}$$

which is known as the *normalized decision coefficient* for object *i* to belong to grey class *k*. Then, accordingly, we construct for object *i* its *normalized decision coefficient vector* $\delta_i=(\delta_i^1,\delta_i^2,\cdots,\delta_i^s)$, $i=1,2,\cdots,n$, and its *normalized decision coefficient matrix* $\prod=(\delta_i^k)_{n\times s}$.

Proposition 7.6. When decision object *i* is respectively clustered using its decision coefficient σ_i^k and its normalized decision coefficient δ_i^k, the outcome is the same.

Within a grey class, after all decision coefficients are normalized, the magnitudes of their coordinates may well determine which vector should be ordered first. For instance, for the given $\delta_1=(0.4,0.35,0.25)$ and $\delta_2=(0.41,0.2,0.39)$, the corresponding objects belong to the first grey class. If we compare directly their comprehensive decision coefficients, from $0.41>0.4$, object 2 will be ordered ahead of object 1. However, if we closely compare the coordinates of the vectors δ_1 and δ_2, we might well be convinced that object 1 should be ordered ahead of object 2. This difference in opinion is caused by the fact that when we compared the normalized decision coefficient vectors, we did not look at the vectors holistically. To resolve this problem, the concept of synthetic measures as proposed in the following two definitions comes to rescue.

Definition 7.1. Assume that there are *s* different grey classes. Then

$\eta_1=(s,s-1,s-2,\cdots,1)$,

$\eta_2=(s-1,s,s-1,s-2,\cdots,2)$,

$\eta_3=(s-2,s-1,s,s-1,\cdots,3)$,

$\cdots$,

$\eta_k = (s-k+1, s-k+2, \cdots, s-1, s, s-1, \cdots, k)$,
$\cdots$,
$\eta_{s-1} = (2,3,\cdots s-1, s, s-1)$, and
$\eta_s = (1,2,3,\cdots s-1, s)$

are referred to as the *adjustment coefficients* of the first, second, …, *s*th grey class, respectively.

Notice that η_k, $k = 1,2,\cdots,s$, is an *s*-dimensional vector, in which the *k*th coordinate is equal to s. If this coordinate is seen as the center, then the coordinates in both directions from this center decrease by 1 consecutively.

Definition 7.2. Assume that there are *n* decision objects, *s* different grey classes, and object *i* belongs to grey class *k*. Then $\omega_i = \eta_k \cdot \delta_i^T$ is referred to as the *synthetic decision measure* of the object *i*.

Definition 7.3. Assume that $\max\limits_{1\le k\le s}\{\delta_{i_1}^k\} = \delta_{i_1}^{k^*}$ and $\max\limits_{1\le k\le s}\{\delta_{i_2}^k\} = \delta_{i_2}^{k^*}$ hold true. If $\omega_{i_1} > \omega_{i_2}$, then in the grey class k^*, decision object i_1 is seen as *superior* to decision object i_2.

Definition 7.4. Assume that $\max\limits_{1\le k\le s}\{\delta_{i_1}^k\} = \delta_{i_1}^{k^*}$, $\max\limits_{1\le k\le s}\{\delta_{i_2}^k\} = \delta_{i_2}^{k^*}$, …, $\max\limits_{1\le k\le s}\{\delta_{i_l}^k\} = \delta_{i_l}^{k^*}$ hold true. That is, the decision objects $i_1, i_2, \cdots, i_l$ all belong to grey class k^*. If $\omega_{i_1} > \omega_{i_2} > \ldots > \omega_{i_l}$ and the decision grey class k^* is only allowed to have l_1 objects, then the objects $i_1, i_2, \cdots, i_{l_1}$ will be admitted into the class k^* with the rest ordered objects being the candidates on the waiting list of the grey class k^*.

In general, a grey cluster decision can be made by going through the following steps:

Step 1: Decide on the *s* grey classes based on the requirements of the comprehensive evaluation; accordingly, also divide the individual fields of the criteria into s grey classes, and define the whitenization weight function $f_j^k(*)$ for the *k*th subclass of the *j*th criterion, $j = 1,2,\cdots,m$, $k = 1,2,\cdots,s$;

Step 2: Determine the clustering weight w_j of the criteria, *j* = 1, 2, …, *m*;

Step 3: Compute the decision coefficient $\sigma_i^k = \sum_{j=1}^{m} f_j^k(x_{ij})w_j$ for object i to belong to grey class k;

Step 4: Calculate the normalized decision coefficient δ_i^k for object i to belong to grey class k;

Step 5: From $\max_{1 \le k \le s} \{\delta_i^k\} = \delta_i^{k^*}$, it is determined that object i belongs to grey class k^*;

Step 6: Compute the synthetic decision measure $\omega_i = \eta_k \cdot \delta_i^T$ of object i with respect to grey class k; and

Step 7: According to the synthetic decision measures, order all the objects that belong to the same grey class k.

Example 7.3. For the problem of selecting leading industries as studies in Example 3.6, if the total number of selected leading industries is limited, how would we decide on the leading industries in Jiangsu province?

From the detailed computational results in Example 3.6, we know that the extraction of petroleum & natural gas, petroleum refinery & coking, chemicals, manufactures using non-metal minerals, metal smelting & rolling processing, metal products, electronic & communication equipment, meters & office appliances, other manufacturing, supply of electric & steam hot water, production & supply of coal gas, construction, transportation & postal service, real estate, societal services, health, sports & welfare, education, arts, radio, TV & movie, service to scientific research & technology, and government & others, are the 20 industries that are considered for the selection of leading industries.

If the relevant government office likes to select no more than 20 leading industries, then we first need to compute the synthetic decision measure $\omega_i = 3\delta_i^1 + 2\delta_i^2 + \delta_i^1$ of each of the candidate industries, leading to the following specific values: 2.268, 2.4805, 2.7226, 2.5818, 2.6458, 2.4901, 2.5573, 2.7782, 2.5661, 2.7192, 2.5653, 2.6468, 2.4655, 2.6628, 2.5969, 2.538, 2.4722, and 2.5105.

By comparing the synthetic decision measures, we can order the candidate industries as follows: other manufacturing, chemicals, production & supply of coal gas, societal services, transportation & postal service, metal products, health, sports & welfare, manufactures using non-metal minerals, supply of electric & steam hot water, construction, meters & office appliances, education, arts, radio, TV & movie, government & others, electronic & communication equipment, petroleum refinery & coking, service to scientific research & technology, real estate, extraction of petroleum & natural gas, and metal smelting & rolling processing. On the basis of this ordering, we can simply select the x-number of industries listed in the front as the industry leaders according to the requirement, and keep the rest industries on the waiting list.

7.2 Decision Makings with Synthesized Targets

In this section, we look at how to make decisions for multiple-target situations to meet a synthesized criterion. In particular, assume that $A=\{a_1,a_2,\cdots,a_n\}$ is a set of events, $B=\{b_1,b_2,\cdots,b_m\}$ a set of countermeasures, $S=\{s_{ij}=(a_i,b_j)|a_i\in A,b_j\in B\}$ the relevant set of situations, $u_{ij}^{(k)}$ the observed effect value of the situation $s_{ij}\in S$ with respect to objective k, $i=1,2,\cdots,n$, $j=1,2,\cdots,m$. Then,

$$r_{ij}^{(k)}=\frac{u_{ij}^{(k)}}{\max_i\max_j\{u_{ij}^{(k)}\}} \tag{7.2}$$

is known as an *upper effect measure*;

$$r_{ij}^{(k)}=\frac{\min_i\min_j\{u_{ij}^{(k)}\}}{u_{ij}^{(k)}} \tag{7.3}$$

a *lower effect measure*; and

$$r_{ij}^{(k)}=\frac{u_{i_0j_0}^{(k)}}{u_{i_0j_0}^{(k)}+\left|u_{ij}^{(k)}-u_{i_0j_0}^{(k)}\right|} \tag{7.4}$$

a *moderate effect measure*, where $u_{i_0j_0}^{(k)}$ is a chosen moderate effect value with respect to objective k.

The concept of upper effect measure reflects the distance of the observed effect value from the maximum observed effect value. The concept of lower effect measure indicates the distance between the observed effect value from the minimum observed effect value. And the concept of moderate effect measure tells the distance of the observed effect value from the pre-fixed moderate effect value. When making situational decisions with a synthesized target, for the kinds of objectives that the greater or the more the effect sample values are the better, one can make use of the concept of upper effect measure. For the kinds of objectives that the smaller or fewer the effect sample values are the better, one can utilize the concept of lower effect measure. As for the kinds of objectives that require "neither too large nor too small" and/or "neither too many nor too few," one can apply the concept of moderate effect measure.

Proposition 7.7. The three effect measures $r_{ij}^{(k)} (i = 1,2,\cdots,n; j = 1,2,\cdots,m)$ defined in equs. (7.2) – (7.4) satisfy the following properties: (1) $r_{ij}^{(k)}$ is non-dimensional; (2) $r_{ij}^{(k)} \in [0,1]$; and (3) the more ideal the effect is, the greater $r_{ij}^{(k)}$ is.

For the given set S of situations, $R^{(k)} = \left(r_{ij}^{(k)}\right)_{n\times m}$ is known as the *matrix of uniform effect measures* of S with respect to objective *k*. For $s_{ij} \in S$, $r_{ij} = (r_{ij}^{(1)}, r_{ij}^{(2)}, , r_{ij}^{(s)})$ is known as the *vector of uniform effect measures* of the situation s_{ij}. If η_k stands for the decision weight of objective *k*, $k = 1,2,\cdots,s$, satisfying $\sum_{k=1}^{s} \eta_k = 1$, then $\sum_{k=1}^{s} \eta_k \cdot r_{ij}^{(k)}$ is known as the *synthetic effect measure of the situation* s_{ij}, which is still denoted as $r_{ij} = \sum_{k=1}^{s} \eta_k \cdot r_{ij}^{(k)}$; and $R = (r_{ij})_{n\times m}$ is known as the *matrix of synthetic effect measures.*

Definition 7.5. (1) If $\max_{1\le j\le m}\{r_{ij}\} = r_{ij_0}$, then b_{j_0} is known as the *optimum countermeasure* of event a_i ; (2) If $\max_{1\le i\le n}\{r_{ij}\} = r_{i_0 j}$, then a_{i_0} is known as the *optimum event* corresponding to countermeasure b_j ; (3) If $\max_{1\le i\le n, 1\le j\le m}\{r_{ij}\} = r_{i_0 j_0}$, then is known as the *optimum situation.*

Each synthesized situation decision can be made by following the steps below:

Step 1: Based on the set $A = \{a_1, a_2, \cdots, a_n\}$ of events and the set $B = \{b_1, b_2, \cdots, b_m\}$ of countermeasures, construct the set of situations $S = \{s_{ij} = (a_i, b_j) | a_i \in A, b_j \in B\}$;

Step 2: Determine the decision objectives $k = 1,2,\cdots,s$;

Step 3: For each objective $k = 1,2,\cdots,s$, compute the corresponding observed effect matrix $U^{(k)} = (u_{ij}^{(k)})_{n\times m}$;

Step 4: Calculate the matrix $R^{(k)} = (r_{ij}^{(k)})_{n\times m}$ of uniform effect measures of objective $k = 1,2,\cdots,s$;

Step 5: Determine the decision weights $\eta_1, \eta_2, \cdots, \eta_s$ of the objectives;

Step 6: From $r_{ij} = \sum_{k=1}^{s} \eta_k \cdot r_{ij}^{(k)}$, compute the matrix of synthetic effect measures $R = (r_{ij})_{n\times m}$; and

Step 7: Determine the optimum situation $s_{i_0 j_0}$.

Example 7.4. Let us look at a grey decision making on the designs of a construction project.

In the process of constructing the infrastructure of a society, selecting the best design determines whether or not the construction project is successful. During the public bidding of designs, the most important task is how to select the best design. For any engineering project, there are problems of multiplicity of designs and that of multi-objectivity. When evaluating designs, there are both quantitative and qualitative objectives. So, whether or not we can comprehensively and correctly reflect the actual situations of all the sub-objectives and the impacts of the sub-objectives on the overall project's objective directly affect the quality of our decision-making.

In particular, during the public bidding for a classroom building (17,000 m^2) of a certain school, it was required that the total cost is economic and reasonable, the appearance can represent the style and spirit of the school, the actual building forms an organic whole with the surrounding environment, it satisfies all requirements of functionality, and its structure is safe and steady. During the public bidding, three proposals were submitted.

Step 1: Establish the set of events and the set of countermeasures. We treated the three designs as our set of events $A = \{a_1, a_2, a_3\}$, where a_1, a_2, a_3 represented the individual designs. For each of the proposals, there were four possible countermeasures $B = \{b_1, b_2, b_3, b_4\}$, where b_1 stood for selection, b_2 selection with slight modification, b_3 selection with major modification, and b_4 rejection. Based on the sets of events A and countermeasures B, we constructed the set of situations: $S = \{s_{ij} = (a_i, b_j) | a_i \in A, b_j \in B\}$, $i = 1,2,3$, $j = 1,2,3,4$.

Step 2: Determine the decision objectives. Based on the opinions of the school administrators and relevant personnel, five objectives are considered in the decision-making: operational functionality, economic and reasonable structure, beautiful appearance, environmental friendliness, and the overall cost.

Step 3: Compute the effect sample matrices for the individual objectives. For each of the objectives, based on the literature, 15 school administrators and relevant personnel evaluated the objectives and produced the following matrices of comprehensive evaluation scores:

$$U^{(1)} = \begin{bmatrix} 2 & 8 & 3 & 2 \\ 12 & 1 & 2 & 0 \\ 1 & 2 & 9 & 3 \end{bmatrix}, U^{(2)} = \begin{bmatrix} 6 & 3 & 5 & 1 \\ 4 & 5 & 6 & 0 \\ 8 & 2 & 4 & 1 \end{bmatrix},$$

$$U^{(3)} = \begin{bmatrix} 2 & 12 & 1 & 0 \\ 1 & 3 & 10 & 1 \\ 13 & 1 & 1 & 0 \end{bmatrix}, U^{(4)} = \begin{bmatrix} 3 & 8 & 2 & 2 \\ 6 & 4 & 3 & 2 \\ 1 & 11 & 2 & 1 \end{bmatrix}.$$

The proposed costs of the bids are respectively 1.95, 2.25, and 2.45 (10,000,000 Yuan).

Step 4: Compute the matrices of uniform effect measures. Based on the evaluation scores on the four objectives of operational functionality, economic and reasonable structure, beautiful appearance, and environmental friendliness, we applied the upper effect measures. For the objective of the overall cost, we employed the lower effect measure. After processing the data in the effect sample matrices, we obtained the matrices of uniform effect measures:

$$R^{(1)} = \begin{bmatrix} 0.1667 & 0.6667 & 0.25 & 0.1667 \\ 0.0833 & 0.0833 & 0.1667 & 0 \\ 0.0833 & 0.1667 & 0.75 & 0.25 \end{bmatrix}$$

$$R^{(2)} = \begin{bmatrix} 0.75 & 0.625 & 0.375 & 0.125 \\ 0.5 & 0.625 & .75 & 0 \\ 1 & 0.25 & 0.5 & 0.125 \end{bmatrix}$$

$$R^{(3)} = \begin{bmatrix} 0.1538 & 0.9231 & 0.0769 & 0 \\ 0.0769 & 0.2308 & 0.7692 & 0.0769 \\ 1 & 0.07692 & 0.07692 & 0 \end{bmatrix}$$

$$R^{(4)} = \begin{bmatrix} 0.2727 & 0.7273 & 0.1818 & 0.1818 \\ 0.5454 & 0.3636 & 0.2727 & 0.1818 \\ 0.0909 & 1 & 0.1818 & 0.0909 \end{bmatrix}$$

and

$$R^{(5)} = \begin{bmatrix} 1 & 1 & 1 & 1 \\ 0.87 & 0.87 & 0.87 & 0.87 \\ 0.79 & 0.79 & 0.79 & 0.79 \end{bmatrix}$$

Step 5: Determine the decision weights of the objectives.

When using weights of the decision objectives to reflect different kinds of buildings, the specific time when the construction is done, and the economic strength of the construction company, there is a great deal of flexibility in terms of the assignment of the weights. So, it was reasonable for us to determine the weights through joint discussion sessions of the personals from both the construction company and the school. Along with the importance of the said classroom building to the school and the then-current state of the school, the decision weights were respective the following: for operational functionality $\eta_1 = 0.23$; for economic and reasonable structure, $\eta_2 = 0.16$; for beautiful appearance, $\eta_3 = 0.20$; for environmental friendliness, $\eta_4 = 0.11$, and for the overall cost, $\eta_5 = 0.30$.

Step 6: From $r_{ij} = \sum_{k=1}^{5} \eta_k \cdot r_{ij}^{(k)}$, we obtained the matrix of synthetic effect measures:

$$R = \begin{bmatrix} 0.5164 & 0.8107 & 0.4511 & 0.3765 \\ 0.6409 & 0.4627 & 0.6005 & 0.2946 \\ 0.6253 & 0.4307 & 0.5231 & 0.3234 \end{bmatrix}$$

Step 7: Make the decision.

Based on the principle of grey decision-making, because $\max_{1 \le j \le 4}\{r_{1j}\} = r_{12} = 0.8107$, $\max_{1 \le j \le 4}\{r_{2j}\} = r_{21} = 0.6409$, and $\max_{1 \le j \le 4}\{r_{3j}\} = r_{31} = 0.6253$, it implies that proposed designs 2 and 3 can be selected without any need for modification, while proposed design 1 needs some slight modification before its being selected.

7.3 Multi-attribute Intelligent Grey Target Decision Models

In this section, we will study a new decision model, which is constructed on the basis of four new functions of uniform effect measures. This new decision model sufficiently considers the two different scenarios of whether or not the effect values of the objectives actually hit the targets with very clear physics significance. Also, the distinguishability of the synthetic effect measures is greatly improved.

Definition 7.6. (1) Let k be a benefit type objective, that is, for k the greater the effect sample value is the better, and the decision grey target of objective k is $u_{ij}^{(k)} \in [u_{i_0 j_0}^{(k)}, \max_i \max_j \{u_{ij}^{(k)}\}]$, that is, $u_{i_0 j_0}^{(k)}$ stands for the threshold effect value of objective k. Then

$$r_{ij}^{(k)} = \frac{u_{ij}^{(k)} - u_{i_0 j_0}^{(k)}}{\max_i \max_j \{u_{ij}^{(k)}\} - u_{i_0 j_0}^{(k)}} \tag{7.5}$$

is referred to as the *effect measure of a benefit-type objective.*

(2) Let k be a cost-type objective, that is, for k the smaller the effect value is the better, and the decision grey target of objective k is $u_{ij}^{(k)} \in [\min_i \min_j \{u_{ij}^{(k)}\}, u_{i_0 j_0}^{(k)}]$, that is, $u_{i_0 j_0}^{(k)}$ stands for the threshold effect value of objective k. Then

$$r_{ij}^{(k)} = \frac{u_{i_0 j_0}^{(k)} - u_{ij}^{(k)}}{u_{i_0 j_0}^{(k)} - \min_i \min_j \{u_{ij}^{(k)}\}} \tag{7.6}$$

is referred to as the *effect measure of cost-type objective.*

(3) Let k be a moderate-value type objective, that is, for k the closer to a moderate value A the effect value is the better, and the decision grey target of objective k is $u_{ij}^{(k)} \in [A - u_{i_0 j_0}^{(k)}, A + u_{i_0 j_0}^{(k)}]$, that is, both $A - u_{i_0 j_0}^{(k)}$ and $A + u_{i_0 j_0}^{(k)}$ are respectively the lower and upper threshold effect values of objective k. Then,

(i) When $u_{ij}^{(k)} \in [A - u_{i_0 j_0}^{(k)}, A]$,

$$r_{ij}^{(k)} = \frac{u_{ij}^{(k)} - A + u_{i_0 j_0}^{(k)}}{u_{i_0 j_0}^{(k)}} \tag{7.7}$$

is referred to as the *lower effect measure of moderate-value type objective.*

(ii) When $u_{ij}^{(k)} \in [A, A + u_{i_0 j_0}^{(k)}]$,

$$r_{ij}^{(k)} = \frac{A + u_{i_0 j_0}^{(k)} - u_{ij}^{(k)}}{u_{i_0 j_0}^{(k)}} \tag{7.8}$$

is referred to as the *upper effect measure of moderate-value type objective.*

The effect measures of benefit-type objectives reflect the degrees of both how close the effect sample values are to the maximum sample values and how far away they are from the threshold effect values of the objectives. Similarly, the effect measures of cost-type objectives represent the degrees of how close the effect sample values are to the minimum effect sample values and how far away the effect sample values are from the threshold effect values of the objectives; and the lower effect measures of moderate-value type objectives indicate how far away the effect sample values that are smaller than the moderate value A are from the lower threshold effect value, and the upper effect measures how far away the effect sample values that are greater than the moderate value A are from the upper threshold effect values of the objectives.

For the situation of missing the target, it can be considered in the following four different possibilities:

(1) The effect value of a benefit-type objective is smaller than the threshold value $u_{i_0 j_0}^{(k)}$, that is, $u_{ij}^{(k)} < u_{i_0 j_0}^{(k)}$;
(2) The effect value of a cost-type objective is greater than the threshold value $u_{i_0 j_0}^{(k)}$, that is, $u_{ij}^{(k)} > u_{i_0 j_0}^{(k)}$;
(3) The effect value of a moderate-value type objective is smaller than the lower threshold effect value $A - u_{i_0 j_0}^{(k)}$, that is, $u_{ij}^{(k)} < A - u_{i_0 j_0}^{(k)}$; and
(4) The effect value of a moderate-value type objective is greater than the upper threshold effect value $A + u_{i_0 j_0}^{(k)}$, that is, $u_{ij}^{(k)} > A + u_{i_0 j_0}^{(k)}$.

In order for the effect measures of each type objective to satisfy the condition of normality, that is, $r_{ij}^{(k)} \in [-1,1]$, without loss of generality, we can assume that

For a benefit-type objective, $u_{ij}^{(k)} \geq -\max_i \max_j \{u_{ij}^{(k)}\} + 2u_{i_0 j_0}^{(k)}$;

For a benefit-type objective, $u_{ij}^{(k)} \leq -\min_i \min_j \{u_{ij}^{(k)}\} + 2u_{i_0 j_0}^{(k)}$;

For the situation that the effect value of a moderate-value type objective is smaller than the lower threshold effect value $A - u_{i_0 j_0}^{(k)}$, $u_{ij}^{(k)} \geq A - 2u_{i_0 j_0}^{(k)}$; and

For the situation that the effect value of a moderate-value type objective is greater than the upper threshold effect value $A + u_{i_0 j_0}^{(k)}$, $u_{ij}^{(k)} \leq A + 2u_{i_0 j_0}^{(k)}$.

With these assumptions, we have the following:

Proposition 7.8. The effect measures $r_{ij}^{(k)} (i = 1, 2, \cdots, n; j = 1, 2, \cdots, m; k = 1, 2, \cdots, s)$, as defined in Definition 7.6, satisfy the following properties: (1) $r_{ij}^{(k)}$ is non-dimensional; (2) the more ideal the effect, the greater $r_{ij}^{(k)}$ is; and (3) $r_{ij}^{(k)} \in [-1,1]$.

For the situation of hitting the target, $r_{ij}^{(k)} \in [0,1]$; and for the situation of missing the target, $r_{ij}^{(k)} \in [-1,0]$.

Similar to the discussions in Section 7.2, we can also establish the concepts of uniform effect measures and synthetic effect measures. In particular, for the case of synthetic effect measures, $r_{ij} \in [-1,0]$ belongs to the situation of missing the target, while $r_{ij} \in [0,1]$ the situation of hitting the target. For the situation of hitting the target, we can further compare the superiorities of events a_i, countermeasures b_j, and situations s_{ij} respectively by using the magnitudes of the synthetic effect measures, $i = 1,2,...,n$, $j = 1,2,...,m$.

Example 7.5. Let us look at the selection of the supplier of a key component used in the production of large commercial aircrafts.

In China, the production of large commercial aircrafts is managed using the model of main manufacturers – suppliers, where a great amount of key components comes from international suppliers. So, the scientificality of the decision-making on the selection of relevant suppliers is a key link that directly determines the success or failure of the operation. As a typical decision-making problem involved in the production process of sophisticated products, the selection of suppliers is generally accomplished through the form of public bidding. What has been done is that the main manufacturer first lists his clear demands, and then each potential supplier puts together his proposal that meets the needs of the manufacturer. After collecting all the proposals, the manufacturer selects the optimum proposal and signs the purchase agreement by comprehensively evaluating all the submitted proposals of the potential suppliers. As for what factors actually affect the manufacturer's decision, it is an extremely complicated matter. In order to arrive at educated and scientifically sound decisions, there is a need to analyze all involved factors closely and holistically.

During the selection of international suppliers for a specific key component of the production of large commercial aircrafts, there were there suppliers accepted into the second round of competition. To decide on the eventual decision, let us go through the following steps.

Step 1: Establish the sets of events, countermeasures, and situations. Let us define event a_1 to be the selection of a supplier for the said component for the production of large commercial aircrafts. So, the set of events is $A = \{a_1\}$. Define selecting supplier 1, supplier 2, or supplier 3 to be our countermeasures b_1, b_2, and b_3 so that the set of countermeasures is $B = \{b_1, b_2, b_3\}$. Therefore, the set of situations of our concern is $S = \{s_{ij} = (a_i, b_j) | a_i \in A, b_j \in B, i = 1; j = 1,2,3\} = \{s_{11}, s_{12}, s_{13}\}$.

Step 2: Determine the decision objectives. Through three rounds of surveys of relevant experts, the following 5 objectives are considered: quality, price, time of delivery, design proposal, and competitiveness.

Among these objectives, competitiveness, quality, and design proposal are qualitative. They are scored by relevant experts' evaluations satisfying that the higher the evaluation scores are the better. That is, they are benefit-type objectives. Let us take the threshold value $u_{i_0 j_0}^{(k)} = 9, k = 1,4,5$. For the objective of cost, the lower the cost is the better. So, it is a cost-type objective. Let us take the threshold value $u_{i_0 j_0}^{(2)} = 15$. The objective of "time of delivery" is one of moderate-value type. The main manufacturer desires the delivery at the end of the 16th month with 2 months deviation allowed. That is, $u_{i_0 j_0}^{(3)} = 2$, the lower threshold effect value is 16 – 2 = 14, and the upper threshold effect value is 16 + 2 = 18.

Step 3: Determine the decision weights of the objectives. To this end, we apply the AHP method to determine the weights, see Table 7.2 for details.

Table 7.2 The objectives' evaluation system

Objective	Quality	Price	Delivery	Design	Competitiveness
unit	qualitative	Million US$	Month	Qualitative	Qualitative
Order #	1	2	3	4	5
Weight	0.25	0.22	0.18	0.18	0.17

Step 4: Determine the effect sample vectors of each of the objectives:

$U^{(1)} = (9.5, 9.4, 9)$, $U^{(2)} = (14.2, 15.1, 13.9)$, $U^{(3)} = (15.5, 17.5, 19)$,
$U^{(4)} = (9.6, 9.3, 9.4)$, $U^{(5)} = (9.5, 9.7, 9.2)$.

Step 5: Assign the threshold effect values for the objectives. Because competitiveness, quality, and design proposal are all benefit-type objectives, let us take the threshold values $u_{i_0 j_0}^{(k)} = 9, k = 1,4,5$. Because price is a cost-type objective, let us take the threshold value $u_{i_0 j_0}^{(2)} = 15$. Because time of delivery is a moderate-value type objective and the main manufacturer desires the delivery at the end of the 16th month with a tolerance of +/- 2 months, we set $u_{i_0 j_0}^{(3)} = 2$, the lower threshold effect value 16 – 2 = 14, and the upper threshold effect value 16 + 2 = 18.

Step 6: Calculate the vectors of uniform effect measures. For the three qualitative objectives, competitiveness, quality, and design proposal, we employ the effect measures of benefit-type. For the objective of price, we utilize the effect measures of cost-type. For the objective of time of delivery, we apply the lower and upper effect measures. So, we obtain the following vectors of uniform effect measures:

$R^{(1)} = [1, 0.8, 0]$, $R^{(2)} = [0.73, -0.09, 1]$, $R^{(3)} = [0.75, 0.25, -0.5]$,
$R^{(4)} = [1, 0.5, 0.67]$, and $R^{(5)} = [0.71, 1, 0.29]$.

Step 7: From $r_{ij} = \sum_{k=1}^{5} \eta_k \cdot r_{ij}^{(k)}$, we compute the following vector of synthetic effect measures:

$$R = [r_{11}, r_{12}, r_{13}] = [0.8463, 0.4852, 0.2999].$$

Step 8: Make the final decision. Because $r_{11} > 0, r_{12} > 0, r_{13} > 0$, it means that all these three suppliers have hit the target. This result implies that it is reasonable for these suppliers to have entered the second round of competition. From $\max_{1 \le j \le 3}\{r_{1j}\} = r_{11} = 0.08463$, it follows that the main manufacturer should sign agreement with supplier 1.

Chapter 8
Grey Game Models

This chapter focuses on a series of specific and complicated problems created by lossening the constraint of complete knowledge, assumed as one of the most fundamental conditions in the conventional game theory, when one investigates problems of grey games. It makes use of the structural system of the conventional game theory as reference and the theoretical need to resolve practical problems as motivation.

The game theory was initiated in 1944 by John von Neumann and Oskar Morgenstern when they jointly published their book, entitled "Game Theory and Economic Behavior." By the end of the 1950s, cooperative game theory had developed to its peak; noncooperative game theory also appeared during the decade. Since after the 1970s, game theory had gradually evolved into a relatively complete theoretical system. Since after the 1980s, game theory has become a component of the mainstream economics. Especially, in the areas of oligopoly theory and information economics, the theory had brought forward magnificent achievements. To a certain extent, it can be said that it has become part of the fundamental materials of microeconomics.

One important assumption of game theory is that both players of the game work from a base of common knowledge. For instance, rationality is the common knowledge to each player. That is, all participants are rational, and all participants know that all participants are rational, and all participants know that all participants know that all participants are rational, and so on. This is an unimaginably infinite process. In terms of the players' cognitive ability of the realistic world, this is a very strick assumption. Obviously, it is extremely difficult for such a strick assumption to hold true in the physical world. And that is one of the greatest confusions the conventional game theory encounters. It is exactly because of this inexciplicable confusing problem of rationality of the conventional game theory that motivated the appearance and investigation of the evolutionary game theory. Speaking from the development history, game theory has been continuously evolved when it tried to address challenges one after another.

However, in the objective world, other than incompleteness in information, limitations in rationality, and others., there are also many such problems as uncertainties of the future, definite boundaries of knowledge, etc. Unfortunately, for these problems, the current studies of game theory have hardly touched upon. It is especially so in terms of the problem of limited knowledge.

This chapter investigates such problems as limited rationality, finite amount of knowledge, and others by using the rich logic of thinking and good amount of methods of grey systems theory.

8.1 Strategic Game Models for Duopolies with Limited Rationality and Knowledge

In this section, we construct a brand new duopolistic strategic output-making decision model based on the assumptions that the main decision-makers' rationality and knowledge are limited, and that the eventual decision depends on how the decision is made, while considering the ultimate purpose of making the decision is to guarantee the optimum strategic profit.

On the basis of this model, we develop the decision algorithm for output-making, investigate the attributes and characteristics of this model, and many interesting problems, such as the concession equilibrium of the later decision-maker, the damping equilibrium of the advanced decision-maker, and the damping loss and total damping cost for the advanced decision-maker to completely control the market.

8.1.1 Duopolistic Strategic Output-Making Models Based on Empirically Ideal Production and Optimal Decision Coefficients

Based on the assumption of Cournot duopoly model, assume that there are two firms 1 and 2 that produce the same product and make their decisions on their productions at the same time. If firm 1's production is q_1, and firm 2's is q_2, then the aggregate supply in the market is $Q = q_1 + q_2$. Let the market clearing price P be a function of the aggregate market supply $P = P(Q) = Q_0 - Q$, where Q_0 is a constant. We further assume that these firms do not have any fixed costs in their productions and their respective marginal unit production costs are c_1 and c_2. Based on the perfect rationality of these parties, the duopolies would choose their individual quantities of production that maximize their profits while reaching the Nash equilibrium.

When comparing these assumptions with the reality of competition in the business world, however, physically existing duopolies may very well be quite a distance away from being completely rational, and any of the competitors does not have any reason to believe that the other party is completely rational. Additionally, when thinking strategically, it is very possible that one side of the competition could take temporary losses by producing beyond the Nash equilibrium quanlity just for the purpose of taking control of more market share in order to squeeze the space of survival of its opponent while developing and expanding its own strategic space. As a matter of fact, if real life game players are not entirely rational, or even if they are completely rational, but for the reason of winning a certain strategic long-term competition, the Nash equilibrium of the Cournot duopolistic output-making competition with the current profit maximized might not be automatically materialized. Due to reasons like these, realistic duopolistic output-making competitive processes can hardly be fully or satisfactorily explained by using the Cournot duopoly model.

Based on the practical, competitive duopolistic output-making processes and considering the before-and-after sequence of when the duopolies determine their individual strategic outputs, we can establish their output-making formulations based on the empirically ideal outputs and the optimum strategic expansion coefficients as follows:

$$\begin{cases} q_1 = q_{01} + \gamma_1 \left(Q - q_{01}\right) \\ q_2 = q_{02} + \gamma_2 \left(Q - q_{02}\right) \\ Q = q_1 + q_2 \end{cases} \tag{8.1}$$

where

(1) Both q_1 and q_2 stand for the respective production outputs of the duopolistic firms 1 and 2; Q the market capability, which is determined by the duopolies' outputs.

(2) Both q_{01} and q_{02} respectively represent the duopolistic firms' ideal production outputs, which are formed based on their direct and/or indirect experiences and are affected by their own and competitors' states of production and the demands of the market, They generally stand for the levels of production that could bring forward the relatively ideal magnitudes of profits under preferential market conditions. If the duopolies are those of the Cournot model, satisfying the assumptions of the model, and based on the fact that a Cournot oligopoly equilibrium is a Nash equilibrium, we can take the solutions of the Cournot model as the duopolies' empirically ideal outputs of production.

(3) Both γ_1 and γ_2 stand for the intended strategic expansions of the duopolistic firms when they made their decisions of output-making. They are known as strategic expansion coefficients. These coefficients are mainly determined by the firms' actual capabilities of production, expansion, value systems, decision-making habits, social and economic backgrounds, psychological characteristics and personalities of the decision-makers, and other relevant factors. If an oligopoly, say *i*, first responds to the market by determining his strategic expansion coefficient $\gamma_i, (i = 1, 2)$, then he is referred to as the first (or advanced) decision-making oligopoly (or initiator). Otherwise, he is referred to as a later (or passive) decision-making oligopoly. The later decision-making oligopoly can somehow acquire the decision-making details of the first deicison-making oligopoly through, say, the already shown behaviors, various hints, past behavioral

habits, tradtion, personality, the current state of actions and the relevant backgrounds, etc.

(4) The decision-maker determines the current production output based on the sum of the empirically ideal output $q_{0i},(i=1,2)$ and the output $\gamma_i\left(Q-q_{0i}\right),(i=1,2)$ designed for strategic expansion. Under the general conditions, q_{0i}, $i=1,2$, are developed by comparing the current situation with the historical past while referencing to the available data. The quantities $\gamma_i\left(Q-q_{0i}\right),(i=1,2)$, reflect how much additional market share the decision-makers like to control with their strategic expansion. Of course, the strategy could also be about market contraction (negative expansion), which is signaled by the signs of γ_i, $(i=1,2)$, where a positive sign stands for a strategic expansion and a negative sign a contraction.

Theorem 8.1. The domains of the strategic expansion coefficients γ_1 and γ_2 of the duopolistic firms 1 and 2, determined when they decide on their respective production outputs, are given by

$$\gamma_1=\left[\frac{-q_{01}}{Q-q_{01}},1\right] \text{ and } \gamma_2=\left[\frac{-q_{02}}{Q-q_{02}},1\right].$$

Proof. Without loss of generality, assume that the production outputs satisfy the following constraints when the two duopolistic firms make their decisions:

$$\begin{cases} 0\le q_1\le Q \\ 0\le q_2\le Q \\ Q=q_1+q_2 \end{cases} \tag{8.2}$$

By solving the system of equs. (8.1) and (8.2), we obtain the domain of the strategic expansion coefficients γ_1 and γ_2 of the two duopolistic firms as follows:

$$\begin{cases} \dfrac{-q_{01}}{Q-q_{01}}\le\gamma_1\le 1 \\ \dfrac{-q_{02}}{Q-q_{02}}\le\gamma_2\le 1 \end{cases} \tag{8.3}$$

This ends the proof. QED.

Corollary 8.1. When the strategic expansion coefficients γ_1 and γ_2 of the two duopolistic firms are respectively 0 and 1, the firms will respectively choose their empirically ideal production outputs or the entire market capacity as their strategies.

Theorem 8.2. If two duopolies bear the same constant unit production cost c, that is, $C_1(q_1)=q_1c$ and $C_2(q_2)=q_2c$, their decision-making process on their production ouputs satisfy equ. (8.1), and their their empirically ideal production outputs q_{01} and q_{02} are equal to the Cournot equilibrium outputs q_1^* and q_2^*, then the duopolies' optimum strategies for the current period production are $\gamma_1 = \gamma_2 = 0$, and their optimum production outputs are $q_1 = q_1^* = q_{01}$, $q_2 = q_2^* = q_{02}$, and $q_1 = q_2 = \frac{1}{3}(Q_0 - c)$.

Proof. According to the same assumption as that in equ. (8.1), from the Cournot model, it follows that the duopolies' production decisions at the Nash equilibrium are $q_1^* = q_2^* = \frac{1}{3}(Q_0 - c)$. Let $k = Q_0 - c$. Thenm we obtain $q_1^* = q_2^* = \frac{k}{3}$. So, following from the given conditions of the theorem, we have $q_{01} = q_{02} = q_0 = \frac{k}{3}$. QED.

This result shows that the Cournot model is a special case of our model. When the relevant parameters of our model take certain specific values, it reduces into the Cournot model.

Assume that the duopolies' strategic decision-making processes on their production outputs satisfy equ. (8.1). If they want to optimize their strategy profits for the current period, then the total production output of these duopolies is given by the following theorem.

Theorem 8.3. Given the duopolies' empirically ideal production outputs q_{0i}, $i=1,2$, if their decision-making processes of production satisfy equ. (8.1), then when faced with different first decision-making oligopolies j, j = 1, 2, the market product supply Q_{2-j}^*, $j=1,2$, which guarantees the maximization of the current period strategy profits, is respectively determined by the expirically ideal output q_{0i}, $i=1,2$, of the first decision-making oligopoly and the unit production cost c_i, $i=1,2$, of the later decision-making oligopoly as shown below:

$$\begin{cases} Q_{2-j}^* = Q_1^* = \dfrac{q_{02} + (Q_0 - c_1)}{2}, j = 2 \\ Q_{2-j}^* = Q_2^* = \dfrac{q_{01} + (Q_0 - c_2)}{2}, j = 1 \end{cases} \tag{8.4}$$

Proof. Without loss of generality, let us assume that oligopoly 2 is the first decision-maker. That is, the first decision-making coefficient γ_{20} is given. Therefore, the optimum decision-making coefficient $\gamma_1^{(0)}$ of oligopoly 1, which maximizes the current period profit, can be determined. Now, from the assumption that the duopolies' decision-making processes satisfy equ. (8.1), the market product supply of the duopolies can be obtained. Similarly, the market equilibrium supply can be obtained for the case that oligopoly 2 is the later decison-maker. QED.

8.1.2 *Concession Equilibrium of the Later Decision-Maker under Nonstrategic Expansion Damping Conditions: Elimination from the Market*

In the practical process of duopolistic decision-making, considering the maximization of the current profit, the first decision-making oligopoly can materialize his strategic goal of taking greater market share by two methods: one is the strategic expansion coefficients $\gamma_i, (i = 1, 2)$; and the other the empirically ideal production outputs $q_{0i}, (i = 1, 2)$. At the same time, the later decision-making oligopoly makes his production output decision.

Each strategy of expanding production output is often limited and affected by many factors, such as the actual production capability, the potential for expansion, the capital ability for expansion, the managerial talent, concepts, personalitites, habits, etc. all these constraining factors are known as the damping conditions of strategic expansions. For the sake of convenience, let us divide our discussion into two scenarios. For the first one, let us assume that for a certain product there is no problem with damping conditions for its strategic expansion. That is when the $\gamma_i, (i = 1, 2)$, values are relatively large, which are near or equal to 1; For the second scenario, a certain product does have a set of damping conditions for its strategic expansion. That is when $\gamma_i, (i = 1, 2)$, values are relatively small, and can be gradually increased through bringing up the actual production capability, investment skills, and the managerial ability by purpsoeful construction and learning.

Theorem 8.4. Asseume that the duopolies' competition process of the strategic output-making problem satisfies equ. (8.1) and potential strategic expansions are not constrained by any damping condition. If the strategic expansion coefficient of the first decision-making oligopoly is $\gamma_i = \gamma_{i0} > 0$, i = 1, 2, then the optimum reaction of the later decision-making oligopoly in terms of maximizing the current profit is the method of withdraw. That is a psychological concession, reduced $\gamma_i^{(0)}$ value, i = 1, 2, and a reduction of its market share.

Proof. Without loss of generaility, let oligopoly 2 is the first decision-making firm. So, oligopolgy 1 knows about the strategic expansion coefficient of oligopoly 2 and responds to its earlier decisions. Equ. (8.4) indicates that if oligopoly 2 proactively takes actions, oligopoly 1 will respond passively, and the market supply is Q_1^* that maximizes the current profits for the duopolies. Now, from Theorem 8.3, it follows that $Q = Q_1^*$ stays the same. When oligopoly 2 becomes greedier, that is when $\gamma_2 = \gamma_{20}$ increases, his market share q_2 increases. Correspondingly, the market share $q_1 = Q_1^* - q_2$ of oligopoly 1 shrinks. As long as both q_{02} and c_1 stay constant, Q_1^* is also a constant. When $\gamma_2 = \gamma_{20}$ increases, it must cause q_2 to increase. In turn, it causes q_1 to reduce. That is, oligopoly 1's market share shrinks and he consequently has to reduce his production.

When $Q = Q_1^*$ stays the same, oligopoly 2 proactively takes actions and becomes greedier. That is, $\gamma_2 = \gamma_{20}$ increases. From the analysis above, it follows that to maximize his current profit, oligopoly 1 has to reduce his production output, leading to decreased q_1 . From equ. (8.1), we can obtain the following equ. (8.5), which indicates that when q_1 decreases, γ_1 also drops. That is to say, at this time, oligopoly is taking a psychological concession.

$$\gamma_1 = \frac{(q_1 - q_{01})}{(Q - q_{01})} . \text{ QED.} \tag{8.5}$$

From Theorem 8.4, it follows that under the assumptions of this current theorem, there is the phenomenon that the first decision-making oligopoly seizes additional control of the market, while the later decision-making oligopoly materializes concession equilibrium by conceding psychologically and productionally.

Corollary 8.2. In Theorem 8.4, there is no damping condition that limits strategic expandions of the first-decision-making oligopoly; and the strategic decisions of both of the duopolies are made to maximize the current profits so that the optimum strategy for the first decision-making oligopoly is: eliminate the later decision-making oligopoly from the market.

Example 8.1. If no damping conditions exist to affect strategic expansions, then along with the increase of the strategetic expansion coefficient γ_{20} of the frist decision-making oligopoly 2, the simulated concession equilibria of oligopoly 1 are given in Table 8.1.

Table 8.1 Simulated concession equilibria of oligopoly 1

γ_{20} (given)	$\gamma_1^{(0)}$	q_2	q_1	$Q_1^* = q_1 + q_2$	u_1	u_2
-1	1	0	4	4	8	0
-0.8	0.6	0.4	3.6	4	7.2	1.2
-0.6	0.2	0.8	3.2	4	6.4	2.4
-0.4	-0.2	1.2	2.8	4	5.6	3.6
-0.2	-0.6	1.6	2.4	4	4.8	4.8
0	-1	2	2	4	4	6
0.2	-1.4	2.4	1.6	4	3.2	7.2
0.4	-1.8	2.8	1.2	4	2.4	8.4
0.6	-2.4	3.2	0.8	4	1.6	9.6
0.8	-2.6	3.6	0.4	4	0.8	10.8
0.999	-2.998	3.998	0.002	4	0.004	11.994

Note: In this table, γ_{10} and γ_{20} are respectively the strategic expansion coefficients of the duopolies. $\gamma_1^{(0)}$ is the reactive strategic expansion coefficient of oligopoly 1 on the basis of maximizing his current profit with given γ_{20}, Q_1^* the market product supply under the condition that oligopoly 2 makes decisions first, $u_1 = q_1 \cdot P(Q) - c_1 \cdot q_1$ and $u_2 = q_2 \cdot P(Q) - c_2 \cdot q_2$ are respectively the strategy profts of the duopolies. Additionally, the initial values of the simulation are $q_{01} = 3, q_{02} = 2, c_1 = 2, c_2 = 1, Q_0 = 8$.

Table 8.1 indicates that under the condition of having no damping conditions to affect strategic expansions, the concession equilibrium of the later decision-making oligopoly is a conditional Nash equilibrium. With the consideration of maximizing the current profits, the later decision-making oligopoly treats the decision of the first-decision-making oligopoly as invariant so that the optimum strategy of the later decision-making oligopoly is to withdraw from the market – being eliminated from the market.

8.1.3 Damping Equilibrium of the Advanced Decision-Maker under Strategic Expansion Damping Conditions: Giving Up Some Market Share

This subsection focuses on the equilibrium problem of the duopolies' output-making strategies under the condition that there are certain damping conditions affecting strategic expansions.

Theorem 8.6. If the competition process of the duopolies' output-making strategy problem satisfies equ. (8.1) and there are damping conditions that constrain strategic expansions, then there is a strategic output-making equilibrium that maximizes the current profits.

Proof. Without loss of generality, assume that oligopoly 2 is the first decision maker, while oligopoly 1 is the later one. Because any strategic expansion of oligopoly 2 is constrained by some damping conditions, that is, his strategic expansion coefficient is equal to a certain constant $\gamma_2 = \gamma_{20}$, the current strategy profit can be described as follows:

$$\begin{aligned} u_2 &= q_2 \cdot P(Q) - c_2 \cdot q_2 = q_2 \cdot \left(Q_0 - Q_1^* - c_2\right) \\ &= \frac{1}{4}\left(2q_{02} + \left(Q_0 - q_{02} - c_1\right)\gamma_{20}\right)\cdot\left(Q_0 - 2c_2 - q_{02} + c_1\right) \end{aligned} \tag{8.6}$$

By computing the partial derivative with respect to the empirically ideal production q_{02}, and letting $\frac{\partial u_2}{\partial q_{02}} = 0$, we obtain the empirically ideal production q_{02} of oligopoly 2 when he considers to maximize the current profit as follows:

$$q_{02}^* = \frac{\left(Q_0 - 2c_2 + c_1\right)\left(2-\gamma_{20}\right) - \left(Q_0 - c_1\right)\gamma_{20}}{2\left(2-\gamma_{20}\right)} \tag{8.7}$$

From equs. (8.1) and (8.7), we obtain the following equ. (8.8), from which we in turn obtain (8.9) and (8.10) as follows:

$$\begin{aligned} Q_1^{**} = Q_1^*\Big|_{q_{02}=q_{02}^*} &= \frac{\left(Q_0 - 2c_2 + c_1\right)\left(2-\gamma_{20}\right) - \left(Q_0 - c_1\right)\gamma_{20}}{4\left(2-\gamma_{20}\right)} + \frac{\left(Q_0 - c_1\right)}{2} \\ &= \frac{\left(3Q_0 - 2c_2 - c_1\right) - \gamma_{20}\left(2Q_0 - c_2 - c_1\right)}{2\left(2-\gamma_{20}\right)} \end{aligned} \tag{8.8}$$

$$q_2^* = q_{02} + \gamma_{20}\left(Q_1^* - q_{02}\right) = \frac{2\left(Q_0 - 2c_2 + c_1\right) + \gamma_{20}\left(4c_2 - 3c_1 - Q_0\right) + \gamma_{20}^2\left(c_1 - c_2\right)}{2\left(2-\gamma_{20}\right)} \quad (8.9)$$

and

$$q_1^* = Q_1^* - q_2^* = \frac{\left(Q_0 + 2c_2 - 3c_1\right) - \gamma_{20}\left(Q_0 - 4c_1 + 3c_2\right) - \gamma_{20}^2\left(c_1 - c_2\right)}{2\left(2-\gamma_{20}\right)} \quad (8.10)$$

Equs. (8.9) and (8.10) are respectively the production outputs of the oligopolies 1 and 2 at the strategic output-making equilibrium with their current profits maximized. QED.

The equilibrium described in Theorem 8.6 is caused by the damping conditions of the first decison-making oligopoly on his potential strategic expansions. So, let us refer this kind of equilibrium to as damping conditions of the first decision-making oligopoly, or damping conditions for short.

By comparing this result with Theorem 8.4, it can be seen that although the first decision-making oligopoly has the supreme advantage of being the first, the damping conditions on his potential strategic expansions slow him down from quickly taking control of the entire market. So, such a damping equilibrium stands for a conditional Nash equilibrium, which is related to the equilibrium of the classical Cournot model, as shown in the following theorem.

Theorem 8.7. In Theorem 8.6, if the strategic expansion coefficients of the duopolies are 0, that is, the damping coefficients of strategic expansions are large, causing both parties unable to expand strategically, then the production output $q_j^*, (j=1,2)$, of the first decision-making oligopoly is 6 times that of the equilibrium production $q_j^{*\prime}, (j=1,2)$, of the classical Cournot model as shown below:

$$q_j^* = 6q_j^{*\prime} \quad (8.11)$$

Proof. According to the theory of the classical Cournot model, we can obtain the equilibrium production outputs of the duopolies under the conditions of the model as follows:

$$\begin{cases} q_1^{*\prime} = \frac{1}{3}\left(Q_0 - 2c_1 + c_2\right) \\ q_2^{*\prime} = \frac{1}{3}\left(Q_0 + c_1 - 2c_2\right) \end{cases} \quad (8.12)$$

Without loss of generality, let us assume that oligopoly 2 is the first decision-maker, while oligopolgy 1 is the later one. From equs. (8.9) and (8.10), it follows that when

$\gamma_{i0} = 0, (i = 1, 2)$, meaning that the damping constraints of any strategic expansion are infinitely large, the damping equilibrium of the oligopolies' productions can be described as

$$\begin{cases} q_1^* = Q_0 - 3c_1 + 2c_2 \\ q_2^* = 2(Q_0 + c_1 - 2c_2) \end{cases}$$

Solving this system leads to

$$q_2^* = 6q_2^{*/}$$

Similarly, we can prove the result for the case when oligopoly 1 is the first decision-maker, while oligopoly 2 is the later one. QED.

Corollary 8.3. In Theorem 8.6, if the duopolies' strategic expansion coefficients are both 0, meaning that neither of the duopolies has any intension for expansion, that is, the damping coefficients for any potential strategic expansion are very large, making the duopolies unable to make any strategic expansion, then the current market product suppus is related to the supply determined by the classical Cournot model as follows:

$$\begin{cases} Q_1^* - Q^{*/} = \dfrac{7Q_0 - 2c_1 - 5c_2}{3} \\ Q_2^* - Q^{*/} = \dfrac{7Q_0 - 5c_1 - 2c_2}{3} \end{cases} \tag{8.13}$$

Proof. We will only look at the proof of $Q_1^* - Q^{*/} = \dfrac{7Q_0 - 2c_1 - 5c_2}{3}$. The detailed argument for the second formula is similar and omitted.

From equ. (8.13), we can obtain the equilibrium product supply of the market under the conditions of the classical Cournot model

$$Q^{*/} = \frac{1}{3}(2Q_0 - c_1 - c_2) \tag{8.14}$$

Without loss of generality, assume that oligopoly 1 is the first decision-maker, while oligopoly 1 the later one. Under this assumption, the equilibrium market supply is

$$Q_1^* = 3Q_0 - 2c_2 - c_1$$

from which, we obtain

$$Q_1^* - Q^{*/} = \frac{7Q_0 - 2c_1 - 5c_2}{3}. \text{ QED.}$$

Example 8.2. Assume that there are damping conditions on any potential strategic expansion. Then the simulated damping equilibria of oligopoly 2 are given in Table 8.2.

Table 8.2 The simulated damping equilibria of oligopoly 2

q_{02} (given)	γ_{20} (given)	$\gamma_1^{(0)}$	q_2	q_1	$Q_1^* = q_1 + q_2$	u_1	u_2
0.200	0.000	0.818	0.200	2.900	3.100	8.410	0.780
0.500		0.600	0.500	2.750	3.250	7.563	1.875
1.000		0.333	1.000	2.500	3.500	6.250	3.500
1.500		0.143	1.500	2.250	3.750	5.063	4.875
2.000		0.000	2.000	2.000	4.000	4.000	6.000
2.500		-0.111	2.500	1.750	4.250	3.063	6.875
3.000		-0.200	3.000	1.500	4.500	2.250	7.500
3.500		-0.273	3.500	1.250	4.750	1.563	7.875
4.000		-0.333	4.000	1.000	5.000	1.000	8.000
4.500		-0.385	4.500	0.750	5.250	0.563	7.875
5.000		-0.429	5.000	0.500	5.500	0.250	7.500
5.500		-0.467	5.500	0.250	5.750	0.063	6.875
6.000		-0.500	6.000	0.000	6.000	0.000	6.000
0.200	0.500	-0.500	1.650	1.450	3.100	4.205	6.435
0.500		-0.500	1.875	1.375	3.250	3.781	7.031
1.000		-0.500	2.250	1.250	3.500	3.125	7.875
1.500		-0.500	2.625	1.125	3.750	2.531	8.531
2.000		-0.500	3.000	1.000	4.000	2.000	9.000
2.500		-0.500	3.375	0.875	4.250	1.531	9.281
3.000		-0.500	3.750	0.7500	4.500	1.125	9.375
3.500		-0.500	4.125	0.625	4.750	0.781	9.281
4.000		-0.500	4.500	0.500	5.000	0.500	9.000
4.500		-0.500	4.875	0.375	5.250	0.281	8.531
5.000		-0.500	5.250	0.250	5.500	0.125	7.875
5.500		-0.500	5.625	0.125	5.750	0.031	7.031
6.000		-0.500	6.000	0.000	6.000	0.000	6.000

Note: In this table, γ_{10} and γ_{20} stand respectively for the strategic expansion coefficients of the duopolies, Q_1^* the market supply under the condition that oligopoly 2 makes his decision first, and $u_1 = q_1 \cdot P(Q) - c_1 \cdot q_1$ and $u_2 = q_2 \cdot P(Q) - c_2 \cdot q_2$ the respective decision profits of oligopolies 1 and 2. Additionally, the initial values of the simulation are $q_{01} = 2, c_1 = 2, c_2 = 1, Q_0 = 8$.

From Table 8.2, it can be seen that when the damping coefficients for strategic expansion are large, there is a damping equilibrium for oligopoly 2, which maximizes the current profit for his most empirically ideal production output. As γ_{20} takes larger values, the curve of u_2 is elevated, see Figure 8.1 for details.

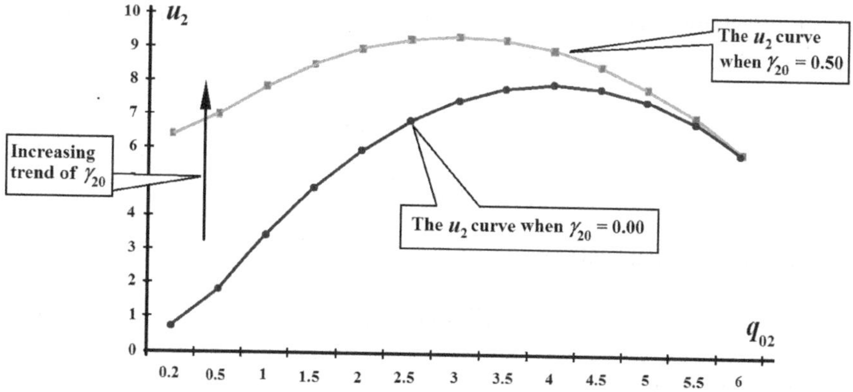

Fig. 8.1 The curve of u_2 in q_{02} with different γ_{20}

8.1.4 Damping Loss and Total Damping Cost for the First Decision-Making Oligopoly to Completely Control the Market

Under the condition of having dampings for any potential strategic expansion, the first decision-making oligopoly could not control the entire market with one leap in order to maximize his profit. In other words, to eliminate his competitor from the market, the first decision-making oligopoly has to sacrifice some of his current profit, a non-maximization strategy of the current profit, and pay a certain price.

Theorem 8.8. If the competition process of the duopolies' production output strategy making satisfies equ. (8.1) and there are damping conditions against any potential strategic expansion, and the first decision-making oligopoly $i,(i=1,2)$, has the absolute advantage over the later decision-maker, then the production output $q_i^{**},(i=1,2)$, at the time when he completely occupies the market with the opponent eliminated, the empirically ideal production output $q_{0i},(i=1,2)$, and the market supply $Q_{2-i}^{*/},(i=1,2)$, with the current profit maximized, are identical and equal to $Q_0-c_{2-i},(i=1,2)$.

Proof. Without loss of generality, assume that oligopoly 2 is the first decision-maker with an absolute advantage over the later decision-making oligopoly 1. From Theorem 8.3, it follows that the market supply, which maximizes the current strategy profit, is $Q_1^* = \dfrac{q_{02}+(Q_0-c_1)}{2}$.

When oligopoly 2 takes the control of the entire market, its production output is $q_2^{**} = Q_1^*$. From equ. (8.1), it follows that

$$\begin{cases} q_2^{**} = q_{02} + \gamma_2 \left(Q_1^* - q_{02}\right) \\ q_2^{**} = Q_1^* \end{cases} \tag{8.15}$$

Solving this equation leads to

$$q_2^{**} = Q_1^* = q_{02} \tag{8.16}$$

Therefore, we have

$$2q_{02} = q_{02} + \left(Q_0 - c_1\right) \tag{8.17}$$

$$q_{02} = Q_0 - c_1$$

$$q_2^{**} = Q_1^* = q_{02} = Q_0 - c_1 \tag{8.18}$$

Similar argument can be employed to show that when oligopoly 1 is the first decision-maker, the following holds true:

$$q_1^{**} = Q_2^* = q_{01} = Q_0 - c_2 \text{. QED.} \tag{8.19}$$

Theorem 8.9. If the competition process of the duopolies' production output strategy making satisfies equ. (8.1) and there are damping conditions against any potential strategic expansion, then the profit of the first decision-making oligopoly $i, (i = 1, 2)$, at the time when the competitor is entirely eliminated from the market is $u_i^{*/} = \left(Q_0 - c_{2-i}\right) \cdot \left(c_{2-i} - c_i\right)$.

Proof. Without loss of generality, assume that oligopoly 2 is the frist decision-maker. From equs. (8.16) and (8.18), we obtain the profit of the time when oligopolgy 2 takes full control of the market with oligopoly 1 eliminated from the market as follows:

$$\begin{aligned} u_2^{*/} &= q_2 \cdot p(Q) - c_2 \cdot q_2 = q_2 \cdot (Q_0 - Q_1^{*/} - c_2) \\ &= (Q_0 - c_1) \cdot (c_1 - c_2) \end{aligned} \tag{8.20}$$

Similarly, we can show that when oligopoly 1 is the first decision-maker, and when he full controls the market, his profit is

$$u_1^{*/} = (Q_0 - c_2) \cdot (c_2 - c_1) \tag{8.21}$$

By combining equs. (8.20) and (8.21), the theorem is shown. QED.

Corollary 8.4. In Theorem 8.9, if the marginal cost c_i of the first decision-making oligopoly $i,(i=1,2)$, is higher than that $c_{2-i},(i=1,2)$, of the later decision-making oligopoly, then the first decision-making oligopoly cannot eliminate the later decision-making oligopoly from the market.

Theorem 8.10. If the competition process of the duopolies' production output strategy making satisfies equ. (8.1) and there are damping conditions against any potential strategic expansion, then the first decision-making oligopoly $i,(i=1,2)$, has to pay certain price to take the full control of the market.

Proof. Without loss of generality, assume that oligopoly 2 is the first decision-maker. Then, his gain at the time of fully controlling the market is given by equ. (8.20). From Theorem 8.8, it follows that the gain of this oligopoly at the damping equilibrium that maximizes the current profit is given by (8.22) below:

$$\begin{aligned} u_2^* &= q_2^* \cdot P(Q) - c_2 \cdot q_2^* = q_2^* \cdot \left(Q_0 - Q_1^* - c_2\right) \\ &= \left(\left(1-\gamma_{20}\right) q_{02}^* + \gamma_{20} Q_1^*\right) \cdot \left(Q_0 - Q_1^* - c_2\right) \end{aligned} \tag{8.22}$$

Because u_2^* stands for the maximum gain of oligopoly 2 when there are damping conditions for any potential strategic expansion, while $u_2^{*/}$ is only the gain of the oligopoly under some specific conditions, we must have $u_2^{*/} \le u_2^*$. So, from equs. (8.21) and (8.22), it follows that the price ΔC_2 oligopoly 2 has to pay for fulling controlling the market is

$$\begin{aligned} \Delta C_2 &= u_2^* - u_2^{*/} \\ &= \left(\left(1-\gamma_{20}\right) q_{02}^* + \gamma_{20} Q_1^*\right) \cdot \left(Q_0 - Q_1^* - c_2\right) - (Q_0 - c_1) \cdot (c_1 - c_2) \ge 0 \end{aligned} \tag{8.23}$$

Similarly we can obtain the price ΔC_1 for oligopoly 1 to pay to fully control the market if he is the first decision-maker as follows:

$$\begin{aligned} \Delta C_1 &= u_1^* - u_1^{*/} \\ &= \left(\left(1-\gamma_{10}\right) q_{01}^* + \gamma_{10} Q_2^*\right) \cdot \left(Q_0 - Q_2^* - c_1\right) - (Q_0 - c_2) \cdot (c_2 - c_1) \ge 0 \end{aligned} \tag{8.24}$$

This ends the proof. QED.

Although the first decision-maker has the absolute advantage to seize the market control, due to the existing damping conditions against any potential strategic expansion, he cannot materialize such control with one leap. And, he has to pay additional price to reach the goal of fully occupying the market. For this reason,

$\Delta C_i, (i=1,2)$, is referred to as the *damping cost* of the first decision-making oligopoly $i, (i=1,2)$, in order for him to take full control of the market.

Throughout the process from starting off at an initial state 0 until entirely controlling the market, let u_{ij} be the strategy gain of the first decision-making oligopoly $i, (i=1,2)$, at the jth period, $j = 0, 1, 2, \ldots, t$; and let u_{ik}^* be the maximum strategy gain at the damping equilibrium achieved in the kth period. Then, the sum of the differences ($u_{ik}^* - u_{ij}$), $j = 0, 1, 2, \ldots, t$, is referred to as the *total damping cost* of the complete strategic market of oligopoly $i, (i=1,2)$.

Theorem 8.11. If the competition process of the duopolies' production output strategy making satisfies equ. (8.1) and there are damping conditions against any potential strategic expansion, then throughout the process of competition that starts off from an initial state 0 until time t, the first decision-making oligopoly $i, (i=1,2)$, has to pay a certain amount of damping cost to fully control the market.

Proof. Without loss of generality, assume that oligopoly 2 is the first decision-maker, who experienced a total of T rounds of decision-making starting at the initial state 0 and ending at time t. For each decision-making, let the gain be $u_{2s} (s = 0,1,2,\cdots,t)$. If oligopoly 2 reaches the damping equilibrium at the kth period with the equilibrium gain u_{2k}^*, then the damping loss of this oligopoly at period j is

$$\Delta d_{2j} = u_{2k}^* - u_{2j}, (j = 0,1,2,\cdots,t) \tag{8.25}$$

Summing up these differences provides the total damping cost D_2 of oligopoly 2 paid throughout the process of competition from the initial period 0 to the last period t, where

$$D_2 = \sum_{j=0}^{t} \Delta d_j = \sum_{j=0}^{t} \left(u_{2k}^* - u_{2s} \right) \tag{8.26}$$

Similarly, the result holds true when oligopoly 1 is the first decision-maker. QED.

Example 8.3. Assume that the competition process of the duopolies' production output strategy making satisfies equ. (8.1) and that there are damping conditions against any potential strategic expansion. Let oligopoly 2 be the first decision-maker. Then the simulated total damping costs of oligopoly 2 for him to fully control the market under different damping conditions against strategic expansion are provided in Table 8.3.

Table 8.3 Simulated total damping costs under different damping conditions

γ_{20} (given)	q_{02}^*	q_2^*	$q_2^{**}=Q_1^*$	u_2^*	u_1^*	$u_2^{*/}$	ΔC_2
0.000	4.000	4.000	6.000	8.000	1.000	6.000	2.000
0.200	3.667	3.900	6.000	8.450	1.089	6.000	2.450
0.300	3.471	3.850	6.000	8.719	1.120	6.000	2.719
0.400	3.250	3.800	6.000	9.025	1.134	6.000	3.025
0.500	3.000	3.750	6.000	9.375	1.125	6.000	3.375
0.600	2.714	3.700	6.000	9.779	1.080	6.000	3.779
0.700	2.385	3.650	6.000	10.248	0.980	6.000	4.248
0.800	2.000	3.600	6.000	10.800	0.800	6.000	4.800
0.900	1.546	3.550	6.000	11.457	0.496	6.000	5.457
1.000	1.000	3.500	6.000	12.250	0.000	6.000	6.250

Note: In this table, γ_{20} stands for the strategic expansion coefficient of oligopoly 2, q_{02}^* and q_2^* respectively the optimum empirical ideal production output and the actual output of oligopoly 2 under the given parameter γ_{20}, $q_2^{**}=Q_1^*$ the entire market supply provided by oligopoly 2, u_2^* and $u_2^{*/}$ the gains of oligopoly 2 when producing the outputs q_2^* and q_2^{**}, respectively, and $\Delta C_2 = u_2^* - u_2^{*/}$. u_1^* represents the gain of oligopoly 1 when producing the output q_1^* that corresponds to q_2^*. The initial values of the simulation are $q_{01}=2, c_1=2, c_2=1, Q_0=8$.

From Table 8.3., it can be seen that under the condition of existing damping conditions against any strategic expansion, as γ_{20} increases, the gain and the damping cost for oligopoly 2 to entirely control the market all grow positively.

Example 8.4. Assume that the competition process of the duopolies' production output strategy making satisfies equ. (8.1) and that there are damping conditions against any potential strategic expansion. Let oligopoly 2 be the first decision-maker and $\gamma_{20}=0.300$. Take $q_{02}^{(1)}=3.000$ as the empirical ideal production output for the first round. If his empirical ideal production output of the *s*th round is equal to the actual production output of the $(s-1)$ round, $q_{02}^{(s)}=q_{2,(s-1)}$, let us iterate this process until the *k*th round when $u_{1k}=0$ and the computation ends. The corresponding simulation is given in Table 8.4.

Table 8.4 Simulated total damping cost of oligopoly 2 with $\gamma_{20} = 0.300$

Game round s	$q_{02}^{(s)}$ (given)	q_{2s}	u_{2s}	u_{1s}	Δd_{2s}
1	1.000	1.750	6.125	4.375	2.594
2	1.750	2.388	7.461	3.161	1.258
3	2.388	2.930	8.221	2.283	0.498
4	2.930	3.391	8.595	1.649	0.124
5	3.391	3.782	8.716	1.191	0.003
6	3.782	4.115	8.678	0.861	0.041
7	4.115	4.398	8.543	0.622	0.176
8	4.398	4.638	8.354	0.449	0.365
9	4.638	4.842	8.140	0.325	0.579
10	4.842	5.016	7.920	0.235	0.799
11	5.016	5.164	7.704	0.169	1.015
12	5.164	5.289	7.500	0.122	1.219
13	5.289	5.396	7.314	0.089	1.405
14	5.396	5.487	7.144	0.064	1.575
15	5.487	5.564	6.991	0.046	1.728
16	5.564	5.629	6.857	0.033	1.862
17	5.629	5.685	6.739	0.024	1.980
18	5.685	5.732	6.635	0.017	2.034
19	5.732	5.772	6.546	0.013	2.173
20	5.772	5.806	6.468	0.009	2.251
21	5.806	5.835	6.401	0.007	2.318
22	5.835	5.860	6.343	0.005	2.376
23	5.860	5.881	6.293	0.003	2.426
24	5.881	5.899	6.250	0.003	2.469
25	5.899	5.914	6.213	0.002	2.506
26	5.914	5.927	6.182	0.001	2.537
27	5.927	5.938	6.155	0.000	2.564
Total damping cost of oligopoly 2 $D_2 = \sum_{j=0}^{t} \Delta d_j = \sum_{j=0}^{t} (u_{2k}^* - u_{2s})$					$D_2 = 40.875$
Given the strategic expansion coefficient of oligopoly 2 $\gamma_{20} = 0.300$, his gain at the damping equilibrium is $u_2^* = 8.719$.					

Note: In this table, γ_{20} stands for the strategic expansion coefficient of oligopoly 2, $q_{02}^{(s)}$ and q_{2s} are respectively the empirically ideal production output and the actual output in the sth period, u_{2s} and u_{1s} respectively the gains of oligopoly 2 and 1 in the sth period, and Δd_{2s} the damping loss of oligopoly 2 in the sth round of decision making. The initial values of simulation are $\gamma_{20} = 0.30, q_{01} = 2, c_1 = 2, c_2 = 1, Q_0 = 8$.

From Table 8.4, it can be seen that under the assumption of existing certain damping conditions against strategic expansions, the first decision-making oligopoly can gradually push the later decision-making oligopoly out from the market by using the method of learning and modifying his empirically ideal production output. However, in terms of his strategies that also maximize the current profit, the damping conditions also make him pay certain prices for his eventual strategic success.

Through the rounds of the game playing, the closer to the time the later decision-making oligopoly is eliminated from the market, the slower the rate of production increase of the first decision-making oligopoly becomes. At the same time, this oligopoly's strategy damping costs of the individual periods go through phases of decreasing to fast increasing and then to slowly decreasing, as shown in Figure 8.2.

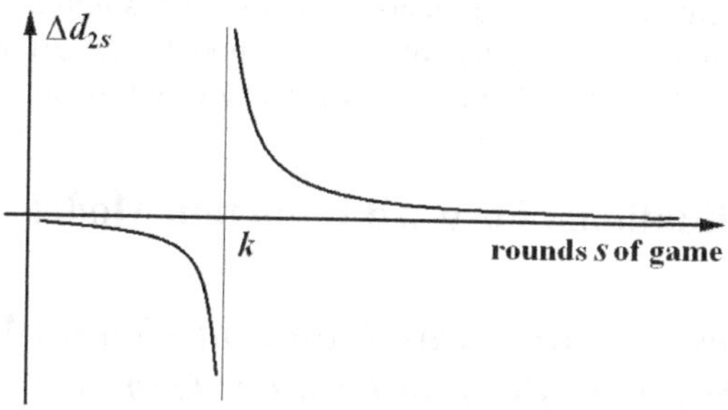

Fig. 8.2 The Δd_{2s} curve of increasing rate for $\gamma_{20} = 0.3$

Our duopolistic strategic output-making model under limited knowledge and limited rationality is essentially different of the classical Cournot model. The difference is mainly reflected in the following four aspects.

(1) The game players' differences in their rationality and knowledge. The classical Cournot model assumes that the players have the complete and symmetric knowledge (or information) relevant to their decision-making and additionally, they are completely rational. In the contrary, our model is established on the assumption that the decision makers have only limited and asymmetric knowledge (information), while they are not entirely rational, either.

(2) The players have different purposes for playing the game. The classical Cournot model believes that the players' purpose of playing the output-making game is to maximize their current profits, while our model is established on the players' purpose of profit maximization of strategic output-making.

(3) There is a difference in time orders of the game play. The classical Cournot model assumes that both players make their decisions at the same time, while our model introduces an order along which the players make their dicisions one after another.

(4) There are differences in the model structures. The classical Cournot model possesses such a structure of optimization models that do not

consider the players' dependence of their decisions on how the decisions are made. In our model, the players' current decisions are heavily dependent on the historical decision-making information; and on the assumption of maximizing the players' strategy profits, we constructed a descriptive model structure that can be widely employed in practical decision-making situations. This new model can be used to describe the actual strategic output making decision processes objectively existing among the dominating firms and other relatively minor firms of some industries. It has the ability to describe the actual scenarios where certain decision makers somehow acquired more information than some others who for reasons are situated in relatively difficult decisoion-making circumstances. Additionally, this model can surely describe the situations the classical Cournot model is initially developed for. It is because when the parameters of this new model take certain specific values, the model will be reduced into the classical Cournot model.

8.2 A New Situational Forward Induction Model

8.2.1 Weaknesses of Backward Induction, Central Mehod of Equilibrium Analysis for Dynamic Games

The method of backward induction of a dynamic game starts with the analysis of the last player's action in the final round. Then from that analysis it determines the choice of action of the previous player in the immediate previous period. Step by step, this method helps to analyze and determine the choice of action of each player in each specific period until reaching the very start of the game. It is a commonly employed method used in the study of game theory and game logic analysis. However, the validity of the backward induction depends on two assumptions. One is that of rationality, that is, every decision maker is rational; and the other that of same expectancy, that is, each player expects correctly how the other player would behave. These assumptions do not agree with the reality of limited rationality and limited ability of making predictions. They can hardly be satisfied in practical applications. Because of these obvious weaknesses, this method has also been criticized. To this end, the most famous paradox is the Centipede Game (Rosenthal, 1981).

What this paradox reveals is the problem existing in the backward induction method that focuses on microscopic logical reasoning while ignoring the underlying holistic, macroscopic thinking. In other words, it places the sole emphasis on the current interest while ignoring the long-term possibilities. In this section, we will employ the philosophical concept of wholeness of the systems theory and the method of holistic thinking and analyzing events and matters to design a new method of "situational" backward induction. By employing this new method, one can conveniently and efficiently compute for the Nash equilibrium solutions of multistage dynamic games.

8.2.2 Backward Derivation of Multi-Stage Dynamic Games' Profits

For a given multistage dynamic game, the first step of designing this new "situational" backward induction method is to determine subgames and backward derivation. For the sake of convenience of our presentation, let us first look at the following three definitions.

Definition 8.1. In a dynamic game, if the subsequent stages of a stage, except the first one, have a set of initial information and all the required information to form a partial game, then the game problem of this part is referred to as a subgame of the original game.

Definition 8.2. In a dynamic game, if there is a subgame B within a given subgame A, then A is referred to as an *upper-layer subgame* of B, while B a *lower-layer subgame* of A. If no subgame can be further extracted from a subgame B, then subgame B is referred to as a *lowest-layer subgame*.

One obvious characteristic of lowest-layer subgames is that there is no longer any subgame of lower layer, and that under any strategy there exists a profict vector of the profits of all players.

Definition 8.3. If a subgame A does have a subgame B, then any strategy in A that directs to B is referred to as an *exogenous conditional strategy* of B, or a *guide strategy* (for lower-layer subgames), and B is referred to as a *subgame under the exogenous (guide) strategy*.

Example 8.5. Figure 8.3 depicts Rosenthal's centipede game (Rosenthal, 1981). Divide this game into subgames, point out the layer-realtionships between the subgames, and list the relevant exogenous (guide) strategies.

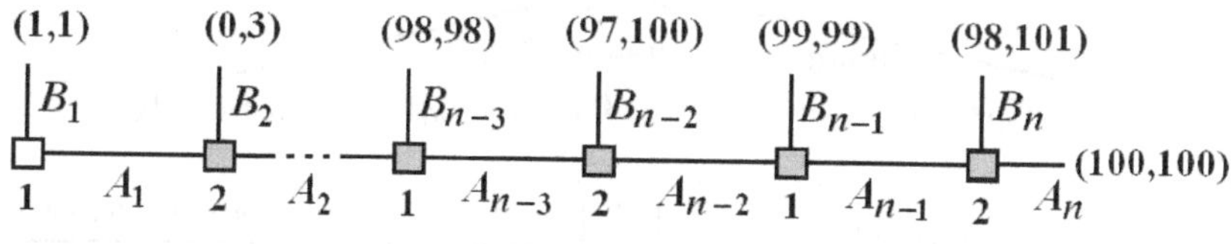

Fig. 8.3 Rosenthal centipede game (Rosenthal, 1981)

Solution. Assume that this game has a total of n stages. Then the lowest-layer subgame contains player 2's choice between strategies A_n and B_n. The immediate upper-layer subgame consists of player 1's choices between strategies A_{n-1} and

B_{n-1}, where A_{n-1} is the exogenous conditional strategy of the lowest-layer subgame. From our reasoning along this line, it can be seen that this overall centipede game consists of many subgames with upper- and lower-layer relationships, where A_i, $i=1,2,\cdots,n$, is the respective exogenous (guide) strategy of the immediate lower-stage subgame.

With the previous three definitions in place, what follows provides us a "situational" backward induction method for multistage games.

Proposition 8.1. Given a dynamic game with K players, one can employ the following "situational" backward induction scheme to solve this game problem.

Step 1: Apply Definitions 8.1 and 8.2 to determine all the subgames.

Step 2: Let v_{ij} be the gain of palyer i under strategy j in the lowest-layer subgame, i = 1, 2, …, K, j = 1, 2, …, S. Define $m_i = \min\{v_{i1}, v_{i2}, \cdots, v_{iS},\}$ and $M_i = \max\{v_{i1}, v_{i2}, \cdots, v_{iS},\}$ so that grey interval numbers $[m_i, M_i]$, $i=1,2,\cdots,K$, are obtained, Figure 8.4.

Step 3: After the prossessing of Step 2 above, the original lowest-layer subgames are simplified, and the original immediate upper-layer subgames where the exogenous conditions exist become new lowest-layer subgames. Repeat Step 2 until all subgames are simplified except the upper most layer game is left.

Step 4: Finish.

The method, described in Proposition 8.1 for multistage dynamic games, starts with the lowest-layer subgames until reaching the upper most layer game. Such a sequence of analysis is exactly the opposite to the order of the game rounds. That is why it is named as a *situational backward induction*.

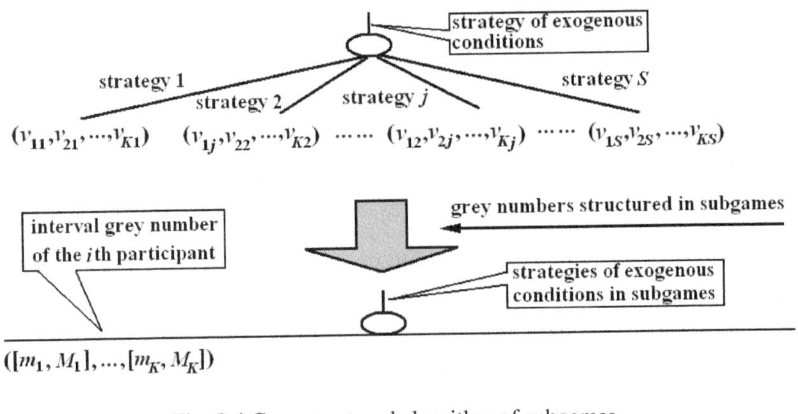

Fig. 8.4 Grey structured algorithm of subgames

Example 8.6. Let us look at how to determine and simplify all the subgames of the centipede game in the previous example.

Solution. We first divide the original centipede game into $(n-1)$ subgames, denoted by G_i, $i = 2, 3, \cdots, n$. Now, let us apply the computational scheme as outlined in Proposition 8.1 to simplify the subgames. The specific simplication processes are shown in Figure 8.5, where the possible future gains of the guide strategies A_1, A_2, …, A_{n-1}, A_n are respectively given as follows: G_0: {[0,100], [1,101]}, G_1: {[0,100], [3,101]}, …, G_{n-3}: {[97,100], [99,101]}, G_{n-2}: {[98,100], [99,101]}, G_{n-1}: {[98,100], [100,101]}.

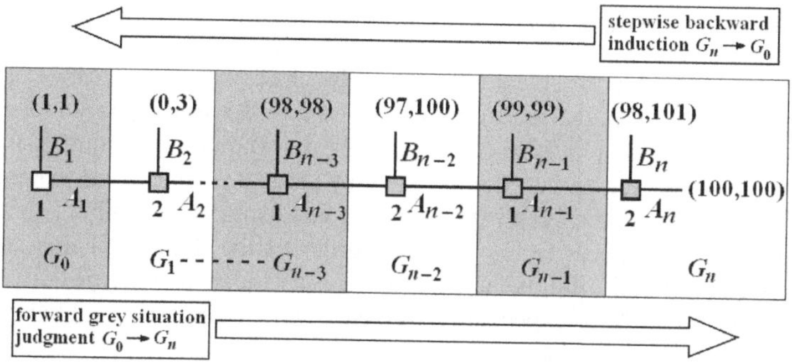

Fig. 8.5 Flow chart of backward induction for analyzing situations

In Proposition 8.1, the situational backward induction applied on a multistage dynamic game constitutes a new form of representation for the structure of dynamic games, see Figure 8.5 for more details.

Definition 8.4. For a multistage dynamic game, if each guide strategy is labeled with possible future gains, then each of the future gains is referred to as the *guide value* of the lower-layer subgame. This structural form of the game is referred to as that of the multistage dynamic game with guide values, or simply the *guide value structure*.

By using the guide value structure of a dynamic game, the players can compare and balance the current gains and possible future gains when they make decisions at any of the time moments. This fact alters the situation of the classical backward induction where at each chosen stage of the game, the current gain can only be compared to that of the next stage, known as *stepwise rationality*. Under certain specific circumstances, such as that of the centipede game, only the worst (irrational) results could be produced. The conventional explanation for this economic phenomenon is the contradiction between individual rationality and group rationality. However, this explanation only reflects one side of the issue without touching on the much deeper and more foundamental problem: There are conflicts and equilibria between the short-term (the current) and long-term (the future) profits (interest) of the players.

8.2.3 Termination of Forward Induction of Multi-Stage Dynamic Games and Guide Nash Equilibrium Analysis

For a game process similar to that of the centipede game, if the stages of the game are clear and few in numbers, people will likely play the game in the way how the centipede game is played. When a game has many stages or does not have a clear structure, people tend to play the game by comparing the current gains with the possible futures gains. To this end, let us first introduce the four assumptions used in the situational forward induction method.

Assumption 1: For a given multistage dynamic game, one can employ the situational backward induction from one step to the one immediately in the front.

Assumption 2: For a multistage dynamic game, when the structure or the future gains are not clear, the players can obtain grey estimates for the current and future gains. In other words, they can express their gains using grey numbers.

Assumption 3: Players follow the time order of the dynamic game and make decisions using certain judgment criteria, such as the potentials or cores of grey numbers, or probabilistic expected values, etc., while comparing and balancing the current and future gains; and

Assumption 4: All other necessary assumptions for game reasoning.

Based on these 4 assumptions, one can analyze the equilibrium solution of a multistage dynamic game along the time order.

Definition 8.5. Given the guide value structure of an arbitrary multistage dynamic game, if a strategic analysis is made along the time order of the game, then this method of equilibrium analysis of the game is referred to as a *situational forward induction*.

Along with this concept of situational forward induction of multistage dynamic games, there is the need to introduce the concepts of termination and guide Nash equilibria.

Definition 8.6. If the occurrence of the equilibrium in the current game at a non-guide value leads to termination of the game, then this game's equilibrium is referred to as a *terminating Nash equilibrium*. If the occurrence of the equilibrium at a guide value does not interrupt the ongoing process of the game, then this game's equilibrium is referred to as a *guide Nash equilibrium*.

Proposition 8.2. If what is actualized by the current player is a terminating Nach equilibrium, then the entire game process ends; if what is actualized is a guide Nash equilibrium, then the current game will be led to the process of a lower layer subgame.

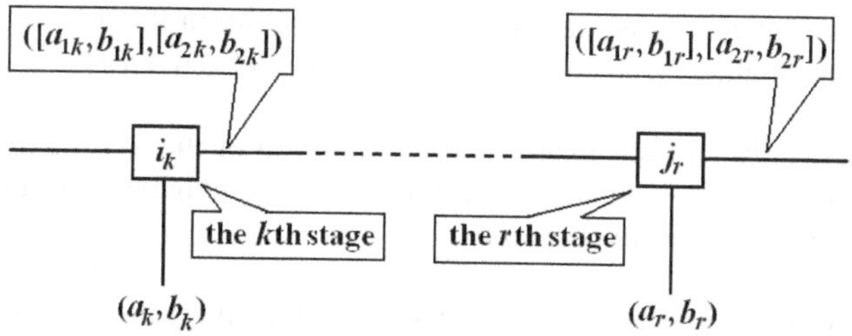

Fig. 8.6 Termination Nash equilibrium analysis based on guide value structure of dynamic games

Proposition 8.3. If the current game player compares the current and future gains to make his decision according to certain judgment criteria, then there must be a terminating Nash equilibrium under the set of judgment criteria.

Proof. For a given multistage dynamic game of two players, according to Proposition 8.1 let us conduct a situational backward induction from one stage to another while labeling the future gains using grey numbers, as shown in Figure 8.6, where the first and second gains are respectively those of player i and j, so that they can be used to compare with the current gains.

In Figure 8.6, assume that player i, $i=1,2$, is the current decision maker at stage k. Then, he will compare the current gain a_k of stage k with the grey number interval $\left[a_{1k}, b_{1k}\right]$ of the possible future gain. Assume that this player cannot acquire any information regarding the eventual gain value in this interval grey number and relevant distribution, then he will make his decision based on the following rules:

(1) When $a_k > \frac{a_{1k}+b_{1k}}{2}$, from the angle of orders of grey potentials, we have $a_k > \left[a_{1k}, b_{1k}\right]$. So, $\left(a_k, b_k\right)$ is a terminating Nash equilibrium of this game. And,

(2) When $a_k < \frac{a_{1k}+b_{1k}}{2}$, again from the angle of orders of grey potentials, we have $a_k < \left[a_{1k}, b_{1k}\right]$. So, $\left(\left[a_{1k}, b_{1k}\right], \left[a_{2k}, b_{2k}\right]\right)$ is the guide Nash equilibrium of the current stage of the game.

If what is actualized in the current game is a guide Nash equilibrium, according to Proposition 8.2, then the current game will be led into the process of a lower-layer subgame.

Continuing this reasoning and derivation, eventually in the process of a certain subgame a terminating Nash equilibrium will be actualized. QED.

Example 8.7. Solve for terminating Nash equilibria of the Rosenthal's centipede game, see Figure 8.5 for details, by using the rule of judgment of either the grey potentials or probabilistic expectations.

Solution. (1) Let us first look at how to solve a dynamic game for its terminating Nash equilibria using the judgment rule of grey potentials.

In the first stage in Figure 8.5, player 1 makes the first decision. He compares the gain 1 and the grey potential of $[0,100]$ of the guide strategy A_1 and the terminating strategy B_1 . Evidently, the former is smaller than the latter. So, player 1 chooses the guide strategy A_1 to actualize a guide Nash equilibrium and turns the decision right of the game to player 2.

We continue to reason similarly until the game proceeds to the process of subgame G_{n-2}, where player 2 is the current decision maker, whose gains of the guide strategy A_{n-2} and the terminating strategy B_{n-2} are respectively $[99,101]$ and 100. The grey potentials of these two grey numbers are the same. That is to say, using this judgment rule, strategies A_{n-2} and B_{n-2} are identical for player 2. So, he can either choose the terminating strategy B_{n-2} to end the game, to collect his gain of 100, while player 1 obtains his gain of 97, and to actualize the terminating Nash equilibrium, or choose strategy A_{n-2} by once again turning the decision right of the game to player 1, while realizing a guide Nash equilibrium.

Similarly, it can be judged that in the period of subgame G_{n-1} for player 1, the terminating Nash equilibrium strategy B_{n-1} is not any different from the guide Nash equilibrium A_{n-1}. If strategy B_{n-1} is taken, then both players materialize their gains of 99. If strategy A_{n-1} is actualized to guide the game into the process of the lower-layer subgame, then the decision right is turned over to player 2.

In the stage G_n, it stands for a determined subgame where player 2 chooses the terminating strategy B_n, which realizes a terminating Nash equilibrium, so that the individual gains of players 1 and 2 are respectively 98 and 101.

Summarizing what is discussed above, under the judgment rule of grey potentials, this game contains three potential terminating Nash equilibria, whose subgame processes and gains of players 1 and 2 are respectively G_{n-2} (97, 100), G_{n-1} (99, 99), and G_n (98, 101).

(2) Next, let us first look at how to solve a dynamic game for its terminating Nash equilibria using the judgment rule of probabilistic expectations.

As a matter of fact, in this game all the interval grey numbers of the possible future gains of the guide strategies are discrete type with certain fixed probabilities. For instance, for the *k*th stage, the probability for a possible value in the interval grey number of the guide value to appear can be estimated by using its frequency of appearance in the future $n-k$ stages.

After having known the respective probability distributions of the interval grey numbers of the guide values, the process of determining the terminating and guide Nash equilibria of each subgame using the expected gains of the strategies becomes a simple matter. Evidently, in the early subgames in Figure 8.5, the equilibria, obtained by using the judgment rule of expected values, are the guide Nash equilibria. In the following, we will only analyze the Nash equilibria for the last a few subgames.

In the stage of subgame G_{n-3}, the decision maker is player 1. The interval grey number of his guide value is [97, 100], where the probability for each of 97, 98, 99, and 100 to appear (not including the current stage) is 0.25. So, the expectation of the guide value is (97 + 98 + 99 + 100) / 4 = 98.5. Since the gain from terminating the game is 98, which is smaller than the expected guide value, player 1 in this case would choose the guide Nash equilibrium and turn the decision right to player 2.

In the stage of subgame G_{n-2}, player 2 is the decision maker. The interval grey number of his guide value is [99, 101], where the probability for each of 99, 100, and 101 to appear (not including the current stage) is 1/3. So, the expectation of the guide value is (99 + 100 + 101) / 3 = 100, which is the same as the gain of terminating the game. So, for player 2, there is no difference for him to choose either the terminating or the guide Nash equilibrium. If he does choose strategy B_{n-2}, then he obtains a terminating Nash equilibrium. in this case, the gains of the players 1 and 2 are respectively 97 and 100. If he pciks strategy A_{n-2}, then he turns the decision right to player 1.

In the stage of subgame G_{n-1}, player 1 is the decision maker. The interval grey number of his guide value is [98, 100], where the probability for each of 98 and 101 to appear (not including the current stage) is 0.5. So, the expectation of the guide value is (98 + 100) / 2 = 99, which is the same as the gain of terminating the game. So, in this case, for player 1 there is no difference between the terminating and guide Nash equilibria. If he chooses strategy B_{n-1}, then he obtains the terminating Nash equilibrium with 99 as the gain for either players. If he picks strategy A_{n-1}, then he turns the decision right to player 2.

In the stage of subgame G_n, because it is the lowest layer subgame, it is easy to see that player 2 would take strategy B_n to end the game. So, the gains of players 1 and 2 are respectively 98 and 101.

Summarizing what is discussed above, under the current judgment rule, this game contains three possible terminating Nash equilibria with their specific processes and gains as follows: G_{n-2} (97, 100), G_{n-1} (99, 99), and G_n (98, 101). That is, we obtained the identical results as in (1) above.

(3) Let us now look at how the classical backward induction works for this game. In the stage of subgame G_{n-1}, if player 1 can definitely determine that in subgame G_n, player 2 will choose strategy B_n, then he will choose strategy B_{n-1} in this stage. If we reason like this stage by stage, then player 1 would simply choose strategy B_1 in the first subgame G_1 to end the game. In this case, the gain of either players is 1.

This analysis indicates that the classical backward induction is a special case of our method with particular probabilities assigned to the gains, where player 1 chooses the terminating Nash equilibrium in the very first stage of the game.

8.3 Chain Structure Model of Evolutionary Games of Industrial Agglomerations and Its Stability

This section studies the evolutionary game problem of industrial agglomerations using grey game theory. This work generalizes the evolutionary games of industrial agglomerations to the area of games with uncertain gains. After introducing the concept of grey gains and losses, we investigate firms' learning mechanism and behavior in the process of forming industrial agglomerations so that the realization process and relevant mechanism of their equlibria can be explored meaningfully.

8.3.1 Chained Evolutionary Game Model for the Development of Industrial Agglomerations

(1) The leaning mechanism of firms

Learning is a way that spreads knowledge. Transfer of knowledge requires all the participants to interact with each other consistently over periods of time so that actual transfer of knowledge occurs between business entities.

(2) Development of model

Given the game problem of an industrial agglomeration, assume that the strategy options the firms can choose include either to agglomerate or not to. All the player firms in the game respond (not necessarily optimal responses) to the current game situation. They either learn from each other, or minic the optimal game strategis of the successful firms. If an evolutionary game is constructed in such a way that all the relevant firms are grouped into two sets, labeled as "agglomerate" and "not agglomerate," respectively, along with arrowed curves that are labeled with the relevant proportion of firms that change their positions and their gains materialized from changing their positions, where the arrow heads stand for the new strategy the firms are taking, while the arrow tails the original stragey the firms took, we obtain the chain structure model of the evolutionary game of the industrial agglomeration, see Figure 8.7. For the sake of convenience of discussion, this section looks at a specific evolutionary game of industrial agglomeration involving uncertain gains. The general form of the matrix of gains and losses is given in Table 8.5.

Table 8.5 The gain and loss matrix of an industrial agglomeration

Group of firms-2		Agglomerate	Not agglomerate
Groupg of firms-1	Agglomerate	$\otimes_1, \otimes_1$	$\otimes_2, \otimes_3$
	Not agglomerate	$\otimes_3, \otimes_2$	$\otimes_4, \otimes_4$

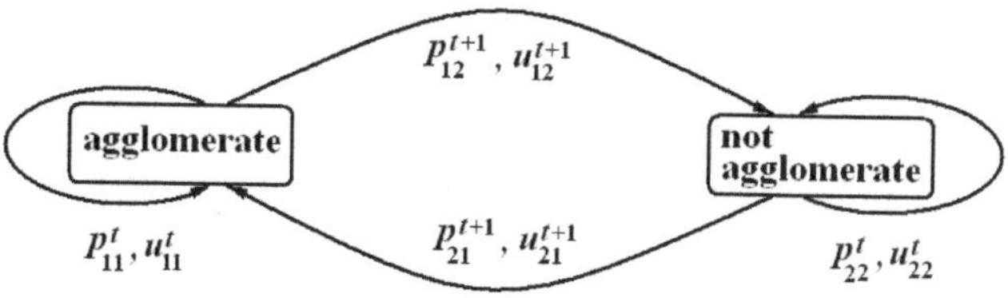

Fig. 8.7 Evolutionary game chain model for the development of industrial agglomeration

Assume that at time moment t the proportion x of the firms choose to agglomerate, and the proportion $1-x$ of the firms choose not to agglomerate. According to Figure 8.7, the state of change of the game from time moment t to t + 2, either maintain the original strategies or minic and learn the different strategies, can be seen.

$$\begin{aligned} &p_{11}^t = x \\ &u_{11}^t = x \cdot \otimes_1 + (1-x) \cdot \otimes_2 = \\ &x \cdot (a_1 + (b_1 - a_1)\gamma_1) + (1-x) \cdot (a_2 + (b_2 - a_2)\gamma_2) \end{aligned} \tag{8.27}$$

$$\begin{aligned} &p_{22}^t = 1-x \\ &u_{22}^t = x \cdot \otimes_3 + (1-x) \cdot \otimes_4 = \\ &x \cdot (a_3 + (b_3 - a_3)\gamma_3) + (1-x) \cdot (a_4 + (b_4 - a_4)\gamma_4) \end{aligned} \tag{8.28}$$

$$\bar{u}_t(\otimes) = x \cdot u_{11}^t + (1-x) \cdot u_{22}^t, \text{ where } 0 \le \gamma_i \le 1, \tag{8.29}$$

$$p_{12}^{t+1} = p_{11}^{t} \cdot \frac{\bar{u}_t(\otimes) - u_{11}^{t}}{\left|\bar{u}_t(\otimes)\right| + \left|u_{11}^{t}\right|} \tag{8.30}$$
$$u_{12}^{t+1} = u_{22}^{t+1}$$

$$p_{21}^{t+1} = p_{22}^{t} \cdot \frac{\bar{u}_t(\otimes) - u_{22}^{t}}{\left|\bar{u}_t(\otimes)\right| + \left|u_{22}^{t}\right|} \tag{8.31}$$
$$u_{21}^{t+1} = u_{11}^{t+1}$$

$$\begin{aligned} &p_{11}^{t+1} = p_{11}^{t} + p_{21}^{t+1} \\ &u_{11}^{t+1} = p_{11}^{t+1} \cdot \otimes_1 + \left(1 - p_{11}^{t+1}\right) \cdot \otimes_2 = \\ &p_{11}^{t+1} \cdot (a_1 + (b_1 - a_1)\gamma_1) + \left(1 - p_{11}^{t+1}\right) \cdot (a_2 + (b_2 - a_2)\gamma_2) \end{aligned} \tag{8.32}$$

$$\begin{aligned} &p_{22}^{t+1} = 1 - p_{11}^{t+1} = p_{22}^{t} + p_{12}^{t+1} \\ &u_{22}^{t+1} = p_{11}^{t+1} \cdot \otimes_3 + \left(1 - p_{11}^{t+1}\right) \cdot \otimes_4 = \\ &p_{11}^{t+1} \cdot (a_3 + (b_3 - a_3)\gamma_3) + \left(1 - p_{11}^{t+1}\right) \cdot (a_4 + (b_4 - a_4)\gamma_4) \end{aligned} \tag{8.33}$$

$$\bar{u}_{t+1}(\otimes) = p_{11}^{t+1} \cdot u_{11}^{t+1} + p_{22}^{t+1} \cdot u_{22}^{t+1} \tag{8.33a}$$

$$p_{12}^{t+2} = p_{11}^{t+1} \cdot \frac{\bar{u}_{t+1}(\otimes) - u_{11}^{t+1}}{\left|\bar{u}_{t+1}(\otimes)\right| + \left|u_{11}^{t+1}\right|} \tag{8.34}$$
$$u_{12}^{t+2} = u_{22}^{t+2}$$

$$p_{21}^{t+2} = p_{22}^{t+1} \cdot \frac{\bar{u}_{t+1}(\otimes) - u_{22}^{t+1}}{\left|\bar{u}_{t+1}(\otimes)\right| + \left|u_{22}^{t+1}\right|} \tag{8.35}$$
$$u_{21}^{t+2} = u_{11}^{t+2}$$

By letting $p_{21}^{t+1} = 0$, we can obtain the equilibrium solution as follows:

$$\begin{aligned} &p_{21}^{t+1} = p_{22}^{t} \cdot \frac{\bar{u}_t(\otimes) - u_{22}^{t}}{\left|\bar{u}_t(\otimes)\right| + \left|u_{22}^{t}\right|} = 0, \\ &x_1 = 0, x_2 = 1, \\ &x_3 = \frac{a_4 + (b_4 - a_4)\gamma_4 - a_2 - (b_2 - a_2)\gamma_2}{a_1 + (b_1 - a_1)\gamma_1 - a_3 - (b_3 - a_3)\gamma_3 + a_4 + (b_4 - a_4)\gamma_4 - a_2 - (b_2 - a_2)\gamma_2} \end{aligned} \tag{8.36}$$

Its stable region is shown in Figure 8.8.

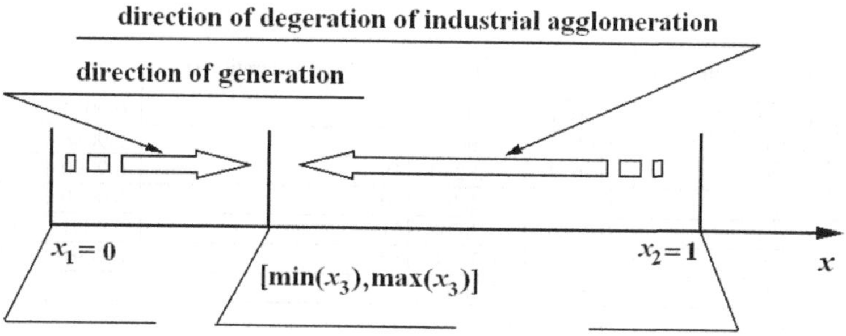

Fig. 8.8 Stable interval of of an industrial agglomeration

8.3.2 Duplicated Dynamic Simulation for the Development Process of Industrial Agglomerations

According to the relevant transfer formulas contained in the evolutionary game model of industrial agglomerations, and considering the specific example given in Table 8.6, we employ Matlab program package to compose our software procedures in order to conduct simulation experiements for duplicating the development dynamics of an industrial agglomeration.

Table 8.6 The gain-loss matrix of an industrial agglomeration

Group of firms-2		Agglomerate	Not agglomerate
Group of firms-1	Agglomerate	$[0.5,1.5],[0.5,1.5]$	0, 2
	Not agglomerate	2, 0	$[-5,-4],[-5,-4]$

From Figure 8.8, it follows that x_1, x_2, x_3 divide the entire range of values into three portions. So, when we simulate this problem, we can take the initial values in the range $[x_1, \min(x_3)]$ or $[\max(x_3), x_2]$. The specific simulations are given in Tables 8.7 – 8.10.

Table 8.7 The industrial agglomeration development $(\min(x_3))$, when the initial value is $p_{11}^0 = 0.0001$

Evolution generation $(t+i)$	p_{11}^{t+i}	$\overline{u}_{t+i}$	Evolution generation $(t+i)$	p_{11}^{t+i}	$\overline{u}_{t+i}$
Initial state (0)	0.0001	-3.9990	20	0.7274	0.3639
5	0.0032	-3.9681	22	0.7273	0.3637
10	0.1024	-3.0337	23	0.7273	0.3636
15	0.7232	0.3554	25	0.7273	0.3636

Table 8.8 The industrial agglomeration development $(\max(x_3))$, when the initial value is $p_{11}^0 = 0.0001$

Evolution generation $(t + i)$	p_{11}^{t+i}	$\overline{u}_{t+i}$	Evolution generation $(t + i)$	p_{11}^{t+i}	$\overline{u}_{t+i}$
Initial state (0)	0.0001	-4.9988	35	0.9084	1.3622
5	0.0032	-4.9617	40	0.9088	1.3631
10	0.1024	-3.8289	45	0.9090	1.3634
15	0.8666	1.2688	50	0.9090	1.3635
20	0.8966	1.3379	51	0.9091	1.3636
25	0.9046	1.3544	52	0.9091	1.3636
30	0.9073	1.3601			

Table 8.9 The industrial agglomeration development $(\min(x_3))$, when the initial value is $p_{11}^0 = 0.9999$

Evolution generation $(t + i)$	p_{11}^{t+i}	$\overline{u}_{t+i}$	Evolution generation $(t + i)$	p_{11}^{t+i}	$\overline{u}_{t+i}$
Initial state (0)	0.9999	0.5001	15	0.7273	0.3638
1	0.9998	0.5002	16	0.7272	0.3636
5	0.9904	0.5091	17	0.7273	0.3637
10	0.7255	0.3600	18	0.7273	0.3636
14	0.7272	0.3634	19	0.7273	0.3636

Table 8.10 The industrial agglomeration development $(\max(x_3))$, when the initial value is $p_{11}^0 = 0.9999$

Evolution generation $(t + i)$	p_{11}^{t+i}	$\overline{u}_{t+i}$	Evolution generation $(t + i)$	p_{11}^{t+i}	$\overline{u}_{t+i}$
Initial state (0)	0.9999	1.4999	45	0.9492	1.4350
1	0.9999	1.4999	50	0.9320	1.4065
5	0.9998	1.4998	55	0.9202	1.3853
10	0.9995	1.4995	60	0.9140	1.3733
15	0.9990	1.4990	65	0.9111	1.3677
20	0.9979	1.4978	70	0.9099	1.3653
25	0.9955	1.4954	75	0.9094	1.3643
30	0.9907	1.4902	80	0.9092	1.3639
35	0.9818	1.4800	83	0.9091	1.3638
40	0.9675	1.4617	84	0.9091	1.3638

From these simulation results, it follows that within this evolutionary game of industrial agglomeration, there is a unique evolutionarily stable strategy grey equilibrium interval [min(x_3), max(x_3)] = [0.7273, 0.9091]. And, when $\gamma_1 = 1, \gamma_4 = 0$,

$$\min(x_3) = \frac{-5+1-0}{-5+1-0+0.5-2} = 0.7273;$$

and when $\gamma_1 = 0, \gamma_4 = 1$,

$$\max(x_3) = \frac{-5-0}{-5-0+0.5+1-2} = 0.9091 \cdot$$

8.3.3 *Stability Analysis for the Formation and Development of Industrial Agglomerations*

Let $x(t)$ be the the number of firms of a region that have agglomerated at time moment t. If we ignore the influence of all external factors, such as recourses, etc., then the rate of increase in the number of agglomerated firms at moment t is only related to $x(t)$. That is,

$$\frac{dx(t)}{dt} = x' = \alpha x(t), \ \alpha \text{ a constant.} \tag{8.37}$$

Solving this equation leads to

$$x(t) = x(0)e^{\alpha t} \tag{8.38}$$

This expression indicates that under the condition that there is an infinite supply of recources, the number of agglomerated firms will grow exponentially and indefinitely. However, for any given region, its available recource has to be limited. So, the growth in the number of agglomerated firms has to be affected by the external conditions, such as recources, and bounded with a maximum possible ceiling. Therefore, let us modify equ. (8.37) as follows. Assume that the maximum number of agglermated firms this specific region could afford to support is k, $k \in [k_1, k_2]$, and $\frac{x(t)}{k}$ stands for the degree of saturation at moment t, where the rate of increase in the number of agglomerated firms is negatively proportional to the degree of saturation. Then the modified formula is

$$\frac{dx(t)}{dt} = x' = \alpha x(t) - \alpha x(t) \cdot \frac{x(t)}{k} = \alpha x(t)(1 - \frac{x(t)}{k}) \tag{8.39}$$

Soliving this equation provides

$$x(t) = \frac{kx(0)e^{\alpha t}}{1 + x(0)e^{\alpha t}} \tag{8.40}$$

So, when $x(0) < k$, $\lim_{t \to \infty} x(t) = k$, meaning that the number of agglomerated firms increases monotomically. For instance, if we choose $x(0) = 0.0001$, it stands for a dynamic process for the development and formation of an industrial agglomeration. During this period of time, the number of agglomerated firms is smaller than the regionally affordable maximum so that there is still space for further growth and more firms will choose to join the agglomeration.

When $x(0) = k$, the initial value is within the equilibrium area of the dynamic process of formation and development of an industrial agglomeration. During this period of time, the number of agglomerated firms stays in a dynamically stable state and the competition between the regional firms are relatively mild.

When $x(0) > k$, $\lim_{t \to \infty} x(t) = k$ and the number of agglomerated firms decreases. If we choose the initial value $x(0) = 0.9999$, it stands for a dynamic duplication process of the formation and development of an industrial agglomeration. During this period of time, the number of agglomerated firms has already gone beyond the region's affordability so that vicious competitions over the limited rcources appear between business entites, causing the number of agglomerated firms to drop.

All the previous three scenarios are respectively depicted by the three curves in Figure 8.9

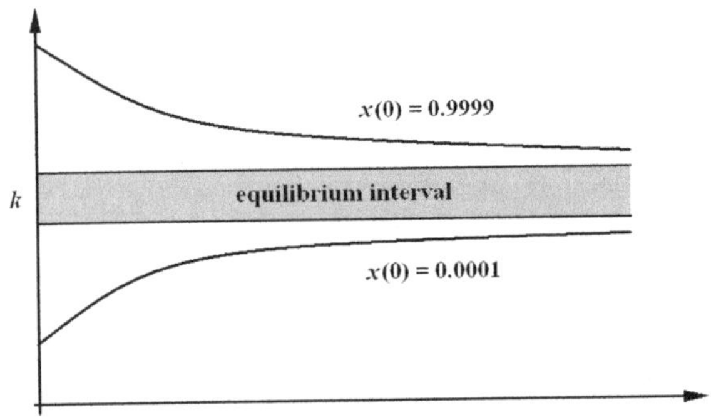

Fig. 8.9 Stability analysis of industrial agglomerations

Due to a lack of information on the system's elements (parameters), structure, boundary, the form of movement, etc., plus the effects of various random and non-random factors, in any game problem of the real world, the decision makers cannot predict accurately about the actual effects of their chosen strategies. However, based on the past experience, they can relatively reliably determine the ranges of the effects. Therefore, problems of grey games exist commonly in the physical world.

Grey game theory is a generalization of the conventional game theory in the area of partial information or limited knowledge. Essentially, the studies of grey games relax the assumption of complete knowledge (information) of the conventional game theory. So, it is not a simpe matter to generalize results of the conventiona game theory to the studies of grey games. On the contrary, solving grey games is much more difficult and complicated than solving traditional games with many more additional factors involved.

Chapter 9
Grey Control Systems

As a scientific concept, the so-called control stands for a special effect a controlling device exerts on controlled equipment. It is a purposeful and selective dynamic activity. A control system contains at least three parts, including a controlling device, controlled equipment, and an information path. A control system that is made up of only these three parts is known as an *open loop control system*, as shown in Figure 9.1. Each open loop control system is quite elementary, in which the input directly controls the output, with the fatal weakness that it does not have any resistance against disturbances.

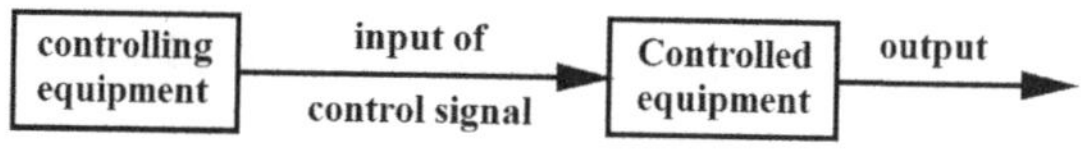

Fig. 9.1

A control system with a feedback return is known as a *closed loop control system*, as shown in Figure 9.2. The closed loop control system materializes its control through the combined effect of the input and the feedback of the output. One of the outstanding characteristics of closed loop systems is its strong ability to assist disturbances with their outputs constantly vibrating around the pre-determined objectives. So, closed loop control systems possess a certain kind of stability.

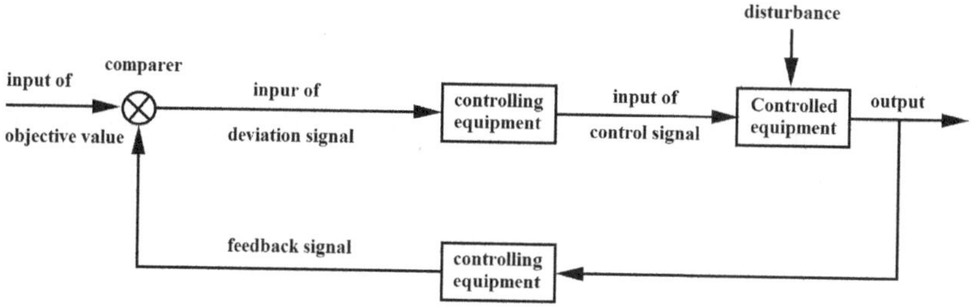

Fig. 9.2

A *grey control system* stands for such a system whose control information is only partially known, and is ordinarily known as a grey system for short. The control of grey systems is different of that of the general white systems. It is mainly because of the existence of grey elements in the systems of concern. Under such conditions, one first needs to understand the possible connection between the systems' behaviors and the parametric matrices of the grey elements, how the systems' dynamics differ from one moment to the next, in particular, how to obtain a white control function to alter the characteristics of the systems and to materialize control of the process of change of the systems. *Grey control* contains not only the general situation of systems involving grey parameters, but also the construction of controls based on grey systems analysis, modeling, prediction, and decision-making. The thinking of grey control can deeply reveal the essence underneath the problems of interest and help materialize the purpose of control.

9.1 Controllability and Observability of Grey Systems

The concepts of controllability and observability are two fundamental structural characteristics of systems seen from the angle of control and observation. This section focuses on the problems of the controllability and observability of grey linear systems.

Assume that $U=[u_1,u_2,\cdots,u_s]^T$ is a control vector, $X=[x_1,x_2,\cdots,x_n]^T$ a state vector, and $Y=[y_1,y_2,\cdots,y_m]^T$ the output vector. Then

$$\begin{cases} \dot{X} = A(\otimes)X + B(\otimes)U \\ Y = C(\otimes)X \end{cases} \tag{9.1}$$

is known as the *mathematical model* of a grey linear control system, where $A(\otimes)\in G^{n\times n}$, $B(\otimes)\in G^{n\times s}$, $C(\otimes)\in G^{m\times n}$. Correspondingly, $A(\otimes)$ is known as the *grey state matrix*, $B(\otimes)$ the *grey control matrix*, and $C(\otimes)$ the *grey output matrix*. In some studies, to emphasize that fact that *U, X*, and *Y* change with time, the dynamic characteristic of the system, we also respectively write the control vector, state vector, and the output vector as $U(t), X(t)$, and $Y(t)$.

The first group of equations

$$\dot{X}(t) = A(\otimes)X(t) + B(\otimes)U(t) \tag{9.2}$$

in the mathematical model of grey linear control systems in equ. (9.1) is known as the state equation, while the second group of equations

$$Y(t) = C(\otimes)X(t) \tag{9.3}$$

is known as the *output equation*.

For a given precision and an objective vector $J = [j_1, j_2, \cdots, j_m]^T$, if there are a controlling device and a control vector $U(t)$ such that through controlling the input the output of the system can reach the objective J while satisfying the required precision, then the system is said to be *controllable*.

For a given time moment t_0 and a pre-determined precision, if there is $t_1 \in (t_0, \infty)$ such that based on the system's output $Y(t)$, $t \in [t_0, t_1]$, one can measure the system's state $X(t)$ within the required precision, then the system is said to be *observable* within the time interval $[t_0, t_1]$. If for any t_0, t_1, the system is observable within the interval $[t_0, t_1]$, then the system is said to be observable.

According to the control theory, it follows that whether or not a grey system is controllable and observable is determined by whether or not the controllability matrix and the observability matrix, made up of $A(\otimes), B(\otimes)$, are of full rank. That is, the following result holds true.

Theorem 9.1. For the system in equ. (9.1), define

$$L(\otimes) = [B(\otimes) \quad A(\otimes)B(\otimes) \quad A^2(\otimes)B(\otimes) \quad \cdots \quad A^{n-1}(\otimes)B(\otimes)]^T$$

$$D(\otimes) = [C(\otimes) \quad C(\otimes)A(\otimes) \quad C(\otimes)A^2(\otimes) \quad \cdots \quad C(\otimes)A^{n-1}(\otimes)]^T$$

Then the following hold true.

(1) When $\text{rank}(L(\otimes)) = n$, the system is controllable, and

(2) When $\text{rank}(D(\otimes)) = n$, the system is observable.

Based on this result, the following four theorems can be established.

Theorem 9.2. For the system in equ. (9.1), if the grey control matrix $B(\otimes) \in G^{n\times n}$ satisfies $B(\otimes)$ = diag[$\otimes_{11}$, $\otimes_{22}$, …, $\otimes_{nn}$], where each grey entry along the diagonal is non-zero, then the system is controllable.

Theorem 9.3. For the system in equ. (9.1), if the grey output matrix $C(\otimes) \in G^{n\times n}$ satisfies $C(\otimes)$ = diag[$\otimes_{11}$, $\otimes_{22}$, …, $\otimes_{nn}$], where each grey entry along the diagonal is non-zero, then the system is observable.

Theorem 9.4. For the system in equ. (9.1), if the control matrix $B(\otimes) \in G^{n\times n}$ satisfies $B(\otimes)$ = diag[$\otimes_{11}, \otimes_{22}$, …, $\otimes_{mm}$, 0, …, 0] with $\text{rank}B(\otimes) = m < n$, and the grey state matrix $A(\otimes)_{n\times n}$ = diag[0, …, 0, $\otimes_{m+1,1}, \otimes_{m+2,2}, \ldots, \otimes_{n,n-m}$] with $\text{rank}A(\otimes) = n - m < n$, then the system is controllable.

Theorem 9.5. For the system in equ. (9.1), if the grey output matrix $C(\otimes) \in G^{m\times n}$ satisfies $C(\otimes)$ = diag[$\otimes_{11}, \otimes_{22}, \ldots, \otimes_{mm}$] with rank$C(\otimes) = m < n$ and the grey state matrix

$$A(\otimes) = \begin{pmatrix} 0 & \cdots & 0 & \otimes_{1,m+1} & 0 & \cdots & 0 \\ 0 & \cdots & 0 & 0 & \otimes_{2,m+2} & \cdots & 0 \\ \vdots & \cdots & \vdots & \vdots & \vdots & \cdots & \vdots \\ 0 & \cdots & 0 & 0 & 0 & \cdots & \otimes_{n-m,n} \\ 0 & \cdots & 0 & 0 & 0 & \cdots & 0 \\ \vdots & \cdots & \vdots & \vdots & \vdots & \cdots & \vdots \\ 0 & \cdots & 0 & 0 & 0 & \cdots & 0 \end{pmatrix}, \ \text{rank}A(\otimes) = n - m < \text{n},$$

then the system is observable.

9.2 Transfer Functions of Grey Systems

The concept of transfer functions stands for a fundamental relationship between the input and output of time invariant, linear grey control systems. Its rich connection with the expressions of the systems' state spaces can be well described by using the concepts of controllability and observability.

9.2.1 Grey Transfer Functions

Assume that the mathematical model of an *n*th order linear system with grey parameters is given as follows:

$$\otimes_n \frac{d^n x}{dt^n} + \otimes_{n-1} \frac{d^{n-1} x}{dt^{n-1}} + \cdots + \otimes_0 x = \otimes \cdot u(t) \tag{9.4}$$

After applying Laplace transform to both sides of this equation, we obtain

$$G(s) = \frac{X(s)}{U(s)} = \frac{\otimes}{\otimes_n s^n + \otimes_{n-1} s^{n-1} + \cdots \otimes_1 s + \otimes_0} \tag{9.5}$$

where $L(x(t)) = X(s)$ and $L(u(t)) = U(s)$. Equ. (9.5) is known as a *grey transfer function*, which is the ratio of the Laplace transform of the response $x(t)$ of the *n*th order grey linear control system and the Laplace transform of the driving term $u(t)$. In fact, the transfer function represents a fundamental relationship between the input and the output of a first order grey linear control system. From the following theorem, it follows that each *n*th order grey linear system can be reduced into an equivalent first order grey linear system.

Theorem 9.6. For an nth order grey linear system as shown in equ. (9.4), there is an equivalent first order grey linear system.

Proof. Assume that the given nth order grey linear system is

$$\otimes_n \frac{d^n x}{dt^n} + \otimes_{n-1} \frac{d^{n-1} x}{dt^{n-1}} + \cdots + \otimes_0 x = \otimes \cdot u(t)$$

Let

$$x = x_1,\ \frac{dx}{dt} = \frac{dx_1}{dt} = x_2,\ \frac{d^2 x}{dt^2} = \frac{dx_2}{dt} = x_3,\ \cdots,\ \frac{d^{n-1} x}{dt^{n-1}} = \frac{dx_{n-1}}{dt} = x_n$$

So, we have

$$\frac{dx_n}{dt} = -\frac{\otimes_0}{\otimes_n} x_1 - \frac{\otimes_1}{\otimes_n} x_2 - \frac{\otimes_2}{\otimes_n} x_3 - \cdots - \frac{\otimes_{n-1}}{\otimes_n} x_n + \frac{\otimes}{\otimes_n} u(t)$$

and the nth order system is reduced into the following first order system

$$\dot{X}(t) = A(\otimes) X(t) + B(\otimes) U(t)$$

where $X(t) = [x_1, x_2, \cdots, x_n]^T$, $U(t) = u(t)$,

$$A(\otimes) = \begin{bmatrix} 0 & 1 & 0 & \cdots & 0 \\ 0 & 0 & 1 & \cdots & 0 \\ \cdots & \cdots & \cdots & \cdots & 0 \\ 0 & 0 & \cdots & \cdots & 1 \\ -\frac{\otimes_0}{\otimes_n} & -\frac{\otimes 1}{\otimes_n} & \cdots & \cdots & -\frac{\otimes_{n-1}}{\otimes_n} \end{bmatrix}, \text{ and } B(\otimes) = \begin{bmatrix} 0 \\ 0 \\ \vdots \\ 0 \\ \frac{\otimes}{\otimes_n} \end{bmatrix}.$$

This ends the proof. QED.

9.2.2 *Transfer Functions of Typical Links*

A grey control system that is symbolically written in an equation is also known as a grey link. When the transfer function of a link is known, from the relationship $X(s) = G(s) \cdot U(s)$ and the Laplace transform of the driving term, one can obtain the Laplace transform of the response, then by using the inverse Laplace transform, he can produce the response $x(t)$. The relationship between the driving term and the response term is depicted in Figure 9.3.

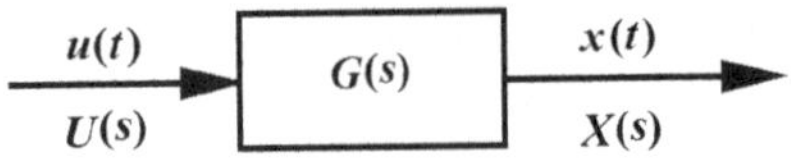

Fig. 9.3

In the following, let us look at the transfer functions of several typical links. The link with the driving term $u(t)$ and the response term $x(t)$ satisfying

$$x(t) = K(\otimes)u(t) \tag{9.6}$$

is known as a *grey proportional link*, where $K(\otimes)$ is the *grey magnifying coefficient* of the link.

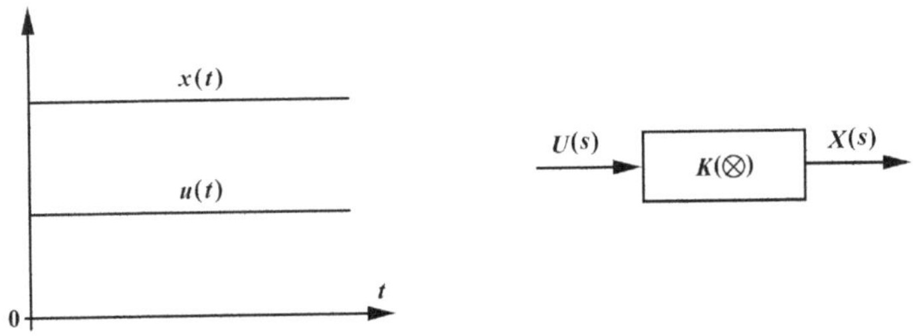

Fig. 9.4

Proposition 9.1. The transfer function of a grey proportional link is

$$G(s) = K(\otimes) \tag{9.7}$$

The characteristics of a grey proportional link are that when a jump occurs in the driving quantity, the response value changes proportionally. This kind of change and relationship between the drive and response are depicted in Figure 9.4.

When driven by a unit jump, if the response is given by

$$x(t) = K(\otimes)(1 - e^{-t/T}) \tag{9.8}$$

then the link is known as a *grey inertia link*, where T stands for a time constant of the link.

Proposition 9.2. The transfer function of a grey inertial link is given by

$$G(s) = \frac{K(\otimes)}{T \cdot s + 1} \tag{9.9}$$

The characteristics of a grey inertia link are that when a jump occurs in the driving quantity, the response can reach a new state of balance only after a period of time.

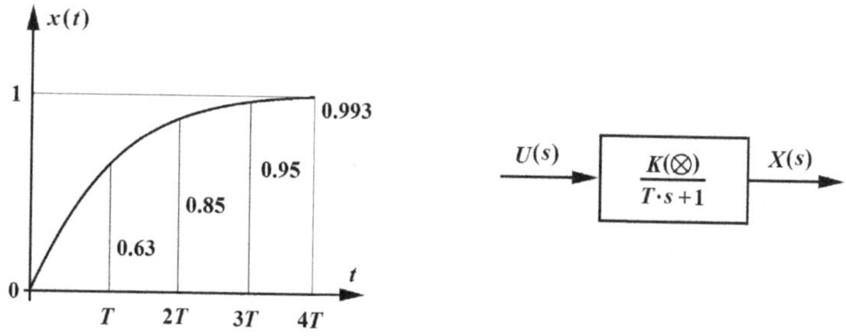

Fig. 9.5

Figure 9.5 provides a block diagram and the curve of change of the response of a grey inertia link when $\widetilde{K}(\otimes)=1$.

When the drive and response are related as follows:

$$x(t)=\int K(\otimes)u(t)dt \tag{9.10}$$

then the link is known as a *grey integral link.*

Proposition 9.3. The transfer function of a grey integral link is given below:

$$G(s)=\frac{K(\otimes)}{s} \tag{9.11}$$

For a grey integral link, when the drive is a jump function, its response is $x(t)=K(\otimes)ut$, as shown in Figure 9.6.

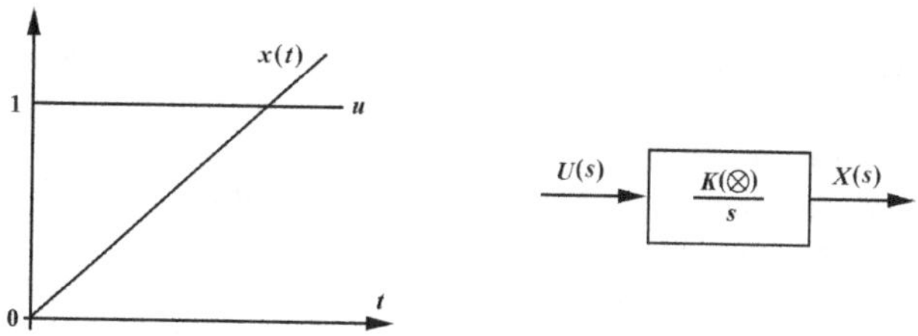

Fig. 9.6

If the response and the drive are related as follows

$$x(t)=K(\otimes)\frac{du(t)}{dt} \tag{9.12}$$

then the link is known as a *grey differential link.*

Proposition 9.4. The transfer function of a grey differential link is given as follows:

$$G(s) = K(\otimes)s \tag{9.13}$$

The characteristics of a grey differential link are that when the drive stands for a jump, the response becomes an impulse with an infinite amplitude.

If the drive and response are related as follows:

$$x(t) = u(t - \tau(\otimes)) \tag{9.14}$$

then the link is known as a *grey postponing link*, where $\tau(\otimes)$ is a grey constant.

Proposition 9.5. The transfer function of a grey postponing link is given below:

$$G(s) = e^{-\tau(\otimes)s} \tag{9.15}$$

For a grey postponing link, when the drive is a jump function, it takes sometime for the response to react accordingly. For details, see Figure 9.7.

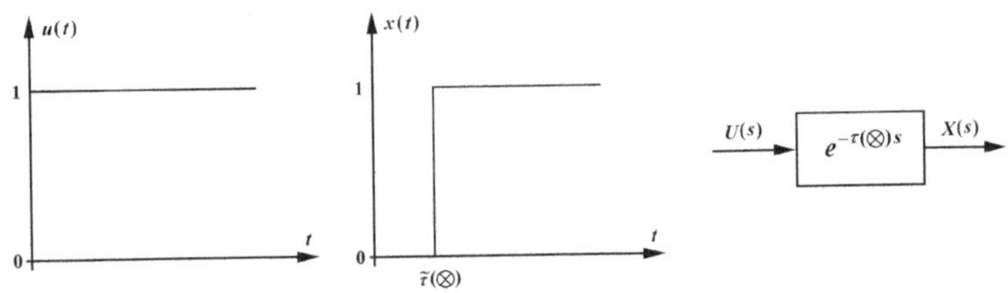

Fig. 9.7

What's given above are some typical links met in practical applications. Many complicated devices and systems can be treated as combinations of these typical links. For instance, when the grey proportional link is combined with a grey differential link, one can obtain a grey proportional differential link. When a grey integral link is connected with grey postponing link, one establishes a grey integral postponing link. Along the same line, multi-layered combinations can be developed for practical purposes. One of the purposes of studying grey transfer functions is that we can investigate the stabilities and other properties of systems by looking at the extreme values of the relevant transfer functions.

9.2.3 Matrices of Grey Transfer Functions

Matrices of grey transfer functions can be employed to express a fundamental relationship between the multi-inputs and multi-outputs of grey linear control systems. In particular, for the following grey linear control system

$$\begin{cases} \dot{X}(t) = A(\otimes)X(t) + B(\otimes)U(t) \\ Y(t) = C(\otimes)X(t) \end{cases}$$

Employing Laplace transforms produces

$$\begin{cases} sX(s) = A(\otimes)X(s) + B(\otimes)U(s) \\ Y(s) = C(\otimes)X(s) \end{cases}$$

and

$$\begin{cases} (sE - A(\otimes))X(s) = B(\otimes)U(s) \\ Y(s) = C(\otimes)X(s) \end{cases}$$

If $(sE - A(\otimes))$ is invertible, then we can further obtain

$$\begin{cases} X(s) = (sE - A(\otimes))^{-1}B(\otimes)U(s) \\ Y(s) = C(\otimes)X(s) \end{cases}$$

That is, we have $Y(s) = C(\otimes)(sE - A(\otimes))^{-1}B(\otimes)U(s)$.

The $m \times n$ matrix

$$G(s) = C(\otimes)(sE - A(\otimes))^{-1}B(\otimes) \tag{9.16}$$

is known as the *matrix of grey transfer functions*.

For an nth order grey linear system, if the state grey matrix $A(\otimes)$ of the corresponding equivalent first order system is non-singular, then

$$\lim_{s\to 0} G(s) = -C(\otimes)A(\otimes)^{-1}B(\otimes) \tag{9.17}$$

is known as a *grey gain matrix*. If the grey gain matrix $-C(\otimes)A(\otimes)^{-1}B(\otimes)$ is used to replace the transfer function $G(s)$, then the system is reduced into a proportional link.

Because $Y(s) = G(s)U(s)$, when $m = s = n$, if $G(s)$ is non-singular, we have the following:

$$U(s) = G(s)^{-1}Y(s) \tag{9.18}$$

The following

$$G(s)^{-1} = B(\otimes)^{-1}(sE - A(\otimes))C(\otimes)^{-1} \tag{9.19}$$

is known as a *grey structure matrix*. When the grey structure matrix is known, to make the output vector $Y(s)$ meet or close to meet a certain expected objective $J(s)$, one can determine the system's control vector $U(s)$ through $G^{-1}(s) \cdot J(s)$.

Additionally, we can also discuss the controllability and observability of systems by using matrices of grey transfer functions.

9.3 Robust Stability of Grey Systems

Stability is a fundamental structural characteristic of systems. It stands for an important mechanism for a system to sustain itself and is a prerequisite for the system to operate smoothly. That explains why it represents an important problem studied in control theory of systems and a key objective in relevant engineering designs. Each physical system has to be stable before it can be employed in practical applications.

The stability of grey systems focuses on the investigations of informational changes or whether or not the grey system of concern stays stable or can recover to its stability when the whitenization value of a grey parameter moves within the field of discourse. The existence of grey parameters makes the study of the stability of grey systems complicated and difficult, assuring that such investigation is a center of attention and represents a hard problem in the area of control theory and control engineering.

In grey systems modeling, there is the distinction between having a postponing term and not having such a term; there is also the difference between having a random term and not having such a term. Ordinarily, the grey systems without involving any random and postponing term are known as grey systems; those involving postponing terms without any random terms grey postponing systems, and those involving random terms grey stochastic systems. In this section, we will study the problem of robust stability of these three kinds of systems.

9.3.1 Robust Stability of Grey Linear Systems

The study of systems' stability is often limited to the systems without the effect of any external input. This kind of system is known as an *autonomous system*. A simple grey linear autonomous system can be written as follows:

$$\begin{cases} \dot{x}(t) = A(\otimes)x(t) \\ x(t_0) = x_0, \forall t \geq t_0 \end{cases} \tag{9.20}$$

where $x \in R^n$ stands for the state vector, and $A(\otimes) \in G^{n\times n}$ is matrix of grey coefficients.

Definition 9.1. If $A(\tilde{\otimes})$ is a whitenization matrix of the grey matrix $A(\otimes)$, then

$$\begin{cases} \dot{x}(t) = A(\tilde{\otimes})x(t) \\ x(t_0) = x_0 \end{cases} \tag{9.21}$$

is referred to as a *whitenization system* of the system in equ. (9.20).

Ordinarily, we assume that the matrix $A(\otimes)$ of grey coefficients of the system in equ. (9.20) has a *continuous matrix cover*:

$$A(D) = [L_a, U_a] = \{A(\tilde{\otimes}) : \underline{a}_{ij} \le \tilde{\otimes} \le \bar{a}_{ij}, i, j = 1,2,\cdots,n\},$$

where $U_a = (\bar{a}_{ij}), L_a = (\underline{a}_{ij})$.

Definition 9.2. If any whitenization system of the system in equ. (9.20) is stable, then the system in equ. (9.20) is referred to as *robust stable.*

The ordinary concept of system's (robust) stability means that of (robust) *asymptotic stability* of the system.

Theorem 9.7. If there is positive definite matrix P such that

$$PL_a + L_a^T P + 2\lambda_{\max}(P) \| U_a - L_a \| I_n < 0$$

then the system in equ. (9.20) is robust stable.

Proof. Let us take the Lyapunov function $V(x) = x^T Px$. For any whitenization matrix $A(\tilde{\otimes}) \in A(D)$, let us compute the derivative of $V(x)$ with respect to t along the trajectory of the whitenization system and obtain

$$\begin{aligned} \dot{V}(x) &= 2x^T PA(\tilde{\otimes})x = x^T \left(PL_a + L_a^T P\right)x + 2x^T P\Delta Ax \\ &\le x^T \left(PL_a + L_a^T P + 2\lambda_{\max}(P)\|U_a - L_a\| I_n\right)x < 0, \forall x \ne 0 \end{aligned}$$

This implies that the system in equ. (9.20) is robust stable. QED.

If in Theorem 9.7 we let $P = I_n$, then we have the following clean result:

Corollary 9.1. If

$$\| U_a - L_a \| < -\lambda_{\max}\left(\frac{L_a + L_a^T}{2}\right) \tag{9.22}$$

holds true, then the system in equ. (9.20) is robust stable.

If we employ another form of decomposition $A(\tilde{\otimes}) = U_a - \Delta A$ of the whitenization matrix $A(\tilde{\otimes})$ to study the robust stability of the system in equ. (9.20), then similar to Theorem 9.7) and Corollary 9.1 we can obtain the following results.

Theorem 9.8. If there is a positive definite matrix P such that

$$PU_a + U_a^T P + 2\lambda_{\max}(P) \| U_a - L_a \| I_n < 0$$

then the system in equ. (9.20) is robust stable.

Corollary 9.2. If

$$\| U_a - L_a \| \le -\lambda_{\max}\left(\frac{U_a + U_a^T}{2}\right) \tag{9.23}$$

holds true, then the system in equ. (9.20) is robust stable.

Both Corollaries 9.1 and 9.2 respectively provide us a practically very meaning result, because $U_a - L_a$ in fact stands for the matrix of disturbance errors of the system in equ. (9.20); and equs. (9.22) and (9.23) indicate that when the norm of the disturbance error matrix varies within the range of $(0, \lambda)$, the system in equ. (9.20) will always be stable, where $\lambda = \max\{-\lambda_{\max}(\frac{L_a + L_a^T}{2}), -\lambda_{\max}(\frac{U_a + U_a^T}{2})\}$.

Theorem 9.9. If $L_a + L_a^T + \lambda_{\max}[(U_a - L_a) + (U_a - L_a)^T]I_n < 0$, then the system in equ. (9.20) is robust stable; if $U_a + U_a^T - \lambda_{\max}[(U_a - L_a) + (U_a - L_a)^T]I_n > 0$, then the system is instable.

Example 9.1. Let us consider the robust stability problem of the following 2-dimensional grey linear system:

$$\dot{x}(t) = \begin{pmatrix} [-2.3,-1.8] & [0.6,0.9] \\ [0.8,1.0] & [-2.5,-1.9] \end{pmatrix} x(t)$$

Solution. Through computations, we have

$$\| U_a - L_a \| = 0.8072 < -\lambda_{\max}\left(\frac{L_a + L_a^T}{2}\right) = 1.3000$$

$$L_a + L_a^T + \lambda_{\max}[(U_a - L_a) + (U_a - L_a)^T]I_n = -1.7759 I_n < 0$$

These inequalities indicate that by using Corollary 9.1 or Theorem 9.9, we can conclude that this given system is robust stable.

9.3.2 Robust Stability of Grey Linear Time-Delay Systems

The phenomena of timely postponing is very common. They are often the main reason for causing instability, vibration, and poor performance in systems. So, it is very important and meaningful to investigate the stability problem of postponing systems. In particular, let us look at the following often seen n-dimensional linear postponing autonomous system:

$$\begin{cases} \dot{x}(t) = Ax(t) + Bx(t-\tau), \forall t \geq 0, \\ x(t) = \varphi(t), \forall t \in [-\tau, 0] \end{cases} \tag{9.24}$$

where $x(t) \in R^n$ stands for the system's state vector, $A, B \in R^{n\times n}$ the known constant matrices, $\tau > 0$ the amount of time of postponing, and $\varphi(t) \in C^n[-\tau, 0]$ (the nth dimensional space of continuous functions).

Definition 9.3. If at least one of the matrices A, B of constants in the linear postponing system in equ. (9.24) is grey, then this system is referred to as a grey linear postponing autonomous system, denoted as

$$\begin{cases} \dot{x}(t) = A(\otimes)x(t) + B(\otimes)x(t-\tau), \forall t \geq 0, \\ x(t) = \varphi(t), \forall t \in [-\tau, 0]. \end{cases} \tag{9.25}$$

In the following, we assume that the constant matrices in the system in equ. (9.25) are all grey and that they have continuous matrix covers, that is, $A(\otimes), B(\otimes)$ respectively have the following form of matrix covers:

$$A(D) = [L_a, U_a] = \{A(\tilde{\otimes}) : \underline{a}_{ij} \leq \tilde{\otimes} \leq \bar{a}_{ij}, i, j = 1, 2, \cdots, n\},$$

$$B(D) = [L_b, U_b] = \{B(\tilde{\otimes}) : \underline{b}_{ij} \leq \tilde{\otimes} \leq \bar{b}_{ij}, i, j = 1, 2, \cdots, n\}.$$

where $U_a = (\bar{a}_{ij}), L_a = (\underline{a}_{ij}), U_b = (\bar{b}_{ij}), L_b = (\underline{b}_{ij})$.

Definition 9.4. If $A(\tilde{\otimes}), B(\tilde{\otimes})$ are respectively whitenization matrices of $A(\otimes), B(\otimes)$, then

$$\begin{cases} \dot{x}(t) = A(\tilde{\otimes})x(t) + B(\tilde{\otimes})x(t-\tau), \forall t \geq 0, \\ x(t) = \varphi(t), \forall t \in [-\tau, 0]. \end{cases} \tag{9.26}$$

is referred to as a *whitenization system* of the system in equ. (9.25).

Definition 9.5. If any whitenization system of the system in equ. (9.25) is stable, the system in equ. (9.25) is referred to as *robust stable*.

Based on whether or not the robust stability condition of a grey postponing system depends on the amount of postponing, the robust stability condition can be divided into two classes: one for postponing independent and the other postponing dependent. In particular, the condition for a robust stable system to be postponing independent is that for any time postponing $\tau > 0$, the system is *robustly asymptotic stable*. Because this condition does not need to know the amount of postponing, it is appropriate for the study of the stability problem of postponing systems whose amounts of postponing are uncertain or unknown.

The condition for a robust stable system to be postponing dependent is that for some values of postponing $\tau > 0$, the system is robust stable, while for some other values of the postponing $\tau > 0$, the system is not stable. That is why the system's stability is dependent on the amount of postponing.

Theorem 9.10. If there are positive definite matrices P, Q and positive constants $\varepsilon_1, \varepsilon_2$ such that the symmetric matrix

$$\begin{pmatrix} \Xi & PL_b & P & P \\ L_b^{\mathrm{T}}P & -Q+\varepsilon_2 \| U_b - L_b \|^2 I_n & 0 & 0 \\ P & 0 & -\varepsilon_1 I_n & 0 \\ P & 0 & 0 & -\varepsilon_2 I_n \end{pmatrix} < 0$$

where $\Xi = L_a^{\mathrm{T}}P + PL_a + Q + \varepsilon_1 \| U_a - L_a \|^2 I_n$, and I_n stands for the identity matrix (the same symbol will be used for the rest of this chapter), then the system in equ. (9.25) is robust stable.

Theorem 9.11. If there are positive definite matrices P, Q, N and positive constants $\varepsilon_1, \varepsilon_2$ such that the symmetric matrix

$$\begin{pmatrix} \Gamma & PL_b & \rho I_n & P & P \\ L_b^{\mathrm{T}}P & -Q+\varepsilon_2 \| U_b - L_b \|^2 I_n & 0 & 0 & 0 \\ \rho I_n & 0 & -\overline{N} & 0 & 0 \\ P & 0 & 0 & -\varepsilon_1 I_n & 0 \\ P & 0 & 0 & 0 & -\varepsilon_2 I_n \end{pmatrix} < 0$$

where $\Gamma = L_a^{\mathrm{T}}P + PL_a + Q + \varepsilon_1 \| U_a - L_a \|^2 I_n$ and $\overline{N} = N^{-1}, \rho = \sqrt{\tau}$, then the system in equ. (9.25) is robust stable.

Example 9.2. Let us look at the following 2-dimensional grey linear postponing system

$$\begin{cases} \dot{x}(t) = A(\otimes)x(t) + B(\otimes)x(t-\tau), \forall t \geq 0, \\ x(t) = \varphi(t), \forall t \in [-\tau, 0]. \end{cases}$$

Assume that the upper and lower bound matrices of the continuous matrix covers of the grey constant matrices $A(\otimes), B(\otimes)$ are respectively give as follows:

$$L_a = \begin{pmatrix} -4.38 & 0.20 \\ 0.19 & -4.33 \end{pmatrix}, U_a = \begin{pmatrix} -4.26 & 0.29 \\ 0.27 & -4.22 \end{pmatrix};$$

$$L_b = \begin{pmatrix} -0.93 & 0.21 \\ 0.23 & -0.86 \end{pmatrix}, U_b = \begin{pmatrix} -0.88 & 0.24 \\ 0.26 & -0.82 \end{pmatrix}.$$

According to Theorem 9.10, by using the solver in the LMI (linear matrix inequality) control toolbox, we obtain the behavioral solution as follows:

$$P = \begin{pmatrix} 9.4642 & 0.6983 \\ 0.6983 & 9.8228 \end{pmatrix}, \quad Q = \begin{pmatrix} 28.0088 & -0.0340 \\ -0.0340 & 27.8605 \end{pmatrix}, \quad \varepsilon_1 = 30.0826, \quad \varepsilon_2 = 30.2461.$$

Now from Theorem 9.11, by using the solver in the LMI (linear matrix inequality) control toolbox, we obtain the behavioral solution below:

$$P = \begin{pmatrix} 6.8592 & 0.5061 \\ 0.5061 & 7.1191 \end{pmatrix}, Q = \begin{pmatrix} 20.2294 & -0.0246 \\ -0.0246 & 20.1920 \end{pmatrix}, N = \begin{pmatrix} 0.0456 & 0 \\ 0 & 0.0456 \end{pmatrix},$$

$\varepsilon_1 = 21.8024$, $\varepsilon_2 = 21.9209$, $\tau = 2.7035$.

These results indicate that the system considered in this example is robust stable. And the maximum allowed postponing length of time as obtained from Theorem 9.11 is 2.7035.

9.3.3 Robust Stability of Grey Stochastic Linear Time-Delay Systems

The mathematical model that describes a stochastic system is generally the Itto stochastic differential equation, where the often seen n-dimensional Itto stochastic differential postponing equation is

$$\begin{cases} dx(t) = Ax(t) + Bx(t-\tau) + [Cx(t) + Dx(t-\tau)]dw(t), \forall t \geq 0, \\ x(t) = \xi(t), \xi(t) \in L^2_{F_0}([-\tau, 0]; R^n), \forall t \in [-\tau, 0]. \end{cases} \quad (9.27)$$

where $x(t) \in R^n$ stands for the system's state vector, $A, B, C, D \in R^{n \times n}$ known constant matrices, $\tau > 0$ the time of postponing, $w(t)$ a 1-dimensional Brownian motion defined on a complete probability space $(\Omega, F, \{F_t\}_{t\geq 0}, P)$, $L^2_{F_0}([-\tau,0];R^n)$ the totality of all F_0-measurable stochastic variables $\xi = \{\xi(t): -\tau \leq t \leq 0\}$ that take values from $C([-\tau,0];R^n)$ satisfying $\sup_{-\tau\leq t\leq 0} E|\xi(\theta)|^2 < \infty$, while $C([-\tau,0];R^n)$ stands for the totality of continuous functions $\varphi:[-\tau,0]\to R^n$. Under the initial condition $x(t)=\xi(t)\in L^2_{F_0}([-\tau,0];R^n)$, the system in equ. (9.27) has an equilibrium point $x(t;\xi)$, and corresponding to the initial value $\xi(t)=0$, $x(t;0)\equiv 0$.

There are several different concepts of stability for the stochastic systems. In the following, we list four of the important stabilities.

Definition 9.6. The equilibrium point $x(t)\equiv 0$ of the system in equ. (9.27) is referred to as *stochastically stable*, if for each $\varepsilon > 0$, $\lim_{x_0\to 0} P(\sup_{t>t_0}|x(t;t_0,x_0)|>\varepsilon)=0$.

Definition 9.7. The equilibrium point $x(t)\equiv 0$ of the system in equ. (9.27) is referred to as *stochastically asymptotically stable*, if it is stochastically stable and $\lim_{x_0\to 0} P(\lim_{t\to +\infty} x(t;t_0,x_0)=0)=1$.

Definition 9.8. The equilibrium point $x(t)\equiv 0$ of the system in equ. (9.27) is referred to as *large-scale stochastically asymptotically stable*, if it is stochastically stable and for any t_0, x_0, $P(\lim_{t\to +\infty} x(t;t_0,x_0)=0)=1$.

Definition 9.9. The equilibrium point $x(t)\equiv 0$ of the system in equ. (9.27) is referred to as *mean square exponential stable*, if there are positive constants $\alpha>0, \beta>0$ such that $E|x(t;t_0,x_0)|^2 \leq \alpha |x_0|^2 \exp(-\beta t)$, $t>t_0$.

A grey system is *stochastic*, if it is such a stochastic system that involves grey parameters. Concepts related to grey stochastic systems are generally introduced based on the relevant concepts of the conventional stochastic systems. Considering the problems we will study, let us provide the following definitions.

Definition 9.10. If at least one of the matrices A, B, C, D of the stochastic linear postponing system in equ. (9.27) is grey, then the system is referred to as a *grey stochastic linear postponing system*, written as follows:

$$\begin{cases} \mathrm{d}x(t) = A(\otimes)x(t) + B(\otimes)x(t-\tau) + [C(\otimes)x(t) + D(\otimes)x(t-\tau)]\mathrm{d}w(t), \forall t \geq 0, \\ x(t) = \xi(t), \xi(t) \in L^2_{F_0}([-\tau,0];R^n), \forall t \in [-\tau,0]. \end{cases} \tag{9.28}$$

In this section, we assume that all the coefficient matrices of the system in equ. (9.28) are grey with continuous matrix covers. That is, the matrix covers of the grey matrices $A(\otimes)$, $B(\otimes)$, $C(\otimes)$, and $D(\otimes)$ are respectively given as follows:

$$A(D) = [L_a, U_a] = \{A(\tilde{\otimes}) = (\tilde{\otimes}_{aij})_{n\times n} : \underline{a}_{ij} \leq \tilde{\otimes}_{aij} \leq \overline{a}_{ij}\},$$

$$B(D) = [L_b, U_b] = \{B(\tilde{\otimes}) = (\tilde{\otimes}_{bij})_{n\times n} : \underline{b}_{ij} \leq \tilde{\otimes}_{bij} \leq \overline{b}_{ij}\},$$

$$C(D) = [L_c, U_c] = \{C(\tilde{\otimes}) = (\tilde{\otimes}_{cij})_{n\times m} : \underline{c}_{ij} \leq \tilde{\otimes}_{cij} \leq \overline{c}_{ij}\},$$

and

$$D(D) = [L_d, U_d] = \{D(\tilde{\otimes}) = (\tilde{\otimes}_{dij})_{n\times n} : \underline{d}_{ij} \leq \tilde{\otimes}_{dij} \leq \overline{d}_{ij}\},$$

where $L_a = (\underline{a}_{ij})_{n\times n}$, $U_a = (\overline{a}_{ij})_{n\times n}$, $L_b = (\underline{b}_{ij})_{n\times n}$, $U_b = (\overline{b}_{ij})_{n\times n}$, $L_c = (\underline{c}_{ij})_{n\times n}$, $U_c = (\overline{c}_{ij})_{n\times n}$, $L_d = (\underline{d}_{ij})_{n\times n}$, and $U_a = (\overline{d}_{ij})_{n\times n}$.

Definition 9.11. If $A(\tilde{\otimes}), B(\tilde{\otimes}), C(\tilde{\otimes})$, and $D(\tilde{\otimes})$ are arbitrary whitenization matrices of the grey matrices $A(\otimes), B(\otimes), C(\otimes)$, and $D(\otimes)$, respectively, then

$$\begin{cases} \mathrm{d}x(t) = A(\tilde{\otimes})x(t) + B(\tilde{\otimes})x(t-\tau) + [C(\tilde{\otimes})x(t) + D(\tilde{\otimes})x(t-\tau)]\mathrm{d}w(t), \forall t \geq 0, \\ x(t) = \xi(t), \xi(t) \in L^2_{F_0}([-\tau,0];R^n), \forall t \in [-\tau,0]. \end{cases} \tag{9.29}$$

is referred to as a *whitenization system* of the system in equ. (9.28).

Definition 9.12. If any whitenization system of the system in equ. (9.28) is large-scale stochastic asymptotic stable, that is,

$$\lim_{t\to\infty} x(t;\xi) = 0 \ a.s.$$

then the system in equ. (9.28) is said to be *large scale stochastic robust asymptotic stable*.

Definition 9.13. If any whitenization system of the system in equ. (9.28) is mean square exponential stable, that is, there are positive constants r_0 and K such that the equilibrium points of whitenization systems of the system in equ. (9.28) satisfy

$$E\,|\,x(t,\xi)\,|^2 \leq Ke^{-r_0 t} \sup_{-\tau\leq\theta\leq 0} E\,|\,\xi(\theta)\,|^2, \quad t \geq 0,$$

or equivalently

$$\limsup_{t\to\infty}\frac{1}{t}\log E\,|\,x(t;\xi)\,|^2\le -r_0\,,$$

then the system in equ. (9.28) is said to be *mean square exponential robust stable.*

Theorem 9.12. For the system in equ. (9.28), if there are a positive definite symmetric matrix Q and positive constants ε_i, $i=1,\cdots,6$, satisfying $M+N<0$, then for any initial condition $\xi\in C^p_{F_0}([-\tau,0];R^n)$ he following holds true

$$\lim_{t\to\infty} x(t;\xi)=0 \ \ a.s.$$

that is, the system in equ. (9.28) is large-scale stochastic robust asymptotic stable, where

$$M=QL_a+L_a^TQ+(\varepsilon_1+\varepsilon_2)Q+\varepsilon_1^{-1}\lambda_{\max}(Q)\cdot\|U_a-L_a\|^2 I_n+(1+\varepsilon_4)(1+\varepsilon_5)L_c^TQL_c$$
$$+(1+\varepsilon_4^{-1})(1+\varepsilon_5)\lambda_{\max}(Q)\|U_c-L_c\|^2 I_n$$

and

$$N=\varepsilon_2^{-1}(1+\varepsilon_3^{-1})\lambda_{\max}(Q)\|U_b-L_b\|^2 I_n+\varepsilon_2^{-1}\cdot(1+\varepsilon_3)L_b^TQL_b+(1+\varepsilon_5^{-1})(1+\varepsilon_6)L_d^TQL_d$$
$$+\ (1+\varepsilon_5^{-1})(1+\varepsilon_6^{-1})\lambda_{\max}(Q)\|U_d-L_d\|^2 I_n.$$

Theorem 9.13. For the system in equ. (9.28), if there are positive definite symmetric matrix Q and positive constants ε_i, $i=1,\cdots,6$, satisfying $K+L<0$, then for any initial condition $\xi\in C^p_{F_0}([-\tau,0];R^n)$, the following holds true:

$$\lim_{t\to\infty} x(t;\xi)=0 \ \ a.s\,.$$

that is, the system in equ. (9.28) is large-scale stochastic asymptotic stable, where

$$K=QL_a+L_a^TQ+(\varepsilon_1+\varepsilon_2)Q+[\varepsilon_1^{-1}\lambda_{\max}(Q)\mathrm{trace}(G_a^TG_a)+(1+\varepsilon_4)(1+\varepsilon_5)\mathrm{trace}(L_c^TL_c)$$
$$+(1+\varepsilon_4^{-1})(1+\varepsilon_5)\lambda_{\max}(Q)\mathrm{trace}(G_c^TG_c)]I_n\,,$$

and

$$L=[\varepsilon_2^{-1}(1+\varepsilon_3^{-1})\lambda_{\max}(Q)\mathrm{trace}(G_b^TG_b)+\varepsilon_2^{-1}(1+\varepsilon_3)\mathrm{trace}(L_b^TL_b)$$
$$+(1+\varepsilon_5^{-1})(1+\varepsilon_6)\mathrm{trace}(L_d^TL_d)+(1+\varepsilon_5^{-1})(1+\varepsilon_6^{-1})\lambda_{\max}(Q)\mathrm{trace}(G_d^TG_d)]I_n.$$

If we let the matrix and constants in Theorems 9.12 and 9.13 be $\varepsilon_1 = \cdots = \varepsilon_6 = 1$ and $Q = I_n$, then we can obtain the following corollaries, respectively.

Corollary 9.3. If the upper and lower bound matrices of the continuous matrix covers of the coefficient matrices of the system in equ. (9.28) satisfy

$$\begin{aligned} &L_a + L_a^T + 2L_b^T L_b + 4L_c^T L_c + 4L_d^T L_d \\ &\quad < -(2\|U_b - L_b\|^2 + \|U_a - L_a\|^2 + 4\|U_d - L_d\|^2 + 4\|U_c - L_c\|^2 + 2)I_n \end{aligned}$$

then the system in equ. (9.28) is large-scale stochastic asymptotic stable.

Corollary 9.4. If the upper and lower bound matrices of the continuous matrix covers of the coefficient matrices of the system in equ. (9.28) satisfy

$$\begin{aligned} &L_a + L_a^T + [2\text{trace}(L_b^T L_b) + 4\text{trace}(L_c^T L_c) + 4\text{trace}(L_d^T L_d)]I_n \\ &\quad < -(\text{trace}(G_a^T G_a) + 2\text{trace}(G_b^T G_b) + 4\text{trace}(G_c^T G_c) + 4\text{trace}(G_d^T G_d) + 2)I_n \end{aligned}$$

then the system in equ. (9.28) is large-scale stochastic asymptotic stable.

Theorem 9.14. For the system in equ. (9.28), if there are positive definite symmetric matrix Q and positive constants ε_i, $i = 1, \cdots, 3$, satisfying

$$\begin{aligned} &QL_a + L_a^T Q + (\varepsilon_1 + \varepsilon_2)Q + \varepsilon_1^{-1}\lambda_{\max}(Q)\|U_a - L_a\|^2 I_n \\ &\quad < -[(1+\varepsilon_3)\lambda_{\max}(Q)\text{trace}(M_c^T M_c) + \varepsilon_2^{-1}\lambda_{\max}(Q)\text{trace}(M_b^T M_b) \\ &\quad\quad + (1+\varepsilon_3^{-1})\lambda_{\max}(Q)\text{trace}(M_d^T M_d)]I_n \end{aligned}$$

then the system in equ. (9.28) is large-scale stochastic robust asymptotic stable.

If in Theorem 9.14 we let $\varepsilon_1 = \varepsilon_2 = \varepsilon_3 = 1$ and $Q = I_n$, then we have

Corollary 9.5. If the upper and lower bound matrices of the matrix covers of the grey coefficient matrices in the system in equ. (9.28) satisfy

$$\begin{aligned} &L_a + L_a^T + 2I_n + \|U_a - L_a\|^2 I_n + 2\text{trace}(M_c^T M_c)I_n \\ &\quad < -[\text{trace}(M_b^T M_b) + 2\text{trace}(M_d^T M_d)]I_n \end{aligned}$$

then the system in equ. (9.28) is large-scale stochastic robust asymptotic stable.

Theorem 9.15. For the system in equ. (9.28), if there are positive definite symmetric matrix Q and positive constants ε_i, $i = 1, \cdots, 6$, satisfying $\lambda_{\max}(M) + \lambda_{\max}(N) < 0$, then for any initial condition $\xi \in C_{F_0}^p([-\tau, 0]; R^n)$, the following holds true:

$$E|x(t,\xi)|^2 \le Ke^{-r_0 t} \sup_{-\tau\le\theta\le 0} E|\xi(\theta)|^2,\ t\ge 0,$$

or equivalently,

$$\limsup_{t\to\infty} \frac{1}{t} \log E|x(t;\xi)|^2 \le -r_0,$$

where the matrices M, N are the same as in Theorem 9.12, $K = \frac{\tau e^{r_0\tau}\lambda_{\max}(N) + \lambda_{\max}(Q)}{\lambda_{\min}(Q)}$, and r_0 is the unique real root of the following equation $r_0\lambda_{\max}(Q) + \lambda_{\max}(M) + e^{r_0\tau}\lambda_{\max}(N) = 0$, then the system in equ. (9.28) is mean square exponential robust stable.

9.4 Several Typical Grey Controls

Grey control stands for the control of essential grey systems, including the situation of general control systems involving grey numbers, by constructing controls through employing the thinking methods of grey systems analysis, modeling, prediction, and decision-making.

9.4.1 Control with Abandonment

The dynamic characteristics of grey systems are mainly determined by the matrices $G(s)$ of grey transfer functions. So, to realize effect control over the systems' dynamic characteristics, one of the effective methods is to modify and correct the matrices of transfer functions and the structure matrices.

Assume that $G^{-1}(s)$ is a system's structure matrix, and $G_*^{-1}(s)$ an objective structure matrix, then

$$\Delta^{-1} = G_*^{-1}(s) - G^{-1}(s) \tag{9.30}$$

is known as a *structural deviation matrix*. From $G^{-1}(s)Y(s) = U(s)$ and $G_*^{-1}(s) = \Delta^{-1} + G^{-1}(s)$, we obtain $(G_*^{-1}(s) - \Delta^{-1})Y(s) = U(s)$, that is

$$G_*^{-1}(s)Y(s) - \Delta^{-1}Y(s) = U(s) \tag{9.31}$$

We will refer $-\Delta^{-1}Y(s)$ to as a *superfluous term*. The control through a feedback of $\Delta^{-1}Y(s)$ to cancel the superfluous term is known as a *control with abandonment*. Through the effect of the feedback of $\Delta^{-1}Y(s)$, the system $G^{-1}(s)Y(s) = U(s)$ is reduced to

$$G^{-1}(s)Y(s)+\Delta^{-1}Y(s)=U(s)\ (G^{-1}(s)+\Delta^{-1})Y(s)=U(s)$$

that is, $G_*^{-1}(s)Y(s)=U(s)$ has already processed the desired objective structure.

The number of entries in the structural deviation matrix Δ^{-1}, used in a control with abandonment, directly affects the number of components in the controlling equipment. So, considering the economics, reliability, and the ease to apply, under the condition of providing the desired dynamical characteristics of the system, one always want to keep the number of elements in the deviation matrix Δ^{-1} as few as possible. That is to say, in the objective structural matrix, one should try to keep the corresponding entries of the original structure matrix.

The idea of control with abandonment is depicted in Figure 9.8.

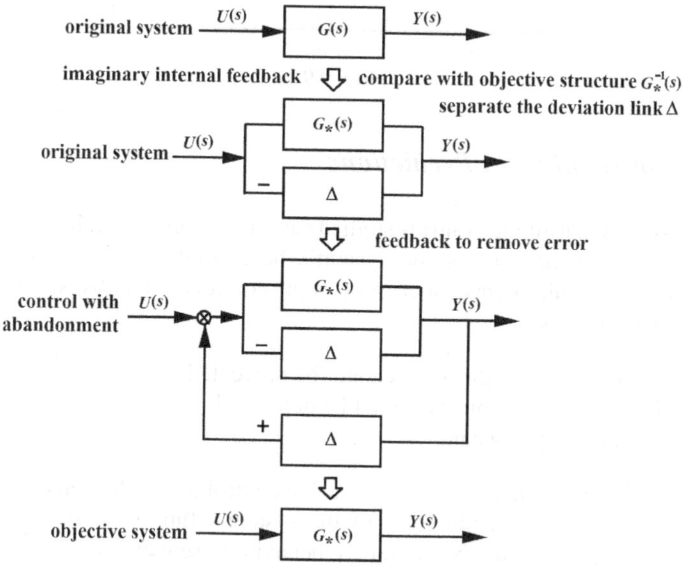

Fig. 9.8

9.4.2 Control of Grey Incidences

Assume that $Y=[y_1, y_2, \cdots, y_m]^T$ stands for the output vector, and $J=[j_1, j_2, \cdots, j_m]^T$ the objective vector. If the components of the control vector $U=[u_1, u_2, \cdots, u_s]^T$ satisfy

$$u_k = f_k(\gamma(J,Y)) \ ; \ k=1, 2, \cdots, s \qquad (9.32)$$

where $\gamma(J,Y)$ is the degree of grey incidence between the output vector Y and the objective vector J, then the system control is known as a *grey incidence control*.

A grey incidence control system is obtained by attaching a grey incidence controller to the general control system. It determines the control vector U through the degree of grey incidence $\gamma(J,Y)$ so that the degree of incidence between the output vector and the objective vector does not go beyond a pre-determined range. The idea of the control systems of grey incidence is depicted in Figure 9.9.

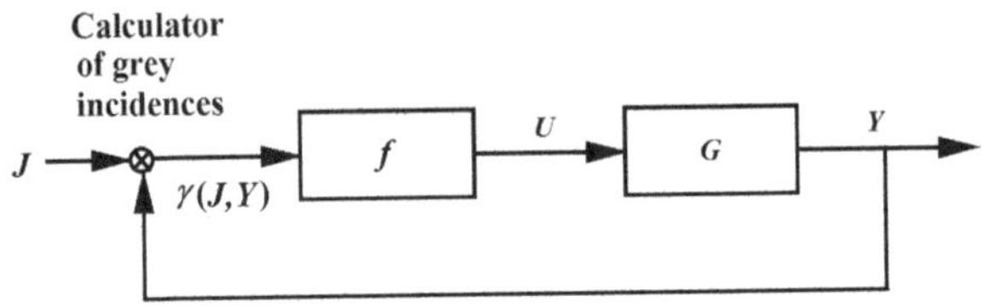

Fig. 9.9

9.4.3 Control of Grey Predictions

All the various kinds of the controls studied earlier are about applying controls after first checking whether or not the system's behavioral sequence satisfies some pre-determined requirements. Such post-event controls evidently suffer from the following weaknesses:

(1) Expected future disasters cannot be prevented;
(2) Instantaneous controls cannot be done; and
(3) Adaptability is weak.

The so-called *grey predictive control* is such a control that is designed based on the control decision made on the basis of the system's future behavioral tendency, which is predicted using the system's behavioral sequences and the patterns discovered from the sequences. This kind of control can be employed to avoid future adverse events from happening, to be implemented in a timely fashion, and possess a wide range of applicability.

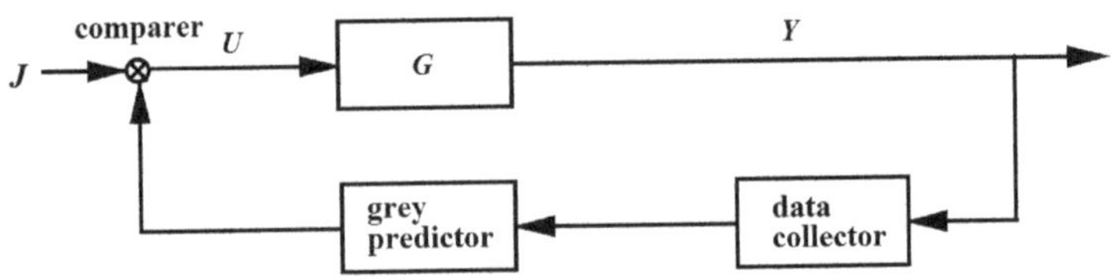

Fig. 9.10

The idea of a grey predictive control system is graphically shown in Figure 9.10. Its working principle is that first collect and organize the device's behavioral sequence of the output vector Y; secondly, use a prediction device to compute the predicted values for the future steps; and lastly, compare the predicted values with the objective and determine the control vector U so that the future output vector Y will be as close to the objective J as possible.

Assume that $j_i(k), y_i(k), u_i(k)$ $(i=1, 2, \cdots, m)$ are respectively the values of the objective component, output component, and control component at time moment k. For $i=1, 2, \cdots, m$, let

$$j_i = (j_i(1), j_i(2), \cdots, j_i(n))$$
$$y_i = (y_i(1), y_i(2), \cdots, y_i(n))$$
$$u_i = (u_i(1), u_i(2), \cdots, u_i(n))$$

For the control operator $f:(j_i(\lambda), y_i(\lambda)) \to u_i(k)$,

$$u_i(k) = f(j_i(\lambda), y_i(\lambda)) \tag{9.33}$$

when $k > \lambda$, the system is known as a *post-event* (or *after-event*) *control*; when $k = \lambda$, the system is known as an *on-time control*; and when $k < \lambda$, the system is known as a *predictive control.*

If the control operator f satisfies

$$f(j_i(\lambda), y_i(\lambda)) = j_i(\lambda) - y_i(\lambda) \tag{9.34}$$

that is,

$$u_i(k) = j_i(\lambda) - y_i(\lambda) \tag{9.35}$$

then when $k > \lambda$, the system is known as an *error-afterward control*; when $k = \lambda$, the system is known as an *error-on-time control*; and when $k < \lambda$, the system is known as an *error-predictive control.*

Let $y_i = (y_i(1), y_i(2), \cdots, y_i(n))(i=1, 2, \cdots, m)$ stand for a sample of the output components, its GM(1,1) response formula be given as follows:

$$\begin{cases} \hat{y}_i^{(1)}(k+1) = (y_i(1) - \dfrac{b_i}{a_i})e^{-a_i k} + \dfrac{b_i}{a_i} \\ \hat{y}_i^{(0)}(k+1) = \hat{y}_i^{(1)}(k+1) - \hat{y}_i^{(1)}(k) \end{cases}$$

If the control operator f satisfies

$$u_i(n+k_0) = f(j_i(k),\, y_i^{(0)}(k)), \quad n+k_0 < k_i,\ i = 1, 2, \cdots, m \tag{9.36}$$

then the system control is known as a *grey predictive control*.

In a grey predictive control system, predictions are often done using metabolic models. So, the parameters of the prediction device vary with time. When a new data value output is produced and accepted by the sampling device, an old data value is removed so that a new model is developed. Accordingly, a series of new predicted values are provided. Doing so guarantees a strong adaptability of the system.

Example 9.3. Let us look at the EDM (electric discharge machining) grey control system.

The investigation on the control systems of EDM machines has been an important effort in the field of electric discharge machining. Each EDM can be seen as a stochastic time-dependent nonlinear system involving many parameters. Applications mainly include those situations when the conventional controls of linear, constant coefficient systems could not produce adequate outcomes. The current commonly employed EDM control systems are established on the modern control theory. Those often applied self-adaptive control systems generally employ the mathematical models of approximation with accompanied high costs without actually realizing the truly significant optimal results. Grey control is not only dislike how to establish the precise mathematical models based on the complete knowledge of the system of concern on the basis of modern control theory, but also unlike fuzzy control where the system is treated as a black box by disregarding all the information about the internal working of the system leading to low accuracy controls. The parameters, structures, and other aspects of grey models vary with time. Such dynamic modeling can be highly appropriate for the study of EDM machines with high degrees of uncertainty and produce relatively more satisfactory control effects.

For EDM control systems, the objects of control are EDM machine tools, where the outputs need the signals from the testing EDM machine tools as well as the control quantity U, the signals about the control of the EDM machine tools. The EDM control systems, in the general sense, mean the control over the systems that serve the EDM machine tools. For instance, let us look at the traditional gap-voltage feedback servo control system. Due to a lack of linear relationship between the gas voltage, gap size, discharge strength, discharge state, and servo reference voltage, the effect of employing only one gap-voltage feedback servo control system is not very good.

In order to make up for the insufficiency of single loop controls, one can employ double-loop controls with the inner loop being the traditional gap-voltage feedback control and the outer loop being an impulse discharge rate feedback control that instantaneously adjusts the inner loop. The block-design chart of this control system is depicted in Figure 9.11. From Figure 9.11, it can be seen that this control design

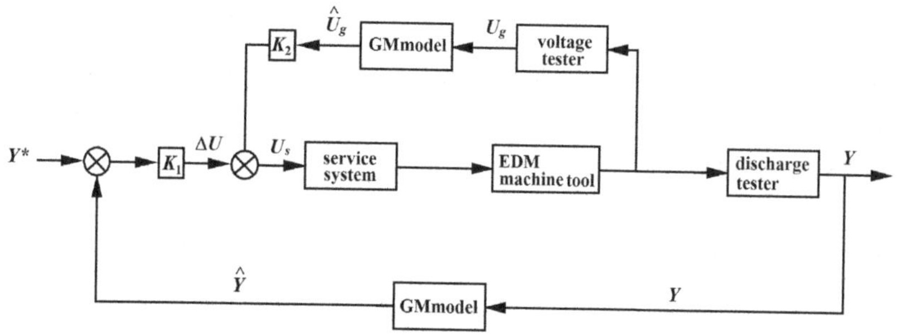

Fig. 9.11

represents a system of two loops. Based on the collected sequence of gap voltage readings $U_g(K)$, the inner loop employs the GM model to predict the next moment $\hat{U}_g(K+i+1)$, where i stands for the prediction steps, which is then fed into the input end to determine the servo reference voltage value U_s, where is a proportionality coefficient. The outer loop establishes a GM model based on a sequence $Y(K)$ of output values to predict the next steps $\hat{Y}(K+i+1)$. When having compared these predicted values with the requirements Y^*, a sequence $e(K)=\hat{Y}-Y^*$ of errors are found. These error values are then fed back into the system to adjust the proportionality coefficient K_1 and the servo reference voltage U_s in order to adjust the inner loop. That is,

$$\Delta U = K_1(Y^* - \hat{Y}), U_s = K_2\hat{U}_g - \Delta U$$

Therefore, $U_s = K_2\hat{U}_g - K_1(Y^* - \hat{Y})$, where the parameters K_1, K_2 are determined by experiments.

Example 9.4. Let us now look at the grey predictive control for the vibration of a rotor system.

The theory and methods for active vibration control of rotors have caught more attention in recent years. Many new control theories, such as neural network theory, time-delay theory, self-learning theory, fuzzy theory, and H^∞ theory have been gradually employed in the research of the active control theory of rotors, leading to some good outcomes. For a Jeffcott symmetric rotor follower system with an electromagnetic damper as its executor, we employ the control theory and methods of grey predictions to investigate an active amplitude control of vibration. We first establish a grey predictive control module with the GM(1,1) as its main component. In the vibration control system of the rotor, let $I^0(k)$ and $x^0(k)$, $k=1,2,\ldots,n$, respectively be electric current inputting into the electromagnetic damper and the corresponding maximum output amplitude of the rotor vibration. By employing the available experimental measurement results from the literature, we obtain a set of data of $I^0(k)$ and the relevant $x^0(k)$, as shown in Table 9.1, when the sensitivity of the transducer is 10^4V/m.

Table 9.1 Sampled data of $I^0(k)$ and $x^0(k)$ when the transducer's sensitivity is 10^4V/m

$I^0(k)$ (A)	0.1	0.125	0.175	0.225	0.325
$x^0(k)$ (dm m)	1.4	1.35	1.2	0.9	0.65

Based on the mechanism of the GM(1,1) modeling, we establish the following modification model of the system based on the errors of the grey predictions:

$$\hat{a}^{(0)}(k+1) = -a[x^{(0)}(1) - \beta]e^{-ak} + \delta(k-i)(-a')[q^{(0)}(1) - \beta']e^{-a'k}$$

where $a = 0.1862; x^{(0)}(1) = 1.4; \beta = 9.3298; a' = 0.14; q^{(0)}(1) = 0.36; \beta' = 3.78$, and

$$\delta(k-i) = \begin{cases} 1 & k \geq i \\ 0 & k < i \end{cases}$$

The design of our grey predictive control of the rotor system is shown in Figure 9.12.

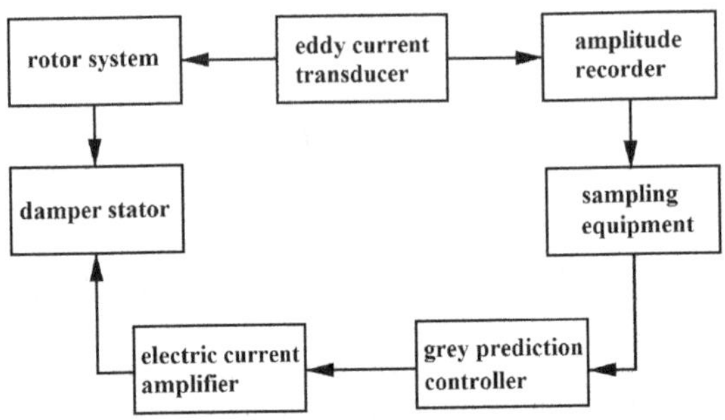

Fig. 9.12

In this control system, the displacement signal of the rotor system is measured by the eddy current transducer, the sampling equipment collects the data from the amplitude recorder, and through the effect of the grey prediction controller, controlling voltage is produced. This voltage is transformed into a controlling electric current through the current amplifier. Then, when this electric current flows through the stator coil of the electromagnetic damper, an electromagnetic force is created, which in turn controls the amplitude of vibration of the rotor within the expected range so that the system's stability is achieved.

For this grey predictive control system developed for the vibration of the said Jeffcott symmetric rotor follower, our computer simulation, when compared to the physical measurements of the amplitudes, indicates that the maximum amplitudes under the control are only about 7% of those physically observed without the control imposed on the rotor system. Table 9.2 respectively provides the results of the maximum amplitudes of two separate computational simulations and the actual measurements X_{1m}, X_{2m}, and X_{3m}, along with the change in the static electricity i of the electromagnetic damper, when the sensitivity of the transducer is $k_1 = 10^4$ V/m, and the corresponding errors e_{12}, e_{13}, and e_{23} between X_{1m} and X_{2m}, X_{1m} and X_{3m}, and X_{2m} and X_{3m}.

Table 9.2 Simulations and actual measurements

$I(A)$	X_{1m} (m)	X_{2m} (m)	X_{3m} (m)	e_{12} (%)	e_{13} (%)	e_{23} (%)
0.1	1.41×10^{-4}	1.4×10^{-4}	1.4×10^{-4}	0.71	0.71	0
0.125	1.27×10^{-4}	1.227×10^{-4}	1.35×10^{-4}	3.5	-5.93	-9.11
0.175	1.03×10^{-4}	0.949×10^{-4}	1.2×10^{-4}	9	-13.8	-20.92
0.225	0.83×10^{-4}	0.745×10^{-4}	0.9×10^{-4}	11.4	-7.78	-17.22
0.325	0.55×10^{-4}	0.46×10^{-4}	0.65×10^{-4}	19.57	-13.38	-29.23

Chapter 10
Introduction to Grey Systems Modeling Software

In 1982, Julong Deng initiated and established the grey systems theory. Currently, grey systems theory has been widely applied in areas like social science, economics, agriculture, meteorology, military science, etc., providing solutions to a large number of practical problems and challenges met in daily lives and day-to-day works. To this end, various versions of grey systems modeling software have played a very important role in such large scale practical applications of grey systems theory. Along with the rapid development of information technology, high level programming languages have gradually matured, applications of computing packages have becoming daily routines, and grey systems modeling programs are also becoming sophisticated.

In 1986, Xuemeng Wang and Jiangjun Luo created their grey systems modeling software using BASIC language and published *Programs of Grey Systems' Prediction, Decision-Making, and Modeling*. In 1991, Xiuli Li and Ling Yang respectively developed grey modeling software using GWBASIC and Turbo C. In 2001, Xuemeng Wang, Jizhong Zhang, and Rong Wang published the book, entitled "Computer Procedures for Grey Systems Analysis and Applications," in which they listed the structure and procedure codes established for grey modeling. All of these computer software packages were developed on the DOS platform and have become obsolete in the more user friendly Windows framework.

In 2003, Dr. Bing Liu developed the first grey systems modeling software for the Windows using VisualBasic6.0. As soon as this package was available, it was most welcomed in the community of scholars and practitioners of the grey systems research, and became the first choice of application in the field of grey systems modeling. With time as the technology of software development evolves, the habit of computer operation changes, and continued progress in the research of the grey theory, some of the weaknesses of the package were found, including mainly the following:

- Data entry was tedious

The single worksheet frame limited the operational flexibility on data sequences. It was especially inconvenient when large amounts of data were dealt with in clustering analysis and grey decision-making. Additionally, the only available way for data entry made users feet tired so that the efficiency and accuracy of data entry were greatly affected.

- The classification of the modules was not scientifically sound

This software system divided modules according to the number of data participating in the modeling process. However, the modules should have been designs according to functions.

- The software system could not show the relevant computational procedures

Most of the users of grey systems theory have been scientists and practitioners. Their purposes of applying the software system were mostly for their scholarly works. So, other than the computational outcomes they are also very interested in knowing the specific procedural details. However the original software package could only provide the final results of computation and was unable to reveal the relevant computational details.

- The system's capability was disconnected with the most recent research progress

Grey systems theory has been an extremely active area of the broader systems science. In particular, in the most recent years, some works with high practical values appeared. However, the software system was not upgraded along with the research progress, causing the capability of the software obsolete in terms of its functions and capability.

- Problems with the choice of the package development

VisualBasic6.0 is a graphics-based software development tool created by Microsoft Company. Due to its simplicity, functionality, versatility, and other strengths, as soon as it was introduced, it was welcome received by many software developers. However, because VisualBasic6.0 is an IDE (integrated development environment) based on BASIC, while BASIC is one typical language with many known weaknesses, it greatly limits its ability of application in scientific computing with relatively high requirements of accuracy. So, the grey systems modeling software, developed using VisualBasic6.0, inherently suffers from many inconveniences.

10.1 Features and Functions

On one hand, an ideal grey systems modeling software package needs to have the computational power to handle practical models, and on the other hand it has to be able to deal with the user confirmation, registration, and other elated affairs. The software system, accompanying this book, sufficiently combines the capabilities of the C/S (client/server) and B/S (Browser/Server) modules, where the C/S part completes computational functions, while the B/S part handles the relevant operations that serve the user and his communication with the server. On the basis of improving the existing systems, the design of this package focuses more on the reliability, practicality, compatibility, upgradability, accuracy, operational convenience, and pleasant appearance to the eye. This package has the following characters:

- Data entry is convenient and fast

For data sequences of the same kind, the package provides a rectangular window, into which the user can simply copy the sequences with one operation. For the module of grey clustering and grey decision-making that involves large amounts of data, it is evidently inconvenient to employ the traditional way to enter the data values. For such instances, the user can enter data in an EXCEL document and then open the data file into this package system. This software system makes use of the powerful data entry ability of EXCEL while making data entry convenient for the user.

- Modules are designed according to functionalities

In software engineering, a module is a relatively independent system unit of procedures. Each such unit of procedures handles and materializes a relatively independent task. Speaking in daily language, it contains a group of independent procedures. Each program module has its own external characteristics, such as its own name, label, interfaces, etc. During the design of this software package, the developer scientifically sorted and organized the contents of grey systems theory, defined the relevant functions and related modules.

- This system provides operational details as well as periodic results

For the modules, in which the computational procedures are complicated and intermediate results are also important, the system provides a textbox that can store and show multi-line operational details. The user can monitor changes of the data in each computational step so that he can further understand how the model operates. Also, to remind the user the relevant formulas employed in the model, the software interface provides the relevant information.

- The functionalities of the modules are greatly expanded

Based on the current situations of practical applications of each part of grey systems theory, combined with the most recent research results, this software system is most up to the date. It includes: weakening operators (mean weakening buffer operators, geometric mean weakening buffer operators weighted mean weakening buffer operators), strengthening operators (mean strengthening buffer operators, geometric mean strengthening buffer operators, weighted mean strengthening buffer operators), grey incidence analysis (relative degree of incidence, closeness degree of incidence), clustering analysis (based on center-point triangular whitenization weight functions), grey prediction (the GM(1, n) model, DGM(1,1) model), greu decision analysis (intelligent grey target decision making), and other contents.

- The degree of accuracy of the computational results can be adjusted

For different systems, their precisions of computation are different. In this software system, there is a ComboBox, which can take selections on the computational precision. So, the user can choose his desired degree of accuracy for his work.

- The operation of the software system is convenient and easy to learn

This software system is based on the Windows interface using pull-down menus, where the commonly employed modeling techniques of grey systems theory are effective gathered together. The user only needs to have an elementary understanding on how a desktop PC works to successfully apply this software system. At the same time, this system has a relative strong ability to locate and correct mistakes. When an illegal operation is performed, the system will provide an accurate and detailed hint.

- This system is developed using Visual C#

C# is an object-oriented programming language Microsoft Company created and an important part of Microsoft's .NET development environment. And, Microsoft Visual C# is an integrated development environment (IDE) constructed on C# by Microsoft. It is designed for the operation of many application software packages created on the .NET framework. C# possesses powerful capabilities, type safety, object orientation, and other superb functions. It is currently the main development tool of C/S software architecture.

10.2 Main Components

On the basis of the currently available achievements of grey systems theory and considering practical applications, the software system modules as shown in Figure 10.1 are constructed.

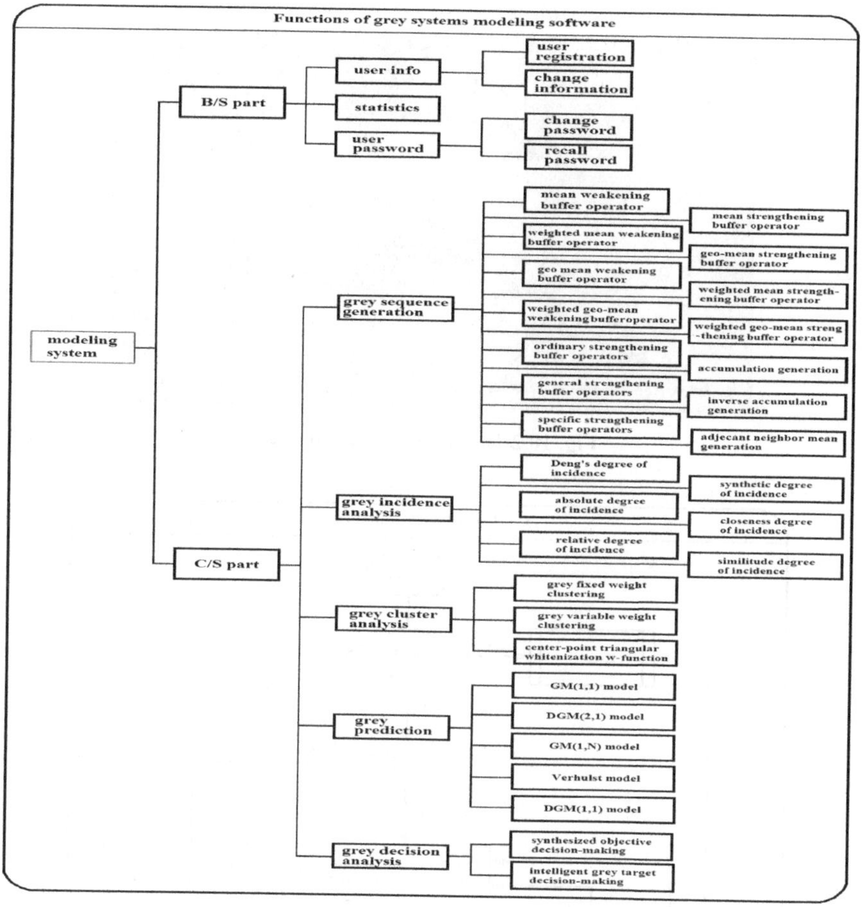

Fig. 10.1

10.3 Operation Guide

10.3.1 The Confirmation System

To verify the legal ownership, the user needs to enter his account number and password before he can actually start using the system. However, if every time the user has to confirm his legal ownership of the software package, it will become an annoyance. So, to guarantee the legal ownership of the user and maintain the simplicity of operation of the system, the system applies the XML-based client programming technique. When the user attempts to run the program the first time, the system will prompt him to provide the needed account number and password.

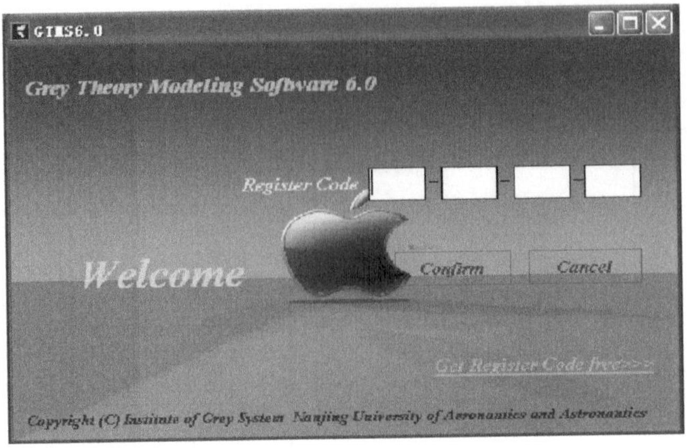

Fig. 10.2 The confirmation window

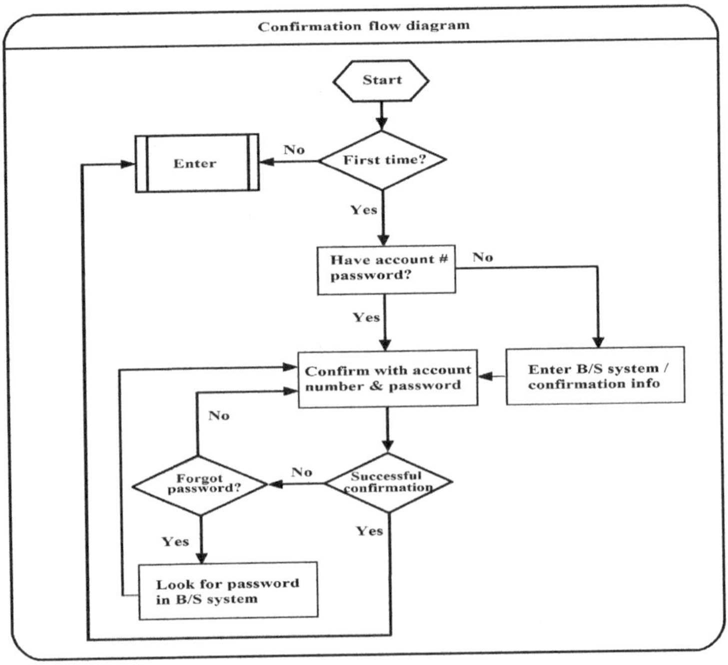

Fig. 10.3 The confirmation flow chart

The provided data will then be delivered through the internet to the data base located at the server to verify their legality. When the user attempts to use the program at different, subsequent occasions, he can directly enter into the main interface window to avoid the annoyance of having to enter account number and password every time.

At the first time of confirmation, if the user does not have any account number or password, then he needs to click at "User registration" button, see Figure 10.2, to do a no-fee registration (B/S). If the user forgot his password, he can click on the "Recall password" button to retrieve his password. Figure 10.3 is the flow chart of confirmation.

10.3.2 Using the Software Package

After the user successfully confirmed his legal ownership, he will enter the system's main interface window, as shown in Figure 10.4. Various modules (and their sub-modules) of grey systems theory are mainly administrated through menus. Figure 10.5 provides the flow chart of various sub-modules of the system.

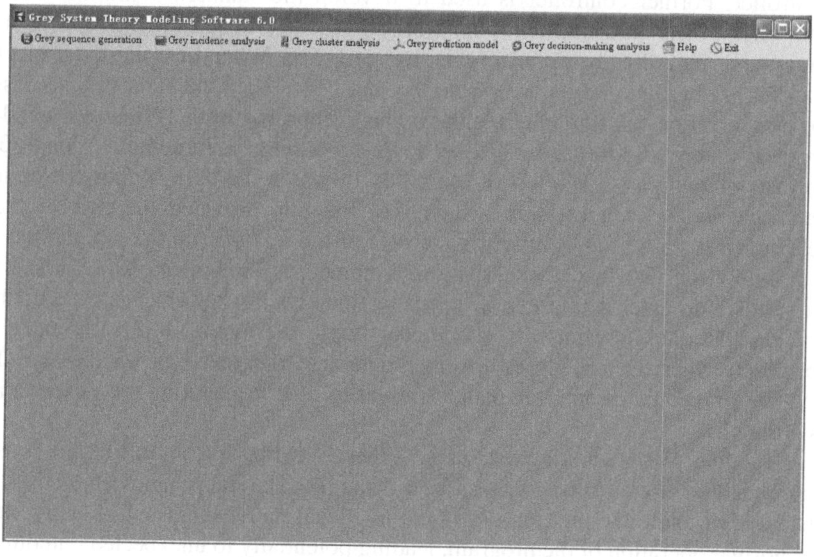

Fig. 10.4 The main interface window

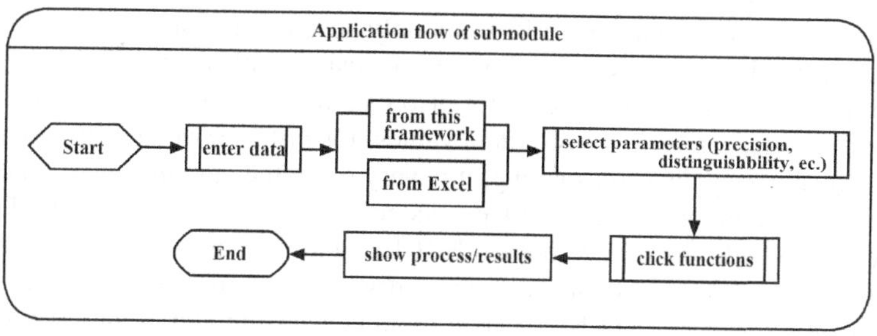

Fig. 10.5 Sub-modules

10.3.2.1 Entering Data

Before running the program, one needs to first enter data into the software package and specify the system parameters. As what has been mentioned earlier, there are two ways to input data. One can directly enter data into the provided text box or import data from an external EXCEL document. For those modules that require large amounts of data, the only way provided for entering data is through importing from EXCEL documents. The following, we will look at the details of these two methods of data entering.

- Enter data directly into the provided text box

With VisualC#, there are two kinds of controllers that are available for directly entering data. One is the TextBox controller, and the other the ComboBox controller. Former controller is used to develop the standard Windows' editing controller of the so-called text-box, which is used to acquire the user's input or show what had already stored in the storage space. When entering data into the text-box, right click the mouse inside the text box. When the cursor blinks inside the text box, one can start entering data. The ComboBox of the Windows' window group is mainly used to show data in down-drop list box. In default, ComboBox consists of two parts: The top is a text box in which the user is allowed to enter data, and below is a list box, which is a list box provided the user to make selections. Because the ComboBox consists of the text box on the top and the list box below the text box, that is why it is named a BomboBox. When using the ComboBox to enter data, the user needs to first check whether or not the list box contains the data he wants. If yes, he can simply use mouse to directly make the selection; otherwise, he needs to enter data into the text box on the top. The detailed procedure of entering data is similar to that of operating the TextBox and is omitted.

Notice: When entering data using either TextBox or ComboBox, the state of entry method needs to be adjusted to half-angle. The data values entered in the state of full angle the program will treat as illegal data, which will directly affect the normal operation of the program, leading potentially to unexpected outcomes.

- Import data from EXCEL document

Both the TextBox and ComboBox can only accept small amount of data values. For entering large sums of information, using either the Textbox or the ComboBox not only is inefficient, but also can also lead to errors. To resolve the problem of entering large sums of data values, for instance when dealing with grey clustering and grey decision-making, it is very often the case that large amounts of information are involved, this software system makes use of the powerful EXCEL. First enter and edit the needed data in EXCEL, and then use the provided interface to import the EXCEL data into the software system. Excel is one of the component of Microsoft Office. It is a tabulated testing and computing software developed for the operating systems of Windows and Apple' Macintosh. Its straightforward interface, excellent capabilities of computation and graphics make it the most widely employed PC software used for data analysis. Through EXECEL, our software package system can conveniently acquire data.

Each EXCEL document generally contains three tables, respectively labeled as Sheet1, Sheet 2, and Sheet3. When an EXCEL document opens, it generally shows Sheet1. When entering data, according to the system's requirements, one can directly type in the corresponding values in the relevant rows and columns. Upon finishing the data entry into the EXCEL document, one can employ our system's input function to import the EXCEL data. When importing an EXCEL document of data, first select the path from which EXCEL file is located. As soon as the importing path and location of the file is confirmed, the data will be successfully imported. In fact, the process of importing data is that through which the EXCEL file is connected to our system through a specific path so that the data in the EXCEL file can be mapped into the data base controller DataGridView.

DataGridView is a data base controller of VisualC#, which can exactly and entirely reveal the data from the source file. Through the DataGridView controller of VisualC#, the acquisition of data from an EXCEL file is materialized. However, our system does not provide any of the editing capabilities of DataGridView. In other words, if it is found in DataGridView that there is an error in the data, this error cannot be corrected directly within DataGridView. Instead, one has to return to the original EXCEL file to make the correction and then reimport the entire corrected file back into the grey systems modeling package.

Notes:

- DataGridView controller does not have any editing capability. To make changes in the data, one has to do it in the original EXCEL file.
- When entering data into an EXCEL document, one needs to do so in the mode of "half-angle."All data entered in the mode of "full-angle" will be treated as illegal entries, which will directly affect the normal operation of the grey systems modeling package, potentially leading to unexpected outcomes.
- The table names of the EXCEL file have to be the default Sheet1, Sheet2 and/or Sheet3 without any modification; otherwise, the import of data will be affected.
- The data entry field of EXCEL is very large. However, generally, one often needs a few rows and columns. So, make sure that there is not symbol. including blank cells, accidently entered in other area of the field. Otherwise, the data transfer will be affected.

10.3.2.2 Model Computations

1. Grey sequence generations

Click on "Sequence generation." From the pull-down menu that appeared select the module according to the practical modeling need. The corresponding detailed modeling interface appears. In the following, let us use "mean weakening buffer operator" as an example to illustrate how to apply grey sequence generations. What is shown in Figure 10.6 is the interface of the mean weakening buffer operator.

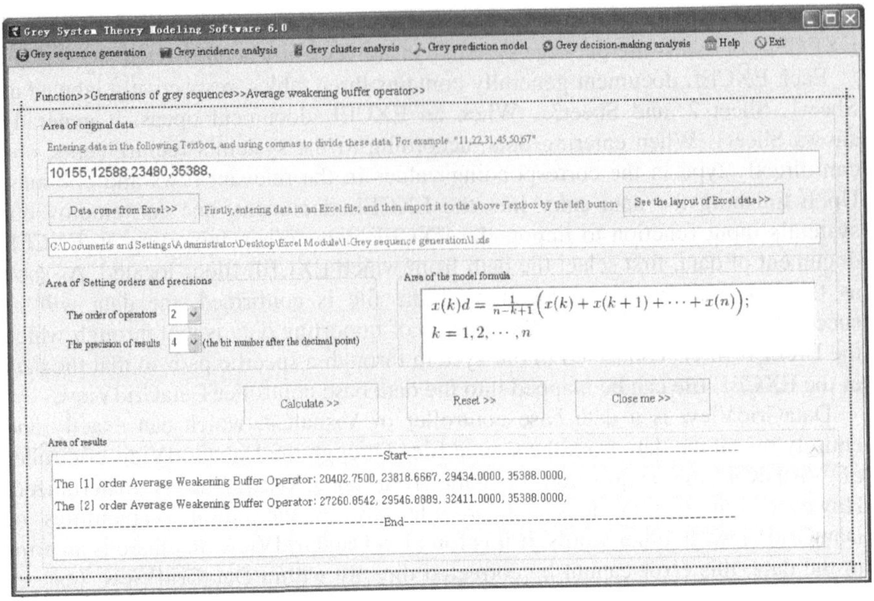

Fig. 10.6 The interface of the mean weakening buffer operator

This interface window contains three main areas: the first shows the original data sequence, which is the area for data entry or importing data; the second area the "order and outcome precision," in which one adjusts the order of the operator being applied and the corresponding precision of the computational outputs based on his need of modeling; and the third the area in which the computational results are shown. After the data entry is completed, click on "mean weakening buffer operator (AWBO)" button. Immediately, the generated sequence will appear in the generated sequence window. Figure 10.7 shows a work sheet of an EXCEL document. When applying this capability of EXCEL and deciding on importing data from EXCEL, the user has to follow this shown format exactly.

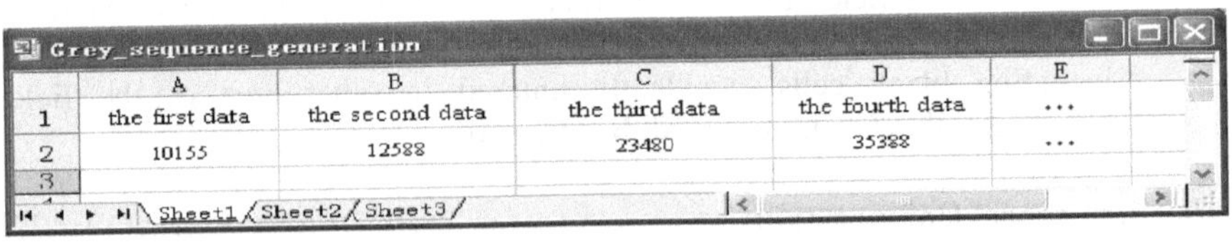

Fig. 10.7 The required EXCEL file format

2. Grey incidence analysis

Similar to the generation of grey sequences, all parts of incidence analysis, except that of Deng's degree of grey incidence, provide two ways to input data, the details of which are omitted here. For Deng's degree of grey incidence, due to the need of large amount of data, this software system provides only the method of data entry through EXCEL documents without making the option of directly data entry available. Figure 10.8 shows the editing format of an EXCEL document, while Figure 10.9 shows the complete work interface.

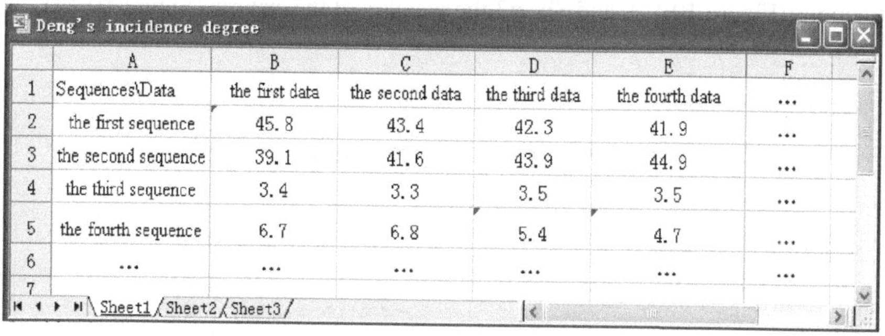

Deng's incidence degree

	A	B	C	D	E	F
1	Sequences\Data	the first data	the second data	the third data	the fourth data	...
2	the first sequence	45. 8	43. 4	42. 3	41. 9	...
3	the second sequence	39. 1	41. 6	43. 9	44. 9	...
4	the third sequence	3. 4	3. 3	3. 5	3. 5	...
5	the fourth sequence	6. 7	6. 8	5. 4	4. 7	...
6	...	...	...	...	...	...
7						

Sheet1 / Sheet2 / Sheet3

Fig. 10.8 The required EXCEL file format

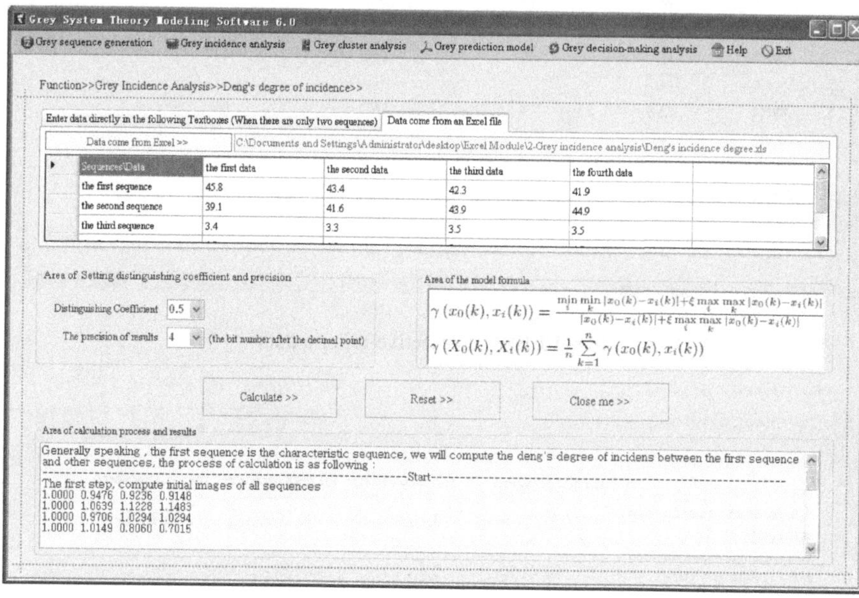

Fig. 10.9 The complete work interface of Deng's degree of grey incidence

3. Grey clustering analysis

Similar to Deng's degree of grey incidence, grey clustering analysis also requires a large amount of original data with the exception that in grey clustering analysis, the data may have several different types, including objective-criterion data, whitenization weight functions, and criteria weights. Therefore, for grey clustering analysis, the system again only provides one way to enter data by importing from EXCEL files. The key of using this part of the functions is to correctly edit the different types of data in the EXCEL documents. Sheet1 contains the objective-criteria data (Figure 10.10), Sheet2 the corresponding whitenization weight functions (Figure 10.11), and Sheet3 the weights of the criteria (Figure 10.12).

Grey fixed clustering [兼容模式]

	A	B	C	D	E	F
1		Parameter values	Parameter values	Parameter values	Parameter values	...
2	Object	22.5	4	0	0	...
3	Object	79.37	6	600	0.75	...
4	Object	144	7	300	0.75	...
5	Object	300	6.1	189	12	...
6	Object	456	12	250	12	...
7	Object	189	8	700	1.5	...
8	Object	369	8	1300	2.25	...
9	Object	1127.11	16.2	550	3	...
10	Object	260	11	600	1	...
11	Object	200	8	600	1.25	...
12	Object	475	10	1000	0.75	...
13	Object	314.1	8	900	0.75	...
14	Object	282.8	7.4	1300	0.5	...
15	Object	240	8	1200	0.5	...
16	Object	160	5	1000	0.25	...
17	Object	270	8	1200	0.25	...
18	Object	9	1	200	0	...
19	...	...	...	...	...	...
20						

Sheet1 Sheet2 Sheet3

Fig. 10.10 The objective-criteria data

Grey fixed clustering [兼容模式]

	A	B	C	D	E	F
1		the first parameter	the second parameter	the third parameter	the forth parameter	...
2	Whitenization weight function of the first grey class	100,300,-,-	3,10,-,-	200,1000,-,-	0.25,1.25,-,-	...
3	Whitenization weight function of the second grey class	50,150,-,250	2,6,-,10	100,600,-,1100	0,0.5,-,1	...
4	Whitenization weight function of the third grey class	-,-,50,100	-,-,4,8	-,-,300,600	-,-,0.25,0.5	...
5	...	...	...	...	...	...

Sheet1 Sheet2 Sheet3

Fig. 10.11 The corresponding whitenization weight functions

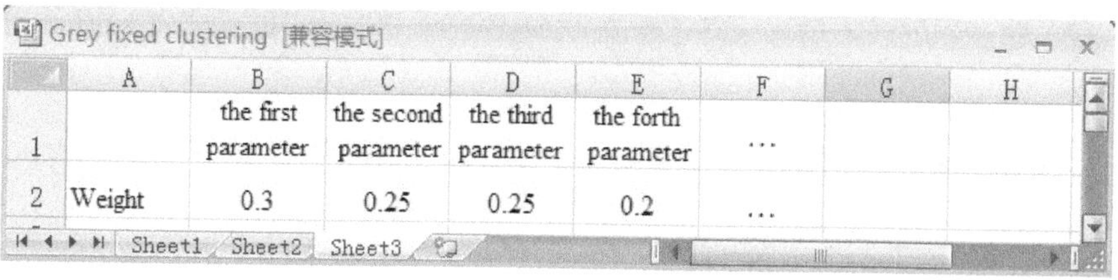

Fig. 10.12 The weights of the criteria

Figure 10.13 shows the operating interface of a grey clustering analysis. As for how to apply grey variable weight clustering and that based on center-point triangular whitenization weight functions, the operational details are similar and omitted.

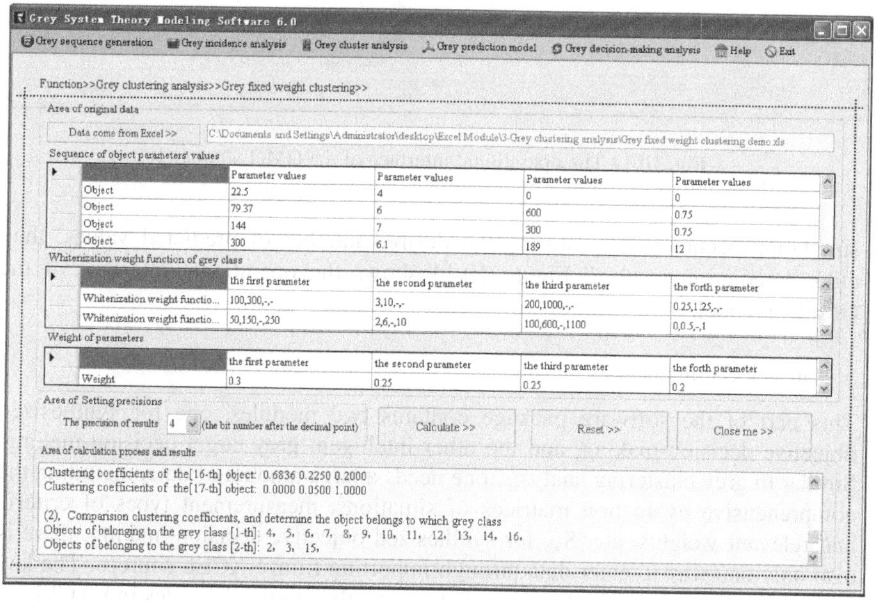

Fig. 10.13 The operating interface of a grey clustering analysis

4. Grey prediction models

Grey prediction models stand for an important part of grey systems theory. The operation of each individual prediction model is roughly the same. So, let us use the GM(1,1) model to illustrate how to use the software system. The main steps include: enter or import data; click the computation, simulation, prediction" button to compute the model parameters and the simulated values, and select the

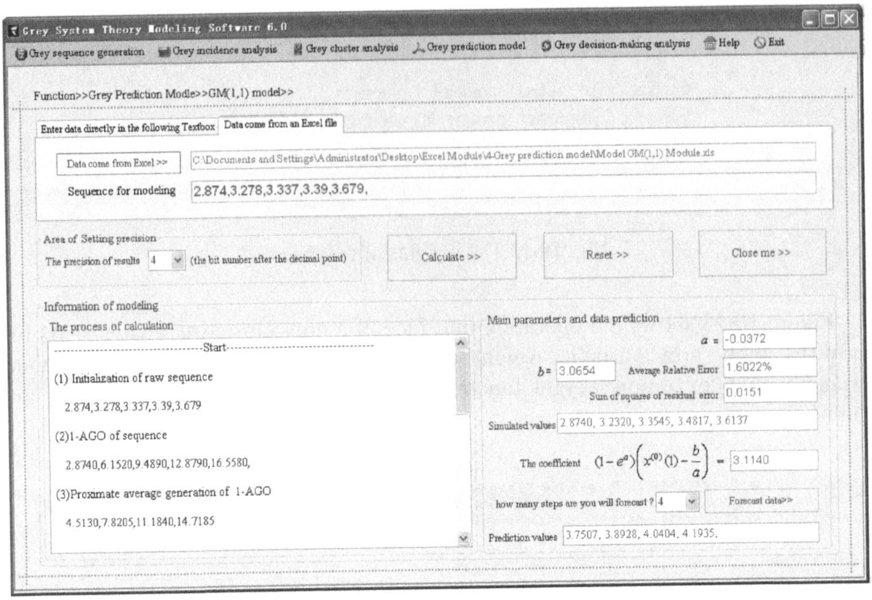

Fig. 10.14 The operational interface of the GM(1,1) model

simulation accuracy; by entering the desired number of predicted values, then click "prediction results." Figure 10.14 shows the operational interface of the GM(1,1) model.

5. Grey decision analysis

This part of the software package contains two modules, one for synthesized objective decision-making, and the other intelligent grey target decision-making. Similar to grey clustering analysis, one needs a large amount of data, including the comprehensive evaluation matrices of situations, measurement types of criteria and relevant weights, etc. So, for synthesized objective decision-making, there is also only one way to enter data through importing from EXCEL sources. The key of applying this part of the functions is to correctly edit the data files in EXCEL.

(1) Synthesized objective decision making

In Sheet1 of the EXCEL file, entered in the first row are the labels indicating the meanings of the values of the columns, while the row values from B – M stand for the matrix of the corresponding comprehensive evaluation scores. Column N provides the measurement types, while the entries in the O column are the corresponding weights of the criteria. When using this part of the functions, one needs to be careful in following the exact layout as shown in Figure 10.15. Figure 10.16 shows the entire operating page.

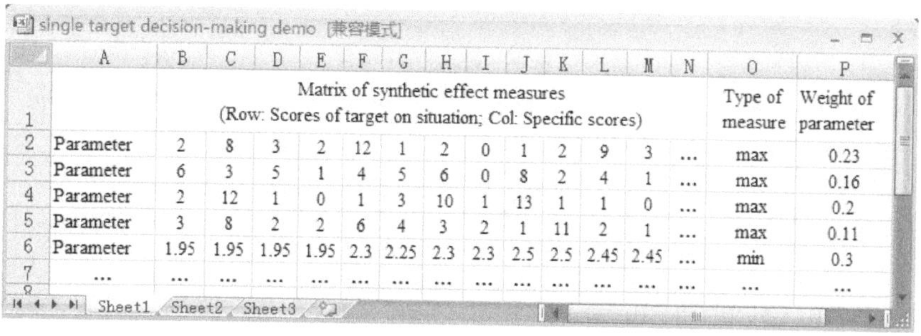

Fig. 10.15 The exact layout of data page for synthesized objective decision making

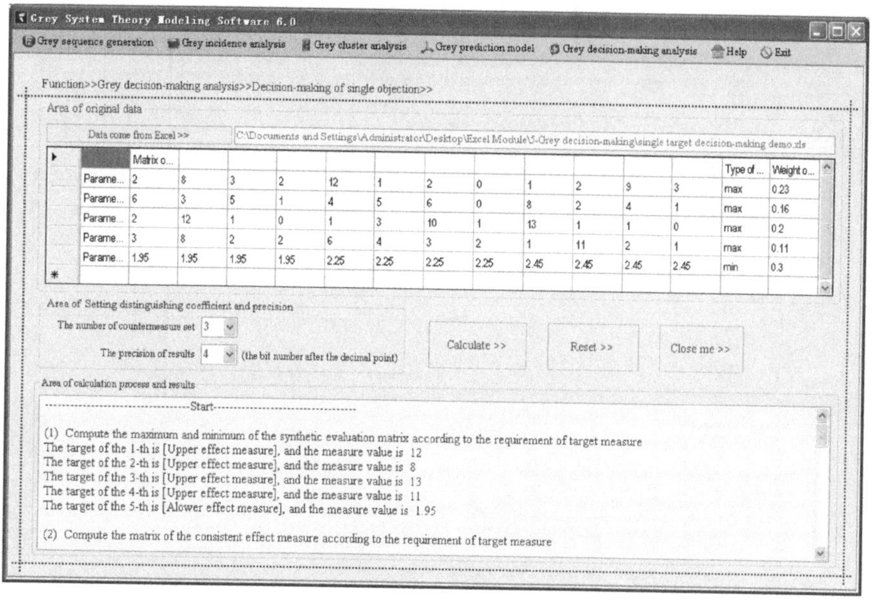

Fig. 10.16 The entire operating page of a synthesized objective decision making

(2) Intelligent grey target decision making

The data layout of an intelligent grey target decision making is the same as that of any synthesized objective decision making except that there is an additional column of threshold value, as shown in Figure 10.17, where the interval [14,18] means that the lower effect threshold value is 14 and the upper effect value is 18. Figure 10.18 shows the entire operating page of an intelligent grey target decision making.

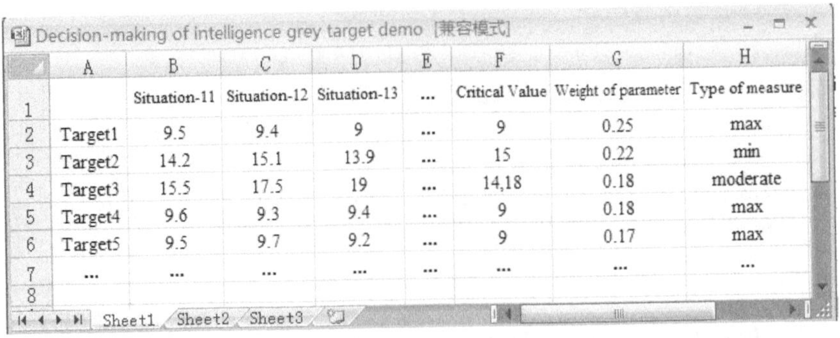

Fig. 10.17 The exact layout of data page for intelligent grey target decision making

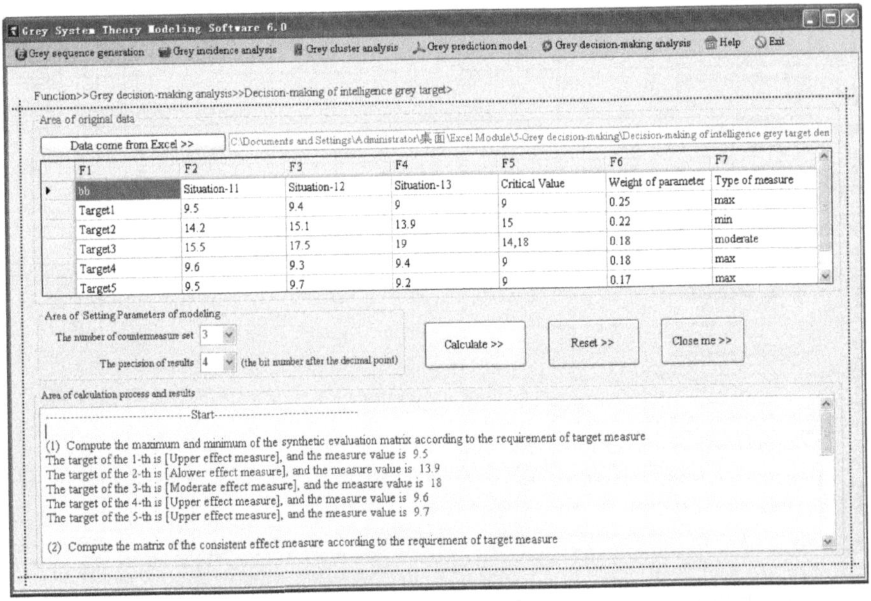

Fig. 10.18 The entire operating page of an intelligent grey target decision making

Appendix A
Interval Analysis and Grey Systems Theory

The purpose of this appendix is to clarify the similarity and difference between interval analysis and grey systems theory. Based on what is presented below and what has been presented earlier in this book, it is quite clear that interval analysis should be seen righteously as a proper sub-portion of grey mathematics. In particular, what is common to these two areas of study, interval analysis and grey systems theory, are described by the contents in Chapter 2 of this book up to equ. (2.8). Beyond that point, both studies have gone their own separate ways.

Due to its constraint on intervals of real numbers and its sole focus on machine computations, the scope of application of interval analysis, in comparison to that of grey systems theory, has been quite limited, considering the fact that it was formally proposed in 1962 by Ramon Moore. Nevertheless, what has been achieved in interval analysis can be a good supplement for furthering the development of grey systems theory. On the other hand, what has been established in grey systems theory can significantly help open doors of generalizing some of the key results of interval analysis to many potential areas of practical applications. This end, by referencing back to the success stories of the scientific history, can be seen as the key for an area of scientific investigation to find its own place in the realm of long lasting theories.

A.1 A Brief Historical Account of Interval Analysis

Interval analysis is a method originated independently by Sunaga (1958) and Moore (1962) and developed ever since the 1950s by a score of mathematicians as an approach to putting bounds on rounding errors and measurement errors in mathematical computations. It represents a numerical quantity as an interval within which the true value of the quantity lies. This end is exactly the same as the concept of grey interval numbers. The arithmetic operations, such as +, −, ×, ÷, of interval numbers are defined exactly the same as those of the interval grey numbers given in the start of Chapter 2 in this book.

As other scientific concepts, the idea of using intervals to estimate numbers has appeared throughout the history in different names. For example, Archimedes estimated the value of π using the interval [223/71, 22/7] in the third century BC. In 1931, Rosalind Cicely Young of University of Cambridge published a paper, entitled "algebra of many-valued quantities," that gives rules for calculating with intervals and other sets of real numbers (Hayes, 2003). Arithmetic work on range numbers to improve reliability of digital systems were then published in a textbook on linear algebra by Paul Dwyer (1951) of University of Michigan, where intervals were used to measure rounding errors associated with floating-point numbers. M. Warmus (1956) established formulas for interval arithmetic operations in both endpoint and midpoint-radius formulations, and the definition of the centered outer multiplication. He then developed an interval space diagram using midpoint-radius coordinates in (Warmus, 1961). (Sunaga, 1958) contained many of the same ideas of interval analysis as later (Moore, 1962) developed in his PhD dissertation. As of this writing, most of the modern literature on interval analysis traces back to (Moore, 1966) as the official start of this area of research. One of the merits of interval analysis is that starting with a set of simple principles, it provides a general method for automated error analysis, not just errors resulting from rounding.

In the following twenty years, groups of German scholars, headed by Götz Alefeld (Alefeld and Herzberger, (1974) and Ulrich Kulisch (1989), carried out pioneering works. In particular, among others, Karl Nickel (1966) explored more effective implementations; and Arnold Neumaier (1990) improved containment procedures for the solution set of systems of equations. In 1968, William Kahan published works on extended arithmetic. Eldon Hansen (1965; 1968; 1969; 1992) dealt with interval extensions for linear equations and then provided crucial contributions to global optimization. Karl Nickel (1966) published in interval analysis. Irene Gargantini and Peter Henrici (1972) investigated complex interval arithmetic. Helmut Ratschek and Jon George Rokne (1984) developed the branch and bound methods, which till then had been only applied to integer values, by using intervals to provide applications for continuous values. Rudolf Lohner (1988, unpublished manuscript) developed Fortran-based software for reliable solutions for initial value problems using ordinary differential equations. Hansen and Walster (2003) studied reliable solution of linear systems and other topics. Since the 1990s, *Reliable Computing*, currently a Springer journal dedicated to the reliability of computer-aided computations, has been published.

A.2 Main Blocks of Interval Analysis

As expected, because the focus of interval analysis is on the reliability of machine computations, most of its contents are centered on the topic of how to carry out traditional computations on machines with more or guaranteed accuracy, so to speak. The rest of this presentation is based on (Moore et al., 2009). For a more detailed account of this subject matter, please consult with (Jaulin et al., 2001).

One of the motivational reasons for studying interval analysis is that rounding errors when accumulated sufficiently can destroy a numerical solution. For instance, one needs to compute x_{75} based on $x_{n+1} = x_n^2$, for n = 0, 1, 2, 3, …, with the initial value being $x_0 = 1 - 10^{-21}$. By performing the computation with 10-place arithmetic leads to the approximated values:

$$x_0 = 1,\ x_1 = 1, \ldots,\ x_{25} = 1.$$

By using 20-place arithmetic, he obtains the same sequence of values. Hence, the two values of x_{75} agree to all 10 places as obtained in the first computation. However, the exact value x_{75} is less than 10^{-10}.

To help control the accumulation of rounding and other relevant errors, interval analysis can be divided roughly into four main blocks: 1) the interval number system and the properties of interval arithmetic; 2) interval functions, interval sequences, and interval matrices; 3) interval Newton methods; and 4) integration of interval functions and related computational problems of integral and differential equations.

A.2.1 Interval Number System and Arithmetic

For the first block of contents of interval analysis, it contains works on the interval number system and the properties of interval arithmetic. In particular, let $X = [\underline{X}, \overline{X}]$ be an interval number with $\underline{X}$ and $\overline{X}$ being respectively the left and right endpoint of the interval. Then, each real number $x = [x, x]$ can be naturally seen as an interval number, known as a degenerate interval. The arithmetic operations +, –, ×, and ÷ are defined the same ways as what is shown in equs. (2.1) – (2.8) in this book. They can be written universally as

$$X * Y = \{x * y : x \in X \ \& y \in Y\}, \text{ where \& can be } +, -, \times, \text{ or } \div.$$

Given interval numbers X and Y, because each interval number is also a set of points of the real number line, one can study $X \cup Y$, $X \cap Y$, and the interval hull $X \,\underline{\cup}\, Y = [\min\{\underline{X}, \underline{Y}\}, \max\{\overline{X}, \overline{Y}\}]$. Corresponding to the geometry of interval X, one can define the width $w(X) = \overline{X} - \underline{X}$, absolute value $|X| = \max\{|\underline{X}|, |\overline{X}|\}$, and the midpoint $m(X)$ $\frac{1}{2}(\underline{X} + \overline{X})$. A partial order < between interval numbers X and Y is defined as follows: $X < Y$ iff $\overline{X} < \underline{Y}$ so that one can speak of positive and negative interval numbers.

An n-tuple of intervals, such as $X = (X_1, \ldots, X_n)$, is known as an n-dimensional interval vector. For a real vector $x = (x_1, \ldots, x_n)$, $x \in X$ means that $x_i \in X_i$, for $i = 1, \ldots, n$. For two interval vectors $X = (X_1, \ldots, X_n)$ and $Y = (Y_1, \ldots, Y_n)$, define $X \cap Y = (X_1 \cap Y_1, \ldots, X_n \cap Y_n)$; $X \subseteq Y$, iff $X_i \subseteq Y_i$, for $i = 1, \ldots, n$; the inner product $X \cdot Y = X_1 \cdot Y_1 + \ldots + X_n \cdot Y_n$; $w(X)$, the width of X, $= \max_i (X_i)$; $m(X)$, the midpoint of X, $= (m(X_1), \ldots, m(X_n))$; and $\|X\|$, the norm of X, serving as the absolute value of X, $= \max_i |X_i|$.

A.2.2 Interval Functions, Sequences and Matrices

The second block of interval analysis establishes interval functions, interval sequences, and interval matrices. Let g: $M_1 \to M_2$ be mapping from set M_1 to set M_2. The so-called united extension of g is the following set-valued mapping $\overline{g}$ from the power set of M_1 to the power set of M_2:

$$\overline{g}(X) = \{g(x) : x \in X\}, \text{ for any } X \in p(M_1). \tag{A.1}$$

In particular, for a $f(x_1, \ldots, x_n)$ real-valued function and interval numbers $X_1, \ldots, X_n$, its united extension is defined as follows:

$$f(X_1, \ldots, X_n) = \{f(x_1, \ldots, x_n) : x_i \in X_i, i = 1, \ldots, n\}. \tag{A.2}$$

A so-called interval extension of f, denoted F, is defined as such an interval-valued function F of n interval variables satisfying that for real arguments $x_1, \ldots, x_n$, $F(x_1, \ldots, x_n) = f(x_1, \ldots, x_n)$. If extensions F of real-valued functions f are obtained by simply replacing (1) the real variables by interval variables, and (2) the real arithmetic operations with corresponding interval operations, then the extensions F are known as natural interval extensions of f. (There are other forms of interval extensions, such as the centered form, the mean value form, the slope form, and the monotonicity form, all of the details of these forms are omitted). For instance, for $f(x) = 1 - x$, one of its interval extension is given by $F(X) = 1 - X$. In general, $F(X) \neq f(X)$. That is, for an arbitrarily given real-valued function f, one does not always have interval extensions identical to the united extension.

An interval function $F = F(X_1, \ldots, X_n)$ is inclusion isotonic, provided $X_i \subseteq Y_i$, for $i = 1, \ldots, n$, imply $F(X_1, \ldots, X_n) \subseteq F(Y_1, \ldots, Y_n)$. Then, one has the following (Moore, 1966):

Theorem A.1 (Fundamental Theorem of Interval Analysis). If F is an inclusion isotonic interval extension of f, then $f(X_1, \ldots, X_n) \subseteq F(X_1, \ldots, X_n)$.

For two intervals X and Y, by considering their distance $d(X,Y)$ as follows: $d(X,Y) = \max\{ |\underline{X} - \underline{Y}|, |\overline{X} - \overline{Y}| \}$, one can study the convergence and continuity of interval mathematics. In particular, a sequence $\{X_k\}$ of intervals is convergent if $\exists$ an interval X^*, $\forall\, \varepsilon > 0$, $\exists$ a natural number $N = N(\varepsilon)$ such that $d(X_k, X^*) < \varepsilon$, $k > N$, written as

$$X^* = \lim_{k \to \infty} X_k \text{ or } X_k \to X^* \tag{A.3}$$

and X^* is referred to as the limit of $\{X_k\}$. For instance, every nested interval sequence $\{X_k\}$, that is, it satisfies $X_{k+1} \subseteq X_k$ for all k, converges to the limit $\bigcap_{k=1}^{\infty} X_k$.

An interval extension F is Lipschitz in X_0, provided $\exists$ a constant L such that $w(F(X)) \le Lw(X)$, $\forall\, X \subseteq X_0$, where X may be an interval or an interval vector.

For any given natural number N, a uniform subdivision of an interval vector X $= (X_1, \ldots, X_n)$ is given by $X_i = \cup_{j=1}^{N} X_{i,j}$ and $X = \cup_{j_i=1}^{N} \left(X_{1,j_1}, \ldots, X_{n,j_n}\right)$, where

$$X_{i,j} = \left[\underline{X}_i + (j-1)w(X_i)/N, \underline{X}_i + jw(X_i)/N\right], j = 1, \ldots, N.$$

If $F(X)$ is an inclusion isotonic, Lipschitz, interval extension for $X \subseteq X_0$, then the interval quantity

$$F_{(N)}(X) = \cup_{j_i=1}^{N} F\left(X_{1,j_1}, \ldots, X_{n,j_n}\right) \tag{A.4}$$

is called a refinement of F over X, which stands for a method for computing arbitrarily sharp upper and lower bounds on the range of values of a real function.

If F is an inclusion isotonic interval extension of f with $F(X)$ defined for $X \subseteq X_0$, then $f(X) \subseteq F(X)$ for $X \subseteq X_0$ so that there is some interval-valued function $E(X)$ such that $F(X) = f(X) + E(X)$. In this case, $w(E(X)) = w(F(X)) - w(f(X))$ is known as the excess width of $F(X)$.

Theorem A.2. If $F(X)$ is an inclusion isotonic, Lipschitz, interval extension for $X \subseteq X_0$, then the excess width of the refinement

$$F_{(N)}(X) = f(X_1, \ldots, X_n) + E_N \tag{A.5}$$

is of order $1/N$, where $w(E_N) \le Kw(X)/N$ for some constant K.

By an interval matrix, it is a matrix with interval number entries. For an interval matrix $A = [A_{ij}]_{n \times m}$ and a real number matrix $B = [b_{ij}]_{n \times m}$, $B \in A$ means that $b_{ij} \in A_{ij}$, for $i = 1, \ldots, n$, and $j = 1, \ldots, m$. The norm of the matrix A is defined by

$$\|A\| = \max_{i} \sum_{j} |A_{ij}| \tag{A.6}$$

as an interval extension of the maximum row rum norm for real matrices. The width $w(A)$ is defined by

$$w(A) = \max_{i,j} w(A_{ij}) \tag{A.7}$$

And, the midpoint of A is the real matrix $m(A)$ whose entries are the midpoints of the corresponding cells in A. For the following system of linear algebraic equations

$$Ax = b \tag{A.8}$$

where $A = [A_{ij}]_{n \times n}$ is an $n \times n$ interval matrix and $b = [B_i]_{n \times 1}$ an n-dimensional interval vector, the set of all solutions may be very complicated and its computation is in general an NP-hard problem. The following result provides a computable test for existence of solutions to this system:

Theorem A.3. If the steps of a direct method for solving the previous system, such as Gaussian elimination, in interval analysis without attempting to divide by an interval containing zero, can be carried out without experiencing any overflow or downflow, then the system has a unique solution for every real matrix in A and every real vector in b, and the solution is contained in the resulting interval vector X.

For using the method of iteration to solve the previous system, multiply both side of the system by a matrix Y and let $E = I - YA$. If $\|E\| < 1$, then the sequence

$$X^{(k+1)} = \{Yb + EX^{(k)}\} \cap X^{(k)},\ k = 0, 1, 2, \ldots \tag{A.9}$$

where $X = [-1,1]\|Yb\|/(1-\|E\|)$, $i = 1, \ldots, n$, is a nested sequence of interval vectors containing the unique solution of the system for every real matrix in A and every vector in b. Because this sequence converges in a finite number of steps to an interval vector containing the set of solutions of the system, the following result holds:

Theorem A.4. The system (A.8) has a unique solution x for every real matrix in A and every real vector in b if $\|I - YA\| < 1$ for some matrix Y (real or interval). And, the solution vector x is contained in the interval vector $X^{(k)}$ of equ. (A.9) for every $k = 0, 1, \ldots$

A.2.3 Interval Newton Methods

The third block of contents of interval analysis deals with interval Newton methods. In particular, for a given continuously differentiable real-valued function $f(x)$, one seeks a solution of the following equation on the interval $[a,b]$:

$$f(x) = 0. \tag{A.10}$$

If such a solution x exists, it should satisfy $f(y) + f'(s)(x - y) = 0$, for any $y \in [a,b]$ in general and $y = m([a,b])$ in particular, where $s \in [x,y]$. So,

$$x = y - f(y)/f'(s) \tag{A.11}$$

Let $F'(X)$ be an inclusion monotonic interval extension of $f'(x)$ and define

$$X^{(k+1)} = X^{(k)} \cap N(X^{(k)}),\ k = 0, 1, 2, \ldots \tag{A.12}$$

where $N(X) = m(X) - f(m(X))/F'(X)$. So, equ. (A.11) implies x and $s \in N(X)$ if $y = m([a,b])$.

Theorem A.5. If an interval $X^{(0)}$ contains a zero x of $f(x)$, then so does $X^{(k)}$ for all $k = 0, 1, 2, \ldots$, as defined in equ. (A.12). And, the nested interval sequence $X^{(k)}$ converges to x if $0 \notin F'(X^{(0)})$.

For a finite system of nonlinear equations

$$f(x) = \begin{bmatrix} f_1(x_1,\ldots,x_n) \\ \vdots \\ f_n(x_1,\ldots,x_n) \end{bmatrix} = 0, \tag{A.13}$$

assume that f_i's are given real-valued continuously differentiable functions in an open interval D. Let F and F' be inclusion isotonic interval functions of f and f', respectively, defined on intervals vectors $X \subseteq D$. Then, the following result provides a computational test for the existence of a solution:

Theorem A.6. Assume that Y is a nonsingular real matrix that approximates the inverse of the real Jacobian matrix $F'(m(X)) = [F'(m(X))_{ij}]_{n \times n} = [\partial f_i(x)/\partial x_j]_{n \times n}$ at $x = m(X)$. For a real vector $y \in$ the interval vector $X \subseteq D$, define

$$K(X) = y - Yf(y) + \{I - YF'(X)\}(X - y) \tag{A.14}$$

If $K(X) \subseteq X$, then equ. (A.13) has a solution in $K(X) \subseteq X$.

Theorem A.7. Assume that $X = (X_1, \ldots, X_n)$ is an n-cube, that is, $w(X_i) = w(X)$, for $i = 1, \ldots, n$, $y = m(X)$, and Y a nonsingular real matrix. If

$$\|K(X) - m(X)\| < w(X)/2 \tag{A.15}$$

then in X the system (A.13) has a unique solution, to which the following algorithm converges:

$$X^{(k+1)} = X^{(k)} \cap K(X^{(k)}),\ k = 1, 2, \ldots \tag{A.16}$$

where $X^{(0)} = X$, $Y^{(0)} = Y$, $x^{(0)} \in X^{(0)}$ is arbitrary, $K(X^{(k)}) = y^{(k)} - Y^{(k)}f(y^{(k)}) + \{I - Y^{(k)}F'(X^{(k)})\}Z^{(k)}$, $y^{(k)} = m(X^{(k)})$, $Z^{(k)} = X^{(k)} - m(y^{(k)})$, and $Y^{(k)}$ is chosen as follows:

$$Y^{(k)} = \begin{cases} Y \left(\approx \left[m(F'(X^{(k)}))\right]^{-1}\right), & if \left\|I - YF'(X^{(k)})\right\| \le \left\|I - Y^{(k-1)}F'(X^{(k-1)})\right\| \\ Y^{(k-1)}, & otherwise \end{cases}$$

A safe starting interval for the interval Newton method is such an $X^{(0)}$ from which equ. (A.16) converges to a solution of the given nonlinear system (A.13). Theorem A.7 implies that $X^{(0)}$ is a safe starting interval if equ. (A.16) holds true.

Similarly, multivariate interval Newton methods can be established.

A.2.4 Integration of Interval Functions

The fourth block of contents of interval analysis investigates integration of interval functions and related computational problems of integral and differential equations. For the sake of communicating effectively, let $FC_n(X_0)$ be the class of all such real-valued function f defined for each $x = (x_1, \ldots, x_n) \in X_0 = (X_{10}, \ldots, X_{n0})$

by a finite sequence of arithmetic operations +, –, ×, ÷, or unary function of one variable of the elementary type: exp( ), ln( ), $\sqrt{\ }$, etc., on the basis of variables $x_1, \ldots, x_n$ and a given set of real coefficients.

Theorem A.8. Assume $f \in FC_1(X_0)$ and that F is an inclusion isotonic, Lipschitz, interval extension of f defined for all $X \subseteq X_0$. For a given natural number N, divide $[a,b] \subseteq X_0$ into N subintervals P: $X_1, \ldots, X_N$ so that $a \le \underline{X}_1 < \overline{X}_1 = \underline{X}_2 < \overline{X}_2 < \ldots < \overline{X}_N = b$. Then,

$$\int_a^b f(t)dt = \int_{[a,b]} f(t)dt \in \sum_{i=1}^{N} F(X_i)w(X_i) \tag{A.17}$$

And, there is a constant L independent of N and P satisfying

$$w\left(\sum_{i=1}^{N} F(X_i)w(X_i)\right) \le L\sum_{i=1}^{N} w(X_i)^2 \tag{A.18}$$

If the subdivision P in Theorem A.8 is uniform, satisfying $w(X_i) = (b-a)/N$, $i = 1, \ldots, N$, then the following result holds true:

Theorem A.9. $\int_{[a,b]} f(t)dt = \cap_{N=1}^{\infty} S_N(F;[a,b]) = \lim_{N\to\infty} S_N(F;[a,b])$, where $S_N(F;[a,b]) = \sum_{i=1}^{N} F(X_i)(b-a)/N$.

Assume that an interval-valued function $F(X)$ is continuous and inclusion isotonic for $X \subseteq X_0$. If for $[a,b] \subseteq X_0$ the sums $S_N(F; [a,b])$, as given in Theorem A.9, are well defined, then the interval integral of F is defined as follows:

$$\int_{[a,b]} F(t)dt = \cap_{N=1}^{\infty} S_N(F;[a,b]) \tag{A.19}$$

where t stands for a real variable although $F(t)$ may not be a degenerate interval value.

By applying the condition of continuity of F, it can be shown that there are continuous real-valued functions $\underline{F}$ and $\overline{F}$ such that for any real t, $F(t) = \left[\underline{F}(t), \overline{F}(t)\right]$; and

$$\int_{[a,b]} F(t)dt = \left[\int_{[a,b]} \underline{F}(t)dt,\ \int_{[a,b]} \overline{F}(t)dt\right]. \tag{A.20}$$

Let T_0 be an interval and $x(t)$ a real-valued function defined on T_0. An interval enclosure of x is an inclusion isotonic interval-valued function $X(T)$ of interval variable defined for all $T \subseteq T_0$, satisfying that $x(t) \in X(t)$, for any $t \in T_0$. If $X(t) = A_0 + A_1 t + \ldots + A_q t^q$, for some fixed $A_i = \left[\underline{A}_i, \overline{A}_i\right]$, for $i = 0, 1, \ldots, q$, then this function $X(t)$ is known as an interval polynomial enclosure of x.

It can be shown that if f is a real-valued function and G an interval-valued, continuous, and inclusion isotonic function satisfying $f(t) \in G(t)$, for any $t \in [a,b]$, then

$$\int_{[a,b]} f(t)dt \in \int_{[a,b]} G(t)dt \tag{A.21}$$

In particular, equ. (A.21) holds true when G is an interval polynomial enclosure of f.

Theorem A.10. For any natural numbers K and N and a real-valued function $f(t)$ with inclusion isotonic interval extensions for itself and its first K derivatives, if both a and b have the same sign, then

$$\int_{[a,b]} f(t)dt \in \sum_{i=1}^{N} \left\{ \sum_{k=1}^{K-1} \frac{1}{k+1} (f)_k (X_i) w(X_i)^{k+1} + \frac{1}{K+1} (f)_K (X_i) w(X_i)^{K+1} \right\} \tag{A.22}$$

where $(f)_k(X)$ is defined as follows:

$$(f)_k(X) = \frac{1}{k!} \frac{d^k f(X)}{dt^k}.$$

Consider some interval methods for establishing the existence of solutions to the operator equations of the following form and relevant computational tests:

$$y(t) = p(y)(t) \tag{A.23}$$

where the operator p: $M_r \to M_r$ is defined for some class M_r of real-valued functions y with common domain $a \le t \le b$, and may include integrals of the function $y(t)$. Assume that the interval operator P: $M \to M$ be defined on a class of interval enclosures of elements of $M_r \subseteq M$. This operator P is known as an interval majorant of p if

$$p(y) \in P(Y), \text{ for } y \in Y \tag{A.24}$$

where for interval- (or interval vector-) valued functions $X(t)$ and $Y(t)$ with a common domain D, $X \subseteq Y$ means that $X(t) \subseteq Y(t)$ for $t \in D$ and for a real-(or real vector-) valued function $x(t)$ defined on D, $x \in X$ means that $x(t) \in X(t)$ for $t \in D$. An interval operator P is inclusion isotonic if

$$X \subseteq Y \to P(X) \subseteq P(Y). \quad \text{A.25)}$$

Theorem A.11. If P is an inclusion isotonic interval majorant of p, and $P(Y^{(0)}) \subseteq Y^{(0)}$, then the nested sequence defined by $Y^{(k+1)} = P(Y^{(k+1)})$, $k = 0, 1, 2, \ldots$, satisfies

1) For every $a \le t \le b$, the limit interval $Y(t) = \cap_{k=0}^{\infty} Y^{(k)}(t)$ exists;
2) Any solution of equ. (A.23) that is in $Y^{(0)}$ is also Y. That is, if $y(t) \in Y^{(0)}(t)$ for $a \le t \le b$, then $y(t) \in Y^{(k)}(t)$ for $k = 0, 1, 2, \ldots, a \le t \le b$.
3) If there is $0 \le c < 1$ such that $\forall\ X \in M, X \subseteq Y^{(0)} \to \sup_t w(P(X)(t)) \le c \sup_t w(X(t))$, $a \le t \le b$, there then is the unique solution $Y(t)$ in $Y^{(0)}$ given by line 1) above.

Example A.1. The initial value problem $\frac{dy}{dy} = \sqrt{y}$, $y \ge 0$, $y(0) = 0$ can be written as the form in equ. (A.23) with $p(y)(t) = \int_0^t \sqrt{y(s)}ds$. Let us define the following interval operator $P(Y)(t) = \int_0^t \sqrt{Y(s)}ds$ with $\sqrt{Y(s)} = \sqrt{[\underline{Y(s)}, \overline{Y(s)}]} = [\sqrt{\underline{Y(s)}}, \sqrt{\overline{Y(s)}}]$, for $0 \le Y(s)$. This operator does not satisfy Theorem A.11 3). Let $Y_0(t) = [0,w]$ with $w > 0$, for $0 \le t \le b$. Then

$$P(Y_0)(t) = \int_0^t \sqrt{[0, w]}ds = [0, \sqrt{w}\,]t.$$

So, $P(Y_0) \subseteq Y_0$ if $\sqrt{w}\, b \le w$. If we take $w = 1$ and $b = 1$, then according to Theorem A.11, we have

$$Y^{(k+1)}(t) = P(Y^{(k)})(t) = \int_0^t \sqrt{Y^{(k)}(s)}ds,\ k = 0, 1, 2, \ldots$$

where $Y^{(0)}(t) = [0,1]t$ with $0 \le t \le 1$. So, according to Theorem A.11, this nested sequence has limit $Y(t) = \cap_{k=0}^{\infty} Y^{(k)}(t)$, and any solution $y(t) \in Y_0(t) = [0,1]$, $0 \le t \le 1$, of the given initial value problem is also in $Y(t)$, $0 \le t \le 1$. Because the initial value problem has infinitely many solutions of the of the form

$$y_a(t) = \begin{cases} 0, & 0 \le t \le a \\ \frac{1}{4}(t-a)^2 & a \le t \end{cases}$$

where a is an arbitrary non-negative real number, we have $y_a(t) \in Y^{(k)}(t)$, $k = 0, 1, 2, \ldots$

References

Alefeld, G., Herzberger, J.: Einführung in die Intervallrechnung. Bibliographisches Institut, Reihe Informatik, Band 12, B.I.-Wissenschaftsverlag, Mannheim - Wien – Zürich (1974)

Dwyer, P.S.: Linear Computations. Chapman & Hall, London (1951)

Gargantini, I., Henrici, P.: Circular arithmetic and the determination of polymonial zeros. Numer. Math. 18(4), 305–320 (1972)

Hansen, E.R.: Interval arithmetic in matrix computations, Part 1. J. Soc. Indust. Appl. Math. Ser. B Numer. Anal. 2(2), 308–320 (1965)

Hansen, E.R.: On solving systems of equations using interval arithmetic. Math. Compt. 22(102), 374–384 (1968)

Hansen, E.R. (ed.): Topics in Interval analysis. Oxford University Press, London (1969)

Hansen, E.R.: Global Optimization Using Interval Analysis. Marcel Dekker, New York (1992)

Hansen, E.R., Walster, G.W.: Global Optimization Using Interval Analysis. Marcel Dekker, New York (2003)

Hayes, B.: A lucid interval. American Scientist 91(6), 484 (2003)

Jaulin, L., Kieffer, M., Didrit, O., Walter, E.: Applied Interval Analysis. Springer, New York (2001)

Kahan, W.M.: A more complete interval arithmetic: Lecture notes for an engineering summer course in numerical analysis at University of Michigan, Technical Report, University of Michigan (1968)

Kulisch, U.: Wissenschaftliches Rechnen mit Ergebnisverifikation. Eine Einführung. Vieweg-Verlag, Wiesbaden (1989)

Moore, R.E.: Interval arithmetic and automatic error analysis in digital computing. Ph. D. Dissertation, Department of Mathematics, Stanford University, Stanford, CA (1962)

Moore, R.E.: Interval Analysis. Prentice Hall, Englewood Cliffs (1966)

Moore, R.E., Kearfott, R.B., Cloud, M.J.: Introduction to Interval Analysis. Society for Industrial and Applied Mathematics, Philadelphia (2009)

Neumaier, A.: Interval Methods for Systems of Equations. Encyclopedia of Mathematics and Its Applications 37. Cambridge Univ. Press, Cambridge (1990)

Nickel, K.L.: Uber die notwendigkeit einer fehlerschranken: Arithmetik fur rechenautomaten. Muner. Math. 9, 69–79 (1966)

Ratschek, H., Rokne, J.G.: Computer methods for the range of functions. Ellis Horwood Ser.: Math. Appl. Chichester. Ellis Horwood, UK (1984)

Sunaga, T.: Theory of interval algebra and its application to numerical analysis. RAAG Memoires 2, 29–46 (1958)

Warmus, M.: Calculus of approximations. Bull. Acad. Polon. Sci., Cl. III IV(5), 253–259 (1956)

Warmus, M.: Approximations and inequalities in the calculus of approximations: Classification of approximate numbers. Bull. Acad. Polon. Sci., Ser. math., astr. et phys IX(4), 241–245 (1961)

Appendix B
Approaches of Uncertainty

(Lin, 1998) and (Lin and OuYang, 2010) showed recently that when predictions of nearly zero probability events are concerned, such as weather forecast or financial market prediction, failure has been the main part of life and no theoretical methods developed so far on the basis of calculus and probability have been successful. Such a lack of success is mainly due to the way of how information and consequent uncertainties are handled. In this appendix, we will show how grey systems theory compares to other concepts of uncertain information.

Because the problem of unsuccessful applications of traditional scientific theories involving uncertain information occurs mainly in the area of predictions of small-probability events, let us first look at the concept of a true historical process. According to George Soros (1998), a legendary investor and financial guru, a true historical process is a process that leads to either unexpected outcomes or non-traditional beliefs, leading to a brand new page in human history. In each true historical process, the past creates expectations for all human participants so that predictions about what will happen in the near future are made. To different participants the expected events are viewed differently. For example, some wish the expected to happen whereas the others might like to avoid it as much as possible. So, consequently, various human participants will behave differently from the past in order to materialize or avoid the expectations and predictions. Consequently to these adjustments in human behaviors the expected and predicted future is in general altered. That is, the expected events in general do not actually occur in the time frame or in the magnitude as expected. That is, in order to be successful in predicting the outcome of a true historical process we must be able to handle uncertainties created by either accurate or inaccurate information. It is because all information, either accurate or inaccurate, brings forward uncertainties.

Historically, the concept of information has been defined in many different ways. To see clearly how grey systems compare with other kinds of uncertain information, we establish various kinds of information on a common ground. In particular, a piece of tidings is meant to be the totality of a special form of objective motion. It is an objective entity, which reduces a human's level of ignorance. For example, the statement that "it will snow today" is a piece of tidings, because it improves our outlook about the weather condition of the day.

Let A be a piece of tidings, then $A \cup A^{-}$ means no tidings, where A^{-} stands for the complement of A, because this union contains the universal description of the tidings A and the opposite A^{-}. For example, let A = the stock market will go up. Then, A^{-} = the market will go down or sideways. Now, $A \cup A^{-}$ = the stock market will go in some direction, which provides no useful information to improve human knowledge. At the same time, the combined tidings $A \cap A^{-}$ = the market will not go in any direction.

We use lowercase letters $x, y, z, \ldots$ to stand for unknowns, which could be variables or statements, and A a piece of tidings. The notation $A\Delta x$ represents that the tidings A can make people know the value of the unknown x. Otherwise, we write $A\nabla x$ to mean that A is a piece of unrelated tidings about x. For example, given two pieces of tidings:

A = everyone has gone to watch a movie in the theater; and
B = the Dow Jones Industrial Average has gone down 400 points.

Now we are concerned with the unknown x = "where is Joe?" So, $A\Delta x$, because A provides an answer to x, even though we do not know whether the answer is true or false. That is, based on A, we know that Joe went to the theater. At the same time, $B\nabla x$, because B does not provide any value and answer to x.

Assume that x is an unknown, U a piece of tidings, and S a set of Cantor type (Kuratowski and Mostowski, 1976). If U makes people realize $x \in S$, U is known as a piece of x-position tidings. Each piece of tidings $A \subseteq U$ is referred to as a piece of information regarding the position tidings U, or just information for short. The totality of all pieces of information of U is referred to as an informational hierarchy.

Each so-called informational uncertainty stands for uncertainties related to information, or the quality of information. As indicated by Soros's reflexivity theory, in a true historic process, all information involved in the formation of predictions about the future could be very certain and definite. However, it is the certainty and the definiteness of the information, and the accuracy and preciseness of the predictions that make the future different and more uncertain. So, informational uncertainties are different from practical uncertainties.

Based on published studies, such as (Liu et al., 2001; Soros, 1998) we have the following types of uncertainties:

1. Grey uncertainty;
2. Stochastic uncertainty;
3. Unascertainty;
4. Ascertainty;
5. Fuzzy uncertainty;
6. Rough uncertainty;
7. Soros reflexive uncertainty; and
8. Blind uncertainty.

B.1 Foundation for a Unified Information Theory

In this section, we will develop all the previously listed uncertainties on a uniform setting so that comprehensive general studies of information theory can be made possible. For relevant literature on unified information theory, please consult with (Hofkirchner, 1998). To this end, let us look at each of the uncertainties.

B.1.1 Grey Uncertainties

A piece of information A is grey, provided it is defined as follows: Let x be an unknown, $S \neq \varnothing$ a set, S' a subset of S, U = "x belongs to S" and A = "x belongs to S'." Then, the so-called grey uncertainty stands for the uncertainty of which specific value of the unknown x should take. For example, suppose that we are given U = "x belongs to S," S = "**R** is the set of all real numbers," S' = the interval [2, 3], and A = "x belongs to S'." Then the piece of grey information A brings about the following uncertainty: We know that x is a number between 2 and 3 inclusive. However, we do not know which value x really assumes.

B.1.2 Stochastic Uncertainties

If x is unknown, S a nonempty set, U = "x belongs to S" and A = "x belongs to S and the possibility for $x = e \in S$ is α_e, where $0 \le \alpha_e \le 1$ and $\sum_{e \in S} \alpha_e = 1$." In this case, A is referred to as a piece of stochastic information. When a piece of stochastic information is given, the consequent uncertainty is referred to as stochastic uncertainty. Such uncertainty is created because the piece of stochastic information A can only spell out how likely the unknown x equals a specific element $e \in S$. This implies that the probability α_e can be very close to 1 or equal to 1. However, the large probability does not guarantee that $x = e$ will definitely be true.

B.1.3 Unascertainties

If in the definition of a piece of stochastic information A, we replace the condition that $\sum_{e \in S} \alpha_e = 1$ by $\sum_{e \in S} \alpha_e \le 1$, then A is referred to as a piece of unascertained information. The main difference between stochastic and unascertained information is that the former concept is developed on the assumption that all possible outcomes of an experiment are known, whereas for unascertained information, we assume that only some possible outcomes of the experiment are known to the researcher.

B.1.4 Fuzzy Uncertainties

A piece A of tidings is referred to as a piece of fuzzy information, if A satisfies the following condition: x is an unknown, S a nonempty set, the position tidings U = "x belongs to S," and A = "x belongs to S and the degree of the membership for $x = e \in S$ is α_e, where $0 \le \alpha_e \le 1$." The so-called fuzzy uncertainty is that for a given piece of fuzzy information A = "x belongs to S and the degree of membership for $x = e \in S$ is α_e, where $0 \le \alpha_e \le 1$," one has no clue on that for a given variable y, should y be considered with the set or not, even though he knows the degree of membership for y to belong to S is α_e.

B.1.5 Rough Uncertainties

Let U be a set of elements. And, a subset $r \subseteq p(U)$, the power set of U, is called a partition of U, if the following conditions hold true:

1. $\cup r = \cup\{x: x \in r\} = U$, and
2. $\forall A, B \in r$, if $A \neq B$, then $A \cap B = \varnothing$.

Let $K = (U, R)$ be a knowledge base over U, where U is the universal set of all objects involved in a study, and R a given set of partitions of the set U, known as a knowledge base over U. A subset $X \subseteq U$ is called exact in K, if there exists a $P \subseteq R$ such that X is the union of some elements in $\cap P$. Otherwise, X is said to be rough in K. Here, no matter how fine the partitions in $\cap P$ are, approximations to a rough set $X \subseteq U$ will never be exact.

B.1.6 Soros Reflexive Uncertainties

Let x be the unknown path a true historic process will eventually take and S = all possible outcomes of this historic process. Then, an Soros reflexive uncertain information is defined as follows: U = "$x \in S$" and $A \subseteq S$ is a piece of information regarding the position of x in S defined by A = "if it is expected $x = e \in S$ with a degree of credence α_e, where $0 \le \alpha_e \le 1$, then $x = e \in S$ has a degree of credence $1 - \alpha_e$." Now the uncertainty associated with a piece of Soros reflexive uncertain information is that the more accurate a prediction about a true historic process is, the more uncertain the expected future will become.

B.2 Relevant Practical Uncertainties

First, let us make it clear about the difference between informational uncertainties and practical uncertainties. In particular, each so-called informational uncertainty stands for uncertainties that relate to information or the quality of information, while as the concept of true historic processes of Soros's reflexivity theory indicates, informational uncertainties are different from practical uncertainties.

Example B.1. On Island Voca, many times in the past the active volcano erupted. And, each eruption destroyed the town below. So, gradually, the people living in the town created canals to channel the lava away from the town and into the ocean. Since the time the project of canals was completed, the active volcano had not caused any damage to the town. So, the residents of the town did all they could to keep up the quality of the canals in order to prevent any potential disaster.

The certainty brought forward by the existence of the canals made the residents feel safe and secure about their futures. However, the next time when the volcano erupted, the lava overflowed the canals and destroyed the town completely.

The morale of this made-to-believe example is that as the residents of the town gained certainty, a major uncertainty appeared. If even after the canals were built and had been successful in terms of protecting the town, the town people were still prepared to protect their lives and economic assets, then even in the case of the canals being over flown, their economic loss and human casualties would still be minimal.

Example B.2. Let us see how grey uncertainty appears in daily lives. In the negotiation process of buying a car, the buyer knows he/she would pay no more than \$30,000. If x stands for the final negotiated price of the car he/she likes, then x is a number between \$0 and \$30,000. Here, the grey uncertainty is that about the final purchase price, which is not known until the very end of the buying process.

Example B.3. In this case, let us look at stochastic uncertainties. First, in the business of commodity trading, we can compute based on the historical price data that the market for S&P500 has a 90% chance to go up on a certain Thursday. So, some traders would buy in to the S&P500 futures contracts on Thursday and sell out on the calculated day, which may be the next Monday or Tuesday. However, the 90% possibility of a rising S&P500 futures market does not guarantee this time when we buy on Thursday the market will actually go up as expected.

As a second example, the current commercial weather forecasting business tends to provide services as follows: The chance of snow for tomorrow is 70%. If it does snow the next day, the weather forecasting service is correct since they said it would. On the other hand, if it does not snow the next day, the weather forecasting service is still correct since it only provided 70% likelihood of a forthcoming snow. Now, if the figure 70% is replaced by 100%, the same thing can be still be said about the weather forecasting service, because service only stated the chance of snowing was 100%, which was not a guarantee.

Example B.4. A group of researchers have a scheduled meeting at 11:30 am on Thursday. However, at around 11:45 am., Genti, as a key member of the group, did not show up. So, the rest of the group needs to decide where to find him for their urgent business decision making. Now, we face two possible situations: 1) The group knows Genti very well. So, the members come up with a definite list of possible places Genti could be at at the very moment. Because they know Genti so well, they could also attach a probability to each place on the list. So, to locate Genti successfully, they only need to check these places in the order from the largest probability to the smallest probability. This is an example of stochastic uncertainties. 2) No member of the group knows Genti well enough to come up with a list of all possible places and relevant probabilities Genti could be at this very moment. This is an example of unascertainties.

The second situation explains that the where about of Genti is at the special moment was certain since as a living being, he must be at some place. However, the decision makers did not know the true state of Genti or relevant information used by Genti in his decision making about where to go and to be at the special moment. That is, the concept of unascertainty deals with the situation that no matter whether an objective event is definite or not, whether it has already occurred or not, it will be "unascertained" as long as the decision maker does not completely understand the essential information.

Example B.5. Now, let us look at an example of fuzzy uncertainties. Assume that Jacklin Ruscitto is an official member of many committees. Due to the nature of these committees, Jacklin does not have enough time to be involved 10% in all of the committee works and relevant decision-makings. Let us look at one committee, say committee A. If Jacklin is listed as a member but did not ever do anything for the committee, then her degree of membership in the committee would be very close to zero. If Jacklin was not listed as a member on the committee and did not do anything for the committee then her degree of membership in the committee is zero. Even though she is not a listed member of the committee, if she has been involved in activities of the committee, then her degree of membership in the committee should be greater than zero.

Now, the so-called fuzzy uncertainty stands for that Jacklin is listed as an official member of Committee A and has been involved in all committee activities. So, her degree of membership in Committee A is 1. However, the fuzzy uncertainty implies that her degree 1 of membership in Committee A does not guarantee her 100% involvement or membership in Committee A in the future. On the other hand, even though John Opalanko is not a listed member of Committee A. It is very likely that since Committee A is involved in a special project, which looks extremely important in John's eyes, John may very well get involved in the project. In this case, John's degree of membership in Committee A should be more than zero even though his previous degree of membership in Committee A was zero.

Example B.6. Let us now look at rough uncertainties. Assume that U is the rectangle area $U = \{(x,y): 0 \le x \le 5, 0 \le y \le 4\}$. A given partition r of U is defined as follows: Each element x in r is a smaller rectangular area as shown in Figure B.1 such that

(1) If x is not on the bottom row, x includes all interior points and points on the upper and left borders;
(2) If x is on the bottom row, then x includes all points as described in (1) above and those on the bottom border;
(3) If x is located on the far right column but not in the bottom, x contains all points as in (1) and the points on the right border; and
(4) If x is located at the lower right corner, then x contains all the points within and on the border of the rectangular area.

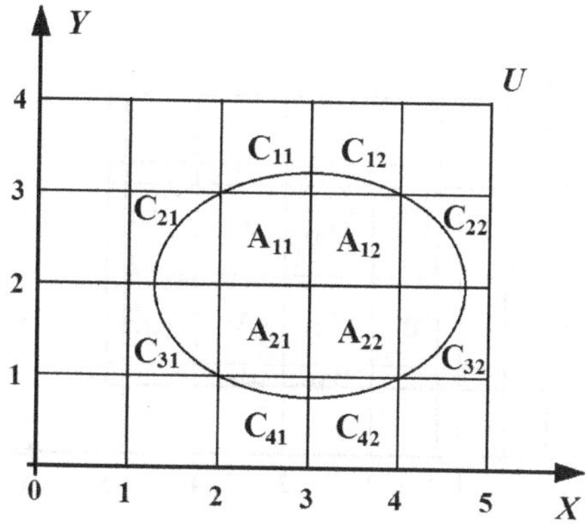

Fig. B.1 The partition r of the area U

Assume that s is another partition of U as shown in Figure B.2 with border points classified in a similar fashion as in partition r above. Then the shaded area of U is an exact subset of U because it is the union of elements A_{11}, A_{12}, A_{21}, and A_{22} of the partition r. Now, the shaded area of A in Figure B.3 is considered a rough subset of U, because it does not equal the union of any combination of elements in r and/or in s. It implies that in order to know more about this rough set A, one has to approximate it in two different ways if he applies the partition s:

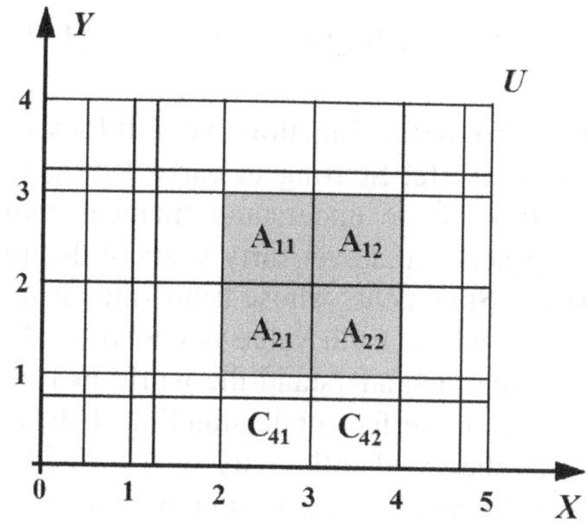

Fig. B.2 The partition s of the area U

1) $\underline{s}A = A_{11} \cup A_{12} \cup A_{21} \cup A_{22} = \cup \{A_{ij}: i, j = 1, 2\}$.
2) $\bar{s}A = \cup \{A_{ij}: i, j = 1, 2\} \cup \{B_{ij}: I = 1, 2, 3, 4, j = 1, 2\}$.

If partition r is applied, the following approximations could be employed:

3) $\underline{r}A = A_{11} \cup A_{12} \cup A_{21} \cup A_{22} = \cup \{A_{ij}: i, j = 1, 2\}$.
4) $\bar{r}A = \cup \{A_{ij}: i, j = 1, 2\} \cup \{C_{ij}: I = 1, 2, 3, 4, j = 1, 2\}$.

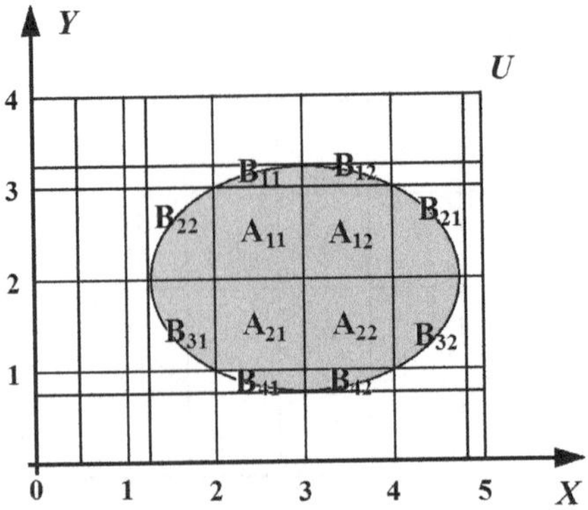

Fig. B.3 Rough approximations of the area A

Since elements in s are finer than those in r, approximations in 1) and 2) of the rough set A are expected to be more accurate than those of 3) and 4). From this example, one can see that no matter how fine a partition is, its approximations to a given rough set always contain uncertainty. This uncertainty is in the area of $\bar{s}A - \underline{s}A$ or $\bar{r}A - \underline{r}A$, where

$$\bar{s}A - \underline{s}A = \cup \{B_{ij}: I = 1, 2, 3, 4, j = 1, 2\}$$

and

$$\bar{r}A - \underline{r}A = \cup \{C_{ij}: I = 1, 2, 3, 4, j = 1, 2\}.$$

Example B.7. (Slater, 1996). In this junction, we will look at how Soros reflexive uncertainty plays out in real life. Starting in early 1960s, George Soros had applied the concept of reflexivity to understand finance, politics, and economics. Based on his success in practice and over forty years of theoretical study, he draws the following conclusion: "Statements whose truth value is indeterminate are more significant than statements whose truth value is known or definite. The later constitute knowledge. They help us understand the world as it is. But the former, expressions of our inherently imperfect understanding, help to shape the world in which we live." With his theory of reflexivity well formed, Soros always looked out for situations where he saw an investment opportunity differently from the prevailing wisdom. He assumes all companies and industries to be flawed and tried to find what the flaws are. As soon as there were signs to show that the flaws were becoming a problem he would take financial positions so that he was putting himself ahead of the game of investment. In particular, when a company has a superior market position, competent management and exceptional profit margins, the stock may be overpriced. Now, the management may become complacent and the competitive or regulatory environment may change. When Soros looks for flaws, he established a hypothesis on which he would invest. Each of his hypotheses satisfied the following conditions:

1) The hypothesis has to differ from the widely accepted wisdom. The greater the difference the greater the profit potential.
2) Each hypothesis does not have to be true in real life to be profitable, as long as it could be generally accepted.
3) The length of time the beneficial effects of the hypothesis can provide depends on whether or not the underlying flaws are recognized and corrected.

For instance, in 1972, Soros sensed that a change was about to occur in the banking industry since banks had the worst reputations at the time. Their employees were considered stodgy and dull, and few believed that the banks would rouse themselves from their deep slumber. That was why investors showed no interest in their stocks. After doing his homework, Soros realized that a new generation of bankers was quietly ready and in place to act aggressively on behalf of their employers. Since the new generation of bank managers were using new financial instruments, the banks' earnings performances were looking up. However, the banks stocks were sold at virtually no premium. Many banks had reached their leveraging limits. So, in order to grow, they needed more equity. When the First National City Bank hosted a dinner for security analysts in 1972, even though Soros was not invited, he sensed the forthcoming aggressive change in the way banks would conduct their business. So, he wrote a brokerage report arguing that because bank shares had been going nowhere they were about to take off, contrary to what many others thought. Timing the publication of his report to coincide with the banks dinner he recommended getting behind some of the better managed banks. As expected bank stocks began to rise and he made a handsome 50% profit in a short period of time. The legendary success of Soros's life long investing has evidenced a wide-spread existence of our so-called Soros reflexive uncertainty in real life situations with human participants.

B.3 Some Final Words and Open Questions

What is presented in this appendix brings up an ambitious research project: Establish a unified information theory (Hofkirchner, 1998). The next questions waiting to be studied and answered immediately include, but not limited to: How can one introduce more analytic or practically more operable methods to handle various types of information and their relevant uncertainties? By definition, a piece of blind information is one piece of such information in which two or more kinds of unascertained information appear simultaneously. How can one tackle the difficulty of describing a piece of blind information and its consequent uncertainties?

References

Hofkirchner, W.: The Quest for a Unified Theory of Information. Gorden and Breach, Cooper Station (1998)

Kuratowski, K., Mostowski, A.A.: Set Theory with an Introduction to Descriptive Set Theory. North-Holland, Amsterdam (1976)

Lin, Y. (guest editor): Mystery of nonlinearity and Lorenz's chaos. Kybernetes: The International Journal of Systems and Cybernetics 27(6 & 7), 605–854 (1998)

Lin, Y., OuYang, S.C.: Irregularities and Prediction of Major Disasters. CRC Press, an imprint of Taylor and Francis, New York (2010)

Liu, K.D., Lin, Y., Yao, L.G.: Informational uncertainties and their mathematical expressions. Kybernetes: The International Journal of Systems and Cybernetics 30(4), 378–396 (2001)

Slater, R.: Soros: The Life, Times and Trading Secrets of the World's Greatest Investor. Irwin, Burr Ridge (1996)

Soros, G.: The Crisis of Global Capitalism: Open Society Endangered. Public Affairs, New York (1998)

Appendix C
How Uncertainties Appear: A General Systems Approach

In recent years, readers of our previously published works from different parts of the world have communicated with us and expressed their interests in learning our opinions regarding grey systems, grey information, and uncertainties in general. Among many open questions they posed are the following: How do uncertainties appear? Are they objective physical existence or artificially created out of the limited knowledge of man? In this appendix, we try to address these and related questions using the systemic yoyo model, developed recently (Lin, 2007; 2008).

This appendix is organized as follows: In Section C-1, we establish the fact that the common form of materials' movement in the universe is eddy motion by employing the results of blown-up theory (Wu and Lin, 2002). In Section C-2, we introduce the systemic yoyo model followed by theoretical and empirical justifications. In Section C-3, we look deeper into the structure of systemic yoyos, interactions of yoyo fields, and then the laws on the state of motion of materials. After establishing the fact that the universe is really an ocean, which is both intelligently abstract and objectively existent, of spinning fields pushing against each other, in Section C-4, we see how uncertainties could be either artificially created or physically existing and appear everywhere we set our eyes on. To this end, we see how uncertainties and inconsistencies even exist in the system of modern mathematics so that the certainty of learned knowledge in general and mathematics in particular and standards of truths are lost forever.

C.1 Evolutionary Transitions

In this section, we learn how general systems of various kinds share the same underlying structure, known as the spinning yoyo field. For a thorough and detailed presentation of this model and intuitive background of general systems, please consult with (Lin, 2008a; Lin and Forrest, 2010).

C.1.1 Blown-Ups: Old Structures Replaced by New Ones

When we study the nature and treat everything we see as a system (Klir, 2001), one fact we can easily see is that many systems in nature evolve in concert. When one thing changes, many other seemingly unrelated matters alter their states of existence accordingly. If we attempt to make predictions regarding an evolving event, which might be from financial markets or about what is forthcoming in the weather system, we may very well find ourselves in such a situation that is somewhere in between that we do not have enough information or that we have too much information. It has been shown (Soros, 2003; Liu and Lin, 2006; Lorenz, 1993) that no matter which situation we are in, too much information or too little, we face with uncertainties. That is why we propose (OuYang, Chen, and Lin, 2009; Lin and Wu, 1998) to look at the evolution of the system or event that we need to predict about its future outlook as a whole. That is, when developments and changes naturally existing in the natural environment are seen as a whole, we have the concept of whole evolutions. And, in whole evolutions, other than continuities, as well studied in modern mathematics and science, what seems to be more important and more common is discontinuity, with which transitional changes (or blown-ups) occur. These blown-ups reflect not only the singular transitional characteristics of the whole evolutions of nonlinear equations, but also the changes of old structures being replaced by new structures.

Thousands of case studies have shown the fact (Lin, 2008a) that reversal and transitional changes are the central issue and extremely difficult open problem of prediction science, because the well-developed method of linearity, which tries to extend the past rise-and-fall into the future, does not have the ability to predict forthcoming transitional changes and what will occur after the changes. In terms of nonlinear evolution models, blown-ups reflect destructions of old structures and establishments of new ones. Although studies of these nonlinear models reveal the limits and weaknesses of calculus-based theories, we can still make use of their analytic forms to describe to a certain degree the realisticity of discontinuous transitional changes of objective events and materials. So, the concept of blown-ups is not purely mathematical; it is also realistic. By borrowing the form of calculus, we can write the concept of blown-ups as follows: For a given (mathematical or symbolic) model, that truthfully describes the physical situation of our concern, if its solution $u = u(t; t_0, u_0)$, where t stands for time and u_0 the initial state of the system, satisfies

$$\lim_{t \to t_0} |u| = +\infty , \tag{C.1}$$

and at the same time moment when $t \to t_0$, the underlying physical system also goes through a transitional change, then the solution $u = u(t; t_0, u_0)$ is called a blown-up solution and the relevant physical movement expresses a blown-up. Our analysis of thousands of real-life cases of various evolutionary systems (Lin, 2008a) indicate that disastrous events appear at the moments of blown-ups in the whole evolutions of systems.

For nonlinear models in independent variables of time (t) and space (x, x and y, or x, y, and z), the concept of blown-ups are defined similarly, where blow-ups in the model and the underlying physical system can appear in time or in space or in both. For instance, if the time-space evolution equation is written as follows:

$$\partial_t u = g(t, u, \partial_x u), \tag{C.2}$$

where u is an $n \times 1$ matrix of the state variables, $g(t,u,\partial_x u)$ an $n \times 1$ matrix of functions in t, u, and $\partial_x u$, ∂_x and ∂_t stand for the differential operations $\frac{\partial}{\partial x}$ and $\frac{\partial}{\partial t}$, respectively. Assume that the initial (or boundary) condition is

$$u(t_0, x) = u_0(x). \tag{C.3}$$

Then, when the solution $u = u(t, x; t_0, u_0)$ or $u_x = u_x(t, x; t_0, u_0)$ changes, for $t \in [t_0, t_b)$, continuously, and when $t \to t_b$, the following holds true

$$\lim_{t \to t_0} |u| = +\infty, \tag{C.4}$$

or

$$\lim_{t \to t_0} |u_x| = +\infty, \tag{C.5}$$

then either u or u_x is called a blown-up solution, if the boundary value problem, consisting of equs. (C.2) and (C.3), truly describes a physical system and at the time moment $t = t_b$, this physical system goes through a transitional change. In the rest of this appendix, we assume that this assumption holds true.

C.1.2 Mathematical Properties of Blown-Ups

To help us understand the mathematical characteristics of blown-ups, let us look at the following constant-coefficient equation:

$$\dot{u} = a_0 + a_1 u + \ldots + a_{n-1} u^{n-1} + u^n = F, \tag{C.6}$$

where u is the state variable, and a_0, a_1, …, a_{n-1} are constants.

For $n = 2$, it can be shown that under different conditions, the solution of the same nonlinear model can either be continuous and smooth or experience either blown-ups or periodic transitional blown-ups. So, even for the simplest nonlinear evolution equations, the well posedness of their evolutions in terms of differential mathematics is conditional. Here, the requirements for well-posedness are: the solution exists; it is unique, and stable (or continuous and differentiable).

For $n = 3$, other than the situation that $F = 0$ has two real solutions, one of which is of multiplicity 2, equ. (C.6) experiences blown-ups. For the general case n, even though the analytical solution of equ. (C.6) cannot be found exactly, the blown-up properties of the solution can still be studied through qualitative means. To this end, rewrite equ. (C.6) as follows:

$$\dot{u} = F = (u - u_1)^{p_1} \dots (u - u_r)^{p_r} (u^2 + b_1 u + c_1)^{q_1} \dots (u^2 + b_m u + c_m)^{q_m}, \quad \text{(C.7)}$$

where p_i and q_j, $i = 1, 2, \ldots, r$ and $j = 1, 2, \ldots, m$, are positive whole numbers, and $n = \sum_{i=1}^{r} p_i + 2\sum_{j=1}^{m} q_j$, $\Delta = b_j^2 - 4c_j < 0$, $j = 1, 2, \ldots, m$. Without loss of generality, assume that $u_1 \geq u_2 \geq \ldots \geq u_r$. Then, the following result has been shown:

Theorem C.1. The condition under which the solution of an initial value problem of equ. (C.3) contains blown-ups is given by

(1) When u_i, $i = 1, 2, \ldots, r$, does not exist, that is, $F = 0$ does not have any real solution; and
(2) If $F = 0$ does have real solutions u_i, $i = 1, 2, \ldots, r$, satisfying $u_1 \geq u_2 \geq \ldots \geq u_r$,
    (a) When n is an even number, if $u > u_1$, then u contains blow-up(s);
    (b) When n is an odd number, no matter whether $u > u_1$ or $u < u_r$, there always exist blown-ups.

C.1.3 The Problem of Quantitative Infinity

One of the features of blown-ups is the quantitative infinity ∞, which stands for indeterminacy mathematically. So, a natural question is how to comprehend this mathematical symbol ∞, which in applications causes instabilities and calculation spills that have stopped each and every working computer.

To address this question, let us look at the mapping relation of the Riemann ball, which is well studied in complex functions (Figure C.1). This so-called Riemann ball, a curved or curvature space, illustrates the relationship between the

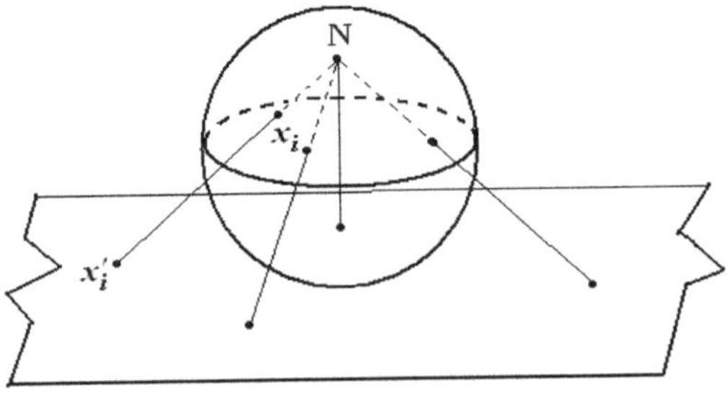

Fig. C.1 The Riemann ball - relationship between planar infinity and three-dimensional North Pole

infinity on the plane and the North Pole N of the ball. Such a mapping relation connects $-\infty$ and $+\infty$ through a blown-up. In other words, when a dynamic point x_i travels through the North Pole N on the sphere, the corresponding image x_i' on the plane of the point x_i shows up as a reversal change from $-\infty$ to $+\infty$ through a blown-up. So, treating the planar points $\pm\infty$ as indeterminacy can only be a product of the thinking logic of a narrow or low dimensional observ-control, since, speaking generally, these points stand implicitly for direction changes of one dynamic point on the sphere at the polar point N. Or speaking differently, the phenomenon of directionless, as shown by blown-ups of a lower dimensional space, represents exactly a direction change of movement in a higher dimensional curvature space.

Therefore, the concept of blown-ups can specifically represent implicit transformations of spatial dynamics. Through blown-ups, problems of indeterminacy of a narrow observ-control in a distorted space are transformed into determinant situations of a more general observ-control system in a curvature space. This discussion shows that the traditional view of singularities as meaningless indeterminacies has not only revealed the obstacles of the thinking logic of the narrow observ-control (in this case, the Euclidean space), but also the careless omissions of spatial properties of dynamic implicit transformations (bridging the Euclidean space to a general curvature space).

Corresponding to the previous implicit transformation between the imaginary plane and the Riemann ball, one can also relate quantitative $\pm\infty$ (symbols of indeterminacy in one dimensional space) to a dynamic movement on a circle (a curved space of a higher dimension) through the modeling of differential equations.

What is shown here is that nonlinearity, speaking mathematically, stands (mostly) for singularities in the Euclidean spaces. In terms of physics, nonlinearity represents eddy motions, the movements on the curvature spaces. Such motions are a problem about structural evolutions, which are a natural consequence of uneven evolutions of materials. So, nonlinearity accidentally describes discontinuous singular evolutionary characteristics of eddy motions (in curvature spaces) from the angle of a special, narrow observ-control system, the Euclidean spaces.

C.1.4 Eddy Motions of the General Dynamic System

In this subsection, let us look at the general dynamic system and how it is related to eddy motions. The following is Newton's second law of motion

$$m\frac{d\vec{v}}{dt}=\vec{F}\,. \tag{C.8}$$

Based on Einstein's concept of uneven time and space of materials' evolutions, we can assume

$$\vec{F} = -\nabla S(t, x, y, z), \tag{C.9}$$

where $S = S(t, x, y, z)$ stands for the time-space distribution of the external acting object. Let $\rho = \rho(t, x, y, z)$ be the density of the object being acted upon. Then, the kinematic equation (C.8) for a unit mass of the object being acted upon can be written as

$$\frac{d\vec{u}}{dt} = -\frac{1}{\rho(t, x, y, z)} \nabla S(t, x, y, z), \tag{C.10}$$

where $\vec{u}$ is used to replace the original $\vec{v}$ in order to represent the meaning that each movement of some materials is a consequence of mutual reactions of materials' structures. Evidently, if ρ is not a constant, then equ. (C.10) becomes

$$\frac{d(\nabla x \vec{u})}{dt} = -\nabla x \left[\frac{1}{\rho} \nabla S \right] \neq 0, \tag{C.11}$$

which stands for an eddy motion because of the nonlinearity involved. In other words, a nonlinear mutual reaction between materials' uneven structures and the unevenness of the external forcing object will definitely produce eddy motions.

On the other hand, since nonlinearity stands for eddy sources, it represents a problem about structural evolutions. This end has essentially ended the particle assumption of Newtonian mechanics and the methodological point of view of "melting shapes into numbers," which has been formed since the time of Newton in natural sciences. What's more important is that the concept of uneven eddy evolutions reveals the fact that forces exist in the structures of evolving objects, and does not exist independently out of objects, as what Aristotle and Newton believed so that the movement of all things had to be pushed first by God. Based on such reasoning, the concept of second stir of materials' movements is introduced. Additionally, the concept of second stir can also be extended into a theory about the evolution of the universe. Since the concept of the second stir assigns forces with materials' structures, it naturally ends Newton's God. The duality of rotations must lead to differences in spinning directions. Such differences surely lead to singular points of singular zones. Through sub-eddies and sub-sub-eddies, breakages (or big bangs) are represented so that evolutionary transitions are accomplished. Evidently, according to the concept of second stir, the number of singular points would be greater than one, and the big bang explosions would also have their individual multiplicities.

C.1.5 Equal Quantitative Effects

Another important concept studied in the blown-up theory is that of equal quantitative effects. Even though this concept was initially proposed in the study of fluid motions, it essentially represents the fundamental and universal characteristics of all movements of materials. The so-called equal quantitative effects stand for the eddy effects with non-uniform vortical vectorities existing naturally in systems of equal quantitative movements due to the unevenness of materials. And, by equal quantitative movements, it is meant to be the movements with quasi-equal acting and reacting objects or under two or more quasi-equal mutual constraints. For example, the relative movements of two or more planets of approximately equal masses are considered equal quantitative movements. In the microcosmic world, an often seen equal quantitative movement is the mutual interference between the particles to be measured and the equipment used to make the measurement. Many phenomena in daily lives can also be considered equal quantitative effects, including such events as wars, politics, economies, chess games, races, plays, etc.

Comparing to the concept of equal quantitative effects, the Aristotelian and Newtonian framework of separate objects and forces is about unequal quantitative movements established on the assumption of particles. On the other hand, equal quantitative movements are mainly characterized by the kind of measurement uncertainty that when I observe an object, the object is constrained by me. When an object is observed by another object, the two objects cannot really be separated apart. At this junction, it can be seen that the Su-Shi Principle of Xuemou Wu's panrelativity theory (1990), Bohr (N. Bohr, 1885 – 1962) principle and the relativity principle about microcosmic motions, von Neumann's Principle of Program Storage, etc., all fall into the uncertainty model of equal quantitative movements with separate objects and forces.

What's practically important and theoretically significant is that eddy motions are confirmed not only by daily observations of surrounding natural phenomena, but also by laboratory studies from as small as atomic structures to as huge as nebular structures of the universe. At the same time, eddy motions show up in mathematics as nonlinear evolutions with wave motions and spraying currents as local characteristics of eddy movements. The birth-death exchanges and the non-uniformity of vortical vectorities of eddy evolutions naturally explain where and how quantitative irregularities, complexities and multiplicities of materials' evolutions, when seen from the current narrow observ-control system, come from. Evidently, if the irregularity of eddies comes from the unevenness of materials' internal structures, and if the world is seen at the height of structural evolutions of materials, then the world is simple. And, it is so simple that there are only two forms of motions. One is clockwise rotation, and the other counter clockwise rotation. The vortical vectority in the structures of materials has very intelligently resolved the Tao of Yin and Yang of the "*Book of Changes*" of the eastern mystery (Wilhalm and Baynes, 1967), and has been very practically implemented in the common form of motion of all materials in the universe.

At this junction, the fundamental problem is why all materials in the universe are in rotational movements. According to Einstein's uneven space and time, we can assume that all materials have uneven structures. Out of these uneven structures, there naturally exist gradients. With gradients, there appear forces. Combined with uneven arms of forces, the carrying materials will have to rotate in the form of moments of forces. That is exactly what the ancient Chinese Lao Tzu, (English and Feng, 1972)) said: "Under the heaven, there is nothing more than the Tao of images," instead of Newtonian doctrine of particles (under the heaven, there is such a Tao that is not about images but sizeless and volumeless particles). The former stands for an evolution problem of rotational movements under stirring forces. Since structural unevenness is an innate character of materials, that is why it is named second stir, considering that the phrase of first push was used first in history (OuYang, et al., 2000). What needs to be noted is that the phrases of first push and second stir do not mean that the first push is prior to the second stir.

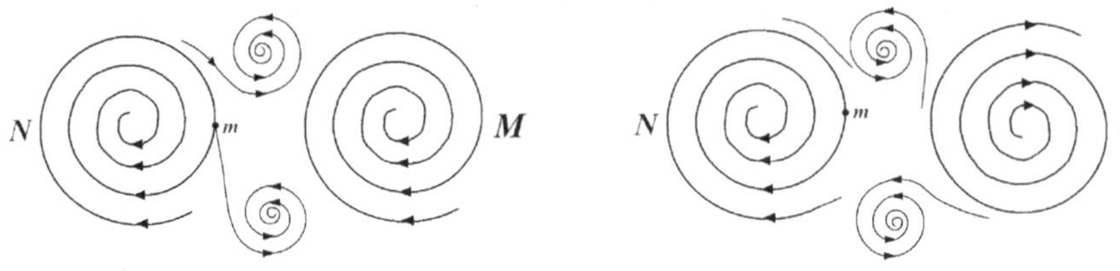

(a) (b)

Fig. C.2 Appearance of sub-eddies

Now, we can imagine that the natural world and/or the universe be entirely composed of eddy currents, where eddies exist in different sizes and scales and interact with each other. That is, the universe is a huge ocean of eddies, which change and evolve constantly. One of the most important characteristics of spinning fluids, including spinning solids, is the difference between the structural properties of inwardly and outwardly spinning pools and the discontinuity between these pools. Due to the stirs in the form of moments of forces, in the discontinuous zones, there exist sub-eddies and sub-sub-eddies (Figure C.2, where sub-eddies are created naturally by the large eddies *M* and *N*). Their twist-ups (the sub-eddies) contain highly condensed amounts of materials and energies. In other words, the traditional frontal lines and surfaces (in meteorology) are not simply expansions of particles without any structure. Instead, they represent twist-up zones concentrated with irregularly structured materials and energies (this is where the so-called small probability events appear and small-probability information is observed and collected so such information (event) should also be called irregular information (and event)). In terms of basic energies, these twist-up zones cannot be formed by only the pushes of external forces and cannot be adequately described by using mathematical forms of separate objects and forces. Since evolution is about changes in materials' structures, it cannot be simply represented by different speeds of movements. Instead, it is mainly about transformations of rotations in the form of moments of forces ignited by irregularities. The enclosed areas in Figure C.3 stand for the potential places for equal quantitative effects to appear, where the combined pushing or pulling is small in absolute terms. However, it is generally difficult to predict what will come out of the power struggles. In general, what comes out of the power struggle tends to be drastic and unpredictable by using the theories and methodologies of modern science.

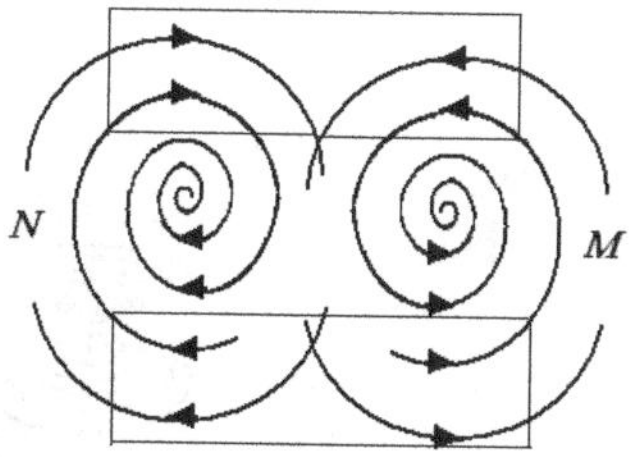

Fig. C.3 Structural representation of equal quantitative effects

C.2 The Systemic Yoyo Structure of General Systems

In this section, we establish the systemic yoyo model for the general system. As what is shown in (Lin, 2008a), this model can be employed as a background and intuition for thinking about systems and manipulating abstract systems. Its role in the research of systems science is expected to be similar to that of Euclidean spaces in modern science. For a thorough and detailed discussion of this model, please consult with (Lin, 2008a).

C.2.1 The Systemic Yoyo Model

On the basis of the blown-up theory and the discussion on whether or not the world can be seen from the point of view of systems (Lin, 1988; Lin, Ma and Port, 1990), the concepts of black holes, big bangs, and converging and diverging eddy motions are coined together in the model shown in Figure C.4. This model was established in (Wu and Lin, 2002) for each object and every system imaginable. In particular, each system or object considered in a study is a multi-dimensional entity that spins about its invisible axis. If we fathom such a spinning entity in our 3-dimensional space, we will have a structure as shown in Figure C.4(a). The side of black hole sucks in all things, such as materials, information, energy, etc. After funneling through the short narrow neck, all things are spit out in the form of a big bang. Some of the materials, spit out from the end of big bang, never return to the other side and some will (Figure C.4(b)). For the sake of convenience of communication, such a structure, as shown in Figure C.4(a), is called a (Chinese) yoyo due to its general shape. More specifically, what this model says is that each physical entity in the universe, be it a tangible or intangible object, a living being, an organization, a culture, a civilization, etc., can all be seen as a kind of realization of a certain multi-dimensional spinning yoyo with an invisible spin field around it. It stays in a constant spinning motion as depicted in Figure C.4(a). If it does stop its spinning, it will no longer exist as an identifiable system. What Figure C.4(c) shows is that due to the interactions between the eddy field, which spins perpendicularly to the axis of spin, of the model, and the meridian field, which rotates parallel to axis of spin (Figure C.4(b)), all the materials returning to the black-hole side travel along a spiral trajectory.

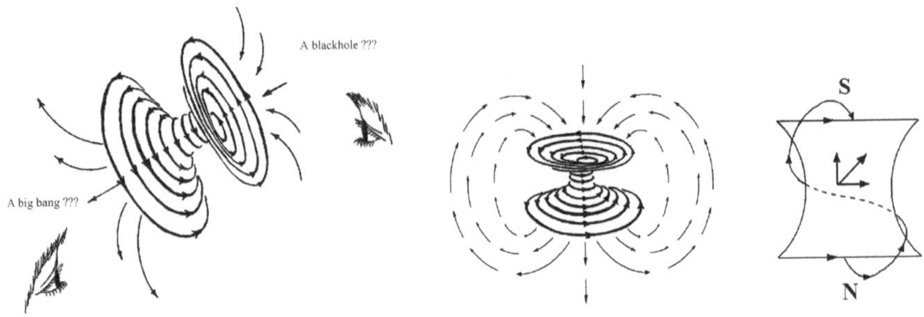

Fig. C.4 The eddy motion model of the general system

C.2.2 Justification Using Conservation Law of Informational Infrastructures

The theoretical justification for such a model of general systems is the blown-up theory, as discussed earlier in the previous section. The model can also be seen as a practical background for the law of conservation of informational infrastructures. More specifically, based on empirical data from particle physics, astronomy, and metoerology, the following law of conservation is proposed (Ren et al., 1998): For each given system, there must be a positive number a such that

$$AT \times BS \times CM \times DE = a \tag{C.12}$$

where A, B, C, and D are some constants determined by the structure and attributes of the system of concern, and T stands for the time as measured in the system, S the space occupied by the system, M and E the total mass and energy contained in the system.

Because M (mass) and E (energy) can exchange to each other and the total of them is conserved, if the system is a closed one, equ. (C.12) implies that when the time T evolves to a certain (large) value, the space S has to be very small. That is, in a limited space, the density of mass and energy becomes extremely high. So, an explosion (a big bang) is expected. Following the explosion, the space S starts to expand. That is, the time T starts to travel backward or to shrink. This end gives rise of the well-known model for the universe as derived from Einstein's relativity theory (Einstein, 1983; Zhu, 1985). In terms of systems, what this law of conservation implies is: Each system goes through such cycles as: … → expanding → shrinking → expanding → shrinking → … Now, the geometry of this model from Einstein's relativity theory is given in Figure C.4.

C.2.3 Justification Using Readily Repeatable Experiments

Empirically, the multi-dimensional yoyo model in Figure C.4 is manifested in different areas of life. For example, each human being, as we now see it, is a

3-dimensional realization of such a spinning yoyo structure of a higher dimension. To illustrate this end, let us consider two simple and easy-to-repeat experiences. For the first one, let us imagine we go to a sport event, say a swim meet. As soon as we enter the pool area, we immediately find ourselves falling into a boiling pot of screaming and jumping spectators, cheering for their favorite swimmers competing in the pool. Now, let us pick a person standing or walking on the pool deck for whatever reason, either for her beauty or for his strange look or body posture. Magically enough, before long, the person from quite a good distance will feel our stare and she or he will be able to locate us in a very brief moment out of the reasonably sized and boiling audience. The reason for the existence of such a miracle and silent communication is because each side is a high dimensional spinning yoyo. Even though we are separated by space and possibly by informational noise, the stare of one side on the other has directed that side's spin field of the yoyo structure into the spin field of the yoyo structure of the other side. That is the underlying mechanism for the silent communication to be established.

As our second example, let us look at the situation of human relationship. When an individual A has a good impression about another individual B, magically, individual B also has a similar and almost identical impression about A. When A does not like B and describes B as a dishonest person with various undesirable traits, it has been clinically proven in psychology that what A describes about B is exactly who A is himself (Hendrix, 2001). Once again, the underlying mechanism for such a quiet and unspoken evaluation of each other is because each human being stands for a spinning yoyo and its rotational field. Our feelings about another person are formed through the interactions of our invisible yoyo structures and their spin fields.

C.3 Laws on State of Motion of Systems

In this section, we look at how materials, social organizations, and events evolve according to certain laws of motions, which are generalizations of Newton's laws of motion by considering the internal structures of matters. For detailed analysis for why these laws hold true in general, please consult with (Lin, 2008a).

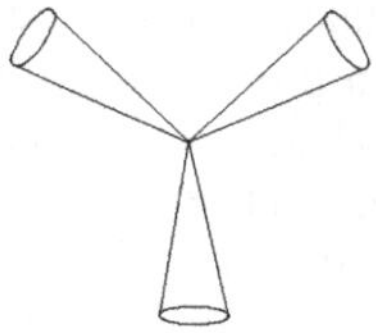

Fig. C.5 A "snapshot" in time and two spatial dimensions of a three-jet event

C.3.1 The Quark Structure of Systemic Yoyos

Let us first look at a well-known laboratory observation from particle physics. The so-called three-jet event is an event with many particles in a final state that appear to be clustered in three jets, each of which consists of particles that travel in roughly the same direction. One can draw three cones from the interaction point, corresponding to the jets, Figure C.5, and most particles created in the reaction appear to belong to one of these cones. These three-jet events are currently the most direct available evidence for the existence of gluons, the elementary particles that cause quarks to interact and are indirectly responsible for the binding of protons and neutrons together in atomic nuclei (Brandelik, et al., 1979). Because jets are ordinarily produced when quarks hadronize, the process of the formation of hadrons out of quarks and gluons, and quarks are produced only in pairs, an additional particle is required to explain such events as the three-jets that contain an odd number of jets. Quantum chromodynamics indicates that this needed particle of the three-jet events is a particularly energetic gluon, radiated by one of the quarks, which hadronizes much as a quark does. What is particularly interesting about these events is their consistency with the Lund string model. And, what is predicted out of this model is precisely what is observed.

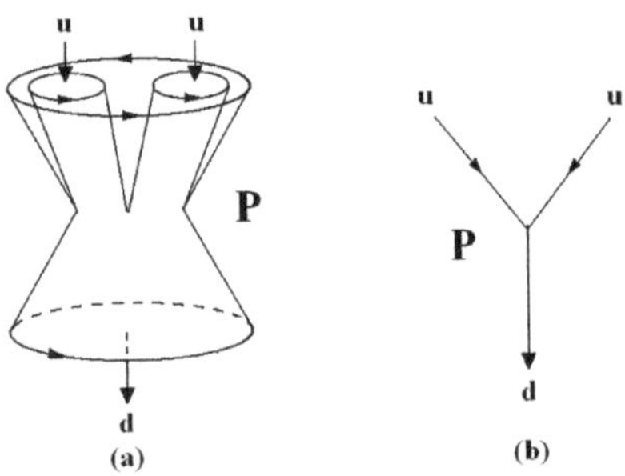

Fig. C.6 The quark structure of a proton P

Now, let us make use of this laboratory observation to study the structure of systemic yoyos and borrow the term of quark structure from (Chen, 2007), where it is argued that each microscopic particle is a whirltron, a similar concept as that of systemic yoyos. In particular, out of the several hundreds of different microscopic particles, other than protons, neutrons, electrons, and several others, most only exist momentarily. That is, it is a common phenomenon for general systemic yoyos to be created and to disappear constantly in the physical microscopic world. According to (Chen, 2007, p. 41) all microscopic systemic yoyos can be classified on the basis of laboratory experiments into two classes using the number of quarks involved. One class contains 2-quark yoyos (or whirltrons), such as electrons, π-, κ-, η-mesons, and others; and the other class 3-quark yoyos (whirltrons), including protons, neutrons, Λ-, Σ-, Ω-, Ξ (Xi) baryons, etc. Here, electrons are commonly seen as whirltrons without any quark. However,

Chen (2007) showed that yes, they are also 2-quark whirltrons. Currently, no laboratory experiment has produced 0-quark or n-quark whirltrons, for natural number $n \geq 4$.

Following the notations of (Chen, 2007), each spinning yoyo, as shown in Figure C.4(a), is seen as a 2-quark structure, where we imagine the yoyo is cut through its waist horizontally in the middle, then the top half is defined as an absorbing quark and the bottom half a spurting quark. Now, let us describe 3-quark yoyos by looking at a proton P and a neutron N. At this junction, the three jet events are employed as the evidence for the structure of 3-quark yoyos, where there are two absorbing and one spurting quarks in the eddy field. The proton P has two absorbing u-quarks and one spurting d-quark (Figure C.6), while the neutron N has two spurting d-quarks and one absorbing u-quark (Figure C.7). In these figures, the graphs (b) are the simplified flow charts with the line segments indicating the imaginary axes of rotation of each local spinning column. Here, in Figure C.6, the absorbing u-quarks stand for local spinning pools while together they also travel along in the larger eddy field in which they are part of. Similarly in Figure C.7, the spurting d-quarks are regional spinning pools. At the same time when they spin individually, they also travel along in the large yoyo structure of the neutron N. In all these cases, the spinning directions of these u- and d-quarks are the same except that each u-quark spins convergently (inwardly) and each d-quark divergently (outwardly).

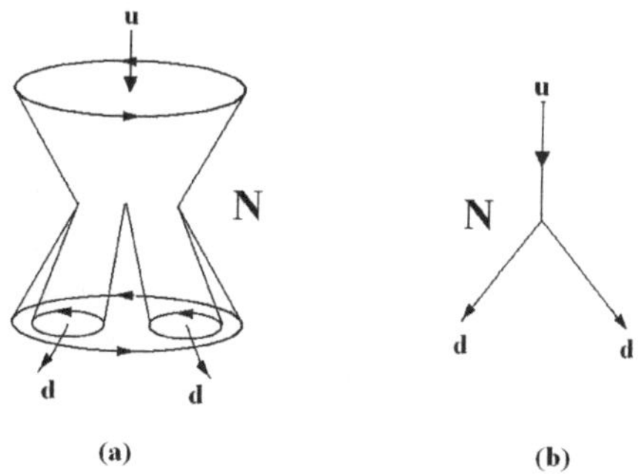

Fig. C.7 The quark structure of a neutron N

Different yoyo structures have different numbers of absorbing u-quarks and d-quarks. And, the u-quarks and d-quarks in different yoyos are different due to variations in their mass, size, spinning speed and direction, and the speed of absorbing and spurting materials. This end is well supported by the discovery of quarks of various flavors, two spin states (up and down), positive and negative charges, and colors. That is, the existence of a great variety of quarks has been firmly established.

Now, if we fit Fultz's (Fultz el al., 1959) dishpan experiment to the discussion above by imagining both the top and the bottom of each yoyo as a spinning dish of fluids, then the patterns as observed in the dishpan experiment suggest that in theory, there could exist such a yoyo structure that it has n u-quarks and

m d-quarks, where $n \geq 1$ and $m \geq 1$ are arbitrary natural numbers, and each of these quarks spins individually and along with each other in the overall spinning pool of the yoyo structure.

From discussions in (Lin, 2007; Lin, 2008a), it can be seen that due to uneven distribution of forces, either internal or external to the yoyo structure, the quark structure of the spinning yoyo changes, leading to different states in the development of the yoyo. This end can be well seen theoretically and has been well supported by laboratory experiments, where, for example, protons and neutrons can be transformed into each other. When a yoyo undergoes changes and becomes a new yoyo, the attributes of the original yoyo in general will be altered. For example, when a two-quark yoyo is split into two new yoyos under an external force, the total mass of the new yoyos might be greater or smaller than that of the original yoyo. And, in one spinning yoyo, no matter how many u-quarks (or d-quarks) exist, although these quarks spin individually, they also spin in the same direction and at the same angular speed. Here, the angular speeds of u-quarks and d-quarks do not have to be the same, which is different of what is observed in the dishpan experiment, because in this experiment everything is arranged with perfect symmetry, such as the flat bottom of the dish and perfectly round periphery.

When the yoyo in Figure C.4(a) models our earth, for details see (Lin and Forrest, 2010), where the 3-dimensional earth ball is a physical realization of this multi-dimensional yoyo, then the spin field of the big bang side corresponding to the visible earth and the black hole side, the invisible side of earth. With this identification in place, one explanation of the meridian field of the yoyo is the earthly geomagnetic field, which is approximately a magnetic dipole. One pole of it is located near the North Pole, and the other near the geographic South Pole. And the imaginary line joining the geomagnetic poles is inclined by approximately 11.3 degrees from the planet's axis of rotation. The geomagnetic field extends several tens of thousands kilometers into space as the magnetosphere (Walt, 2005; Comins, 2006). It is shown in (Lin and Forrest, 2010) that this yoyo field model provides a nice explanation for why and when the geomagnetic pole reversals appear.

C.3.2 Interactions between Systemic Yoyos

When two or more yoyo structures are concerned with, the interactions between these entities, a multi-body problem, become impossible to describe analytically, This end is witnessed by the difficulties scholars have faced in the investigation of the three-body problem since about three hundred years ago when Newton introduced his laws of motion regarding two-body systems. So, to present interactions of yoyo structures meaningfully, figurative expressions are used.

First of all, due to their field structures, as electric or magnetic fields, systemic yoyos also share the property that the like-kind ends repel and the opposite attract. Secondly, for the relative positioning of two same scale yoyo structures, assume that two spinning yoyos X and Y are given as shown in Figure C.8(a). Then the meridian field A of X fights against C of Y so that both X and Y have the

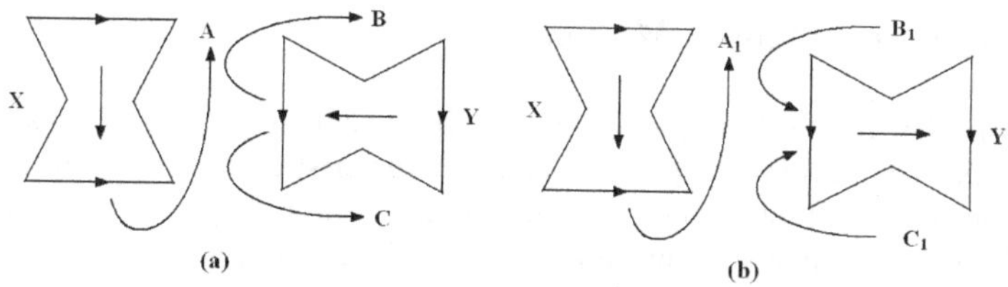

Fig. C.8 The tendency for yoyos to line up

tendency to realign themselves in order to reduce the conflicts along the meridian directions. Similarly in Figure C.8(b), the meridian field A_1 of yoyo X fights against B_1 of Y. So, the yoyos X and Y also have the tendency to realign themselves as in the previous case.

Of the two yoyos X and Y above, if one of them is mighty and huge and the other so small that it can be seen as a tiny particle. Then, the tiny particle yoyo m will be forced to line up with the much mightier and larger spinning yoyo M in such a way that the axis of spin of the tiny yoyo m is parallel to that of M and that the polarities of m and M face the same direction. For example, Figure C.9 shows how the particle yoyo m has to rotate and reposition itself under the powerful influence of the meridian field of the much mightier and larger yoyo structure M. In particular, if the two yoyos M and m are positioned as in Figure C.9(a), then the meridian field A of M fights against C of m so that m is forced to realign itself by rotate clockwisely in order to reduce the conflicts with the meridian direction A of M. If the yoyos M and m are positioned as in Figure C.9(b), the meridian field A_1 of yoyo M fights against B_1 of m so that the particle yoyo m is inclined to readjust itself by rotating once again clockwisely. If the yoyos M and m are positioned as in Figure C.9(c), then the meridian field A_2 of yoyo M fights against B_2 of m so that the tiny particle yoyo m has no choice but to reorient itself clockwisely to the position as in Figure C.9(b). As what has been just analyzed, in this case, yoyo m will further be rotated until its axis of spin is parallel to that of M and its polarities face the same directions as M.

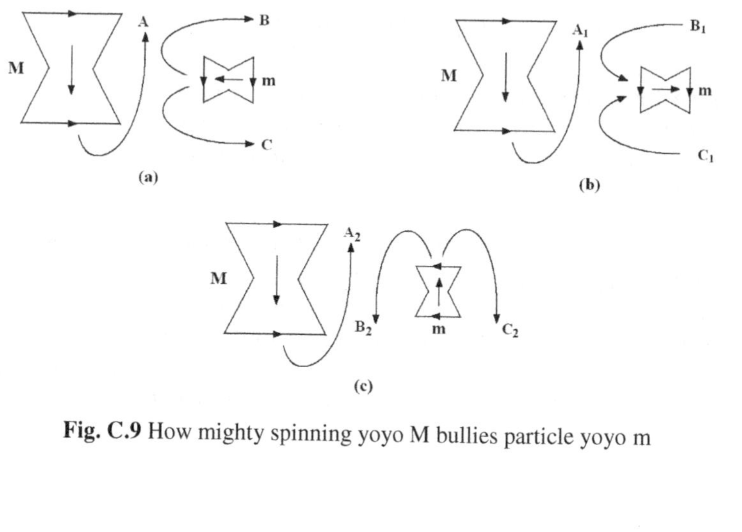

Fig. C.9 How mighty spinning yoyo M bullies particle yoyo m

C.3.3 Laws on State of Motion

Based on the discussion above. it has been concluded that spins are the fundamental evolutionary feature and characteristic of materials, in this subsection, we will look at the laws on state of motion by generalizing Newton's laws of motion so that external forces are no longer required for these laws to work. As what's known, these laws are one of the reasons why physics is an "exact" science. In (Lin and Forrest, 2010), it is shown that these generalized forms of the original laws of mechanics are equally applicable to social sciences and humanity areas as their classic forms in natural science.

The First Law on State of Motion (Lin, 2007): *Each imaginable and existing entity in the universe is a spinning yoyo of a certain dimension. Located on the outskirt of the yoyo is a spin field. Without being affected by another yoyo structure, each particle in the said entity's yoyo structure continues its movement in its orbital state of motion.*

The Second Law on State of Motion (Lin, 2007): *When a constantly spinning yoyo structure M does affect an object m, which is located in the spin field of another object N, the velocity of the object m will change and the object will accelerate. More specifically, the object m experiences an acceleration* $\vec{a}$ *toward the center of M such that the magnitude of* $\vec{a}$ *is given by*

$$a = \frac{v^2}{r} \tag{C.13}$$

where r is the distance between the object m and the center of M and v the speed of any object in the spin field of M about distance r away from the center of M. And, the magnitude of the net pulling force $\vec{F}_{net}$ *that M exerts on m is given by*

$$F_{net} = ma = m\frac{v^2}{r} \tag{C.14}$$

The Third Law on State of Motion: *When the spin fields of two yoyo structures N and M act and react on each other, their interaction falls in one of the six scenarios as shown in Figure C.10(a) – (c) and Figure C.11(a) – (c). And, the following are true:*

(1) For the cases in (a) of Figures C.10 – C.11, if both N and M are relatively stable temporarily, then their action and reaction are roughly equal but in opposite directions during the temporary stability. In terms of the whole evolution involved, the divergent spin field (N) exerts more action on the convergent field (M) than M's reaction peacefully in the case of Figure C.10(a) and violently in the case of Figure C.11(a).

(2) *For the cases (b) in Figures C.10 – C.11, there are permanent equal, but opposite, actions and reactions with the interaction more violent in the case of Figure C.10(b) than in the case of Figure C.11(b).*

(3) *For the cases in (c) of Figure C.10 – C.11, there is a permanent mutual attraction. However, for the former case, the violent attraction may pull the two spin fields together and have the tendency to become one spin field. For the later case, the peaceful attraction is balanced off by their opposite spinning directions. And, the spin fields will coexist permanently.*

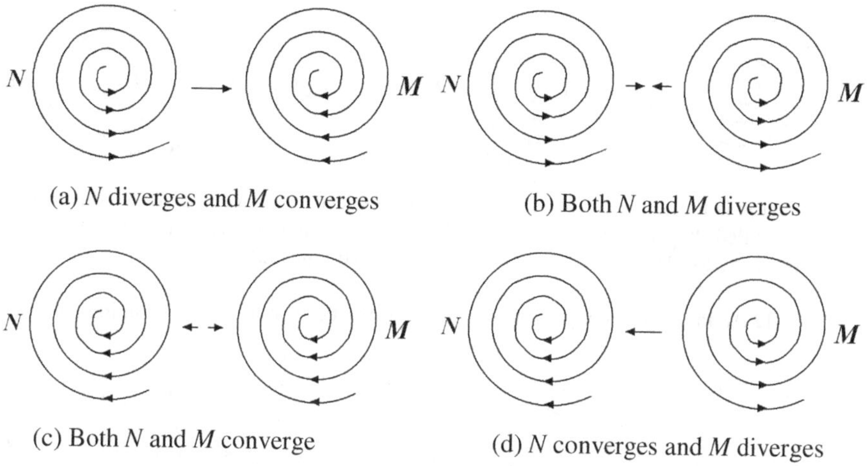

(a) *N* diverges and *M* converges (b) Both *N* and *M* diverges

(c) Both *N* and *M* converge (d) *N* converges and *M* diverges

Fig. C.10 Same scale acting and reacting spinning yoyos of the harmonic pattern

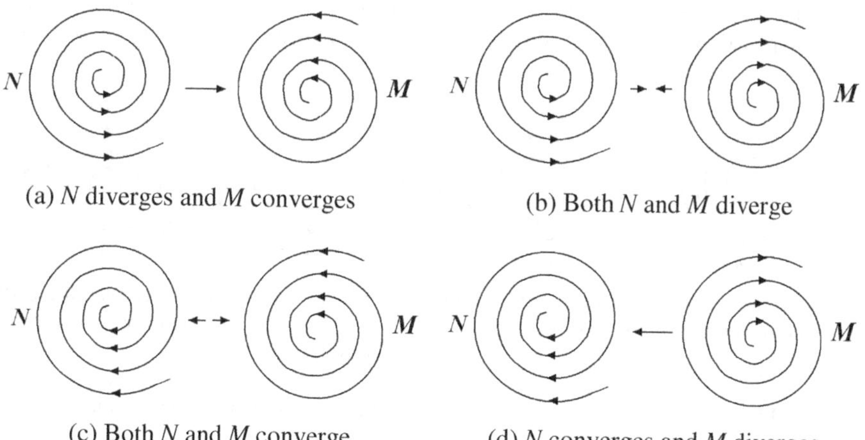

(a) *N* diverges and *M* converges (b) Both *N* and *M* diverge

(c) Both *N* and *M* converge (d) *N* converges and *M* diverges

Fig. C.11 Same scale acting and reacting spinning yoyos of inharmonic patterns

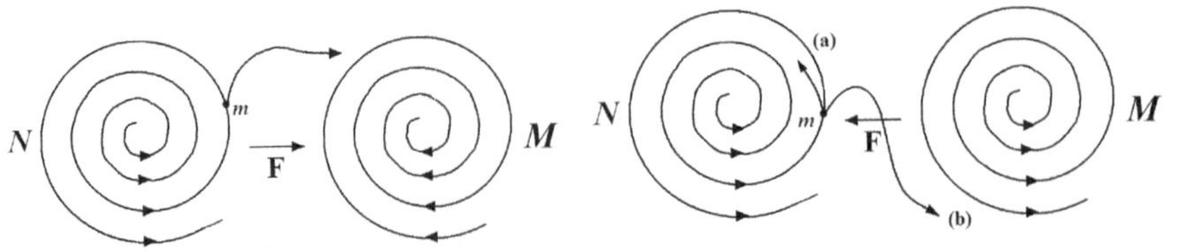

(a) Object m is located in a diverging eddy and pulled by a converging eddy M

(b) Object m is located in a diverging eddy and pulled or pushed by a diverging eddy M

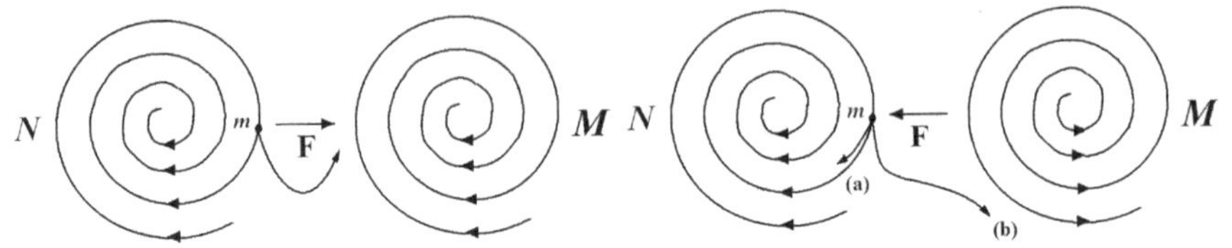

(c) Object m is located in a converging eddy and pulled by a converging eddy M diverging eddy M

(d) Object m is located in a converging eddy and pulled or pushed by a diverging eddy M

Fig. C.12 Acting and reacting models with yoyo structures of harmonic spinning patterns

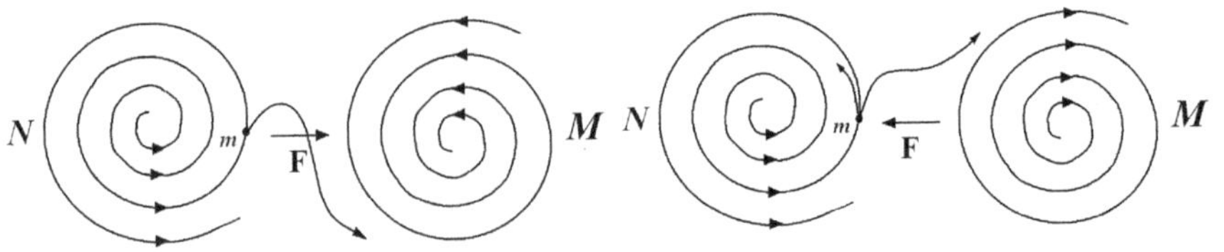

(a) Object m is located in a diverging eddy and pulled by a converging eddy M

(b) Object m is located in a diverging eddy and pushed or pulled by a diverging eddy M

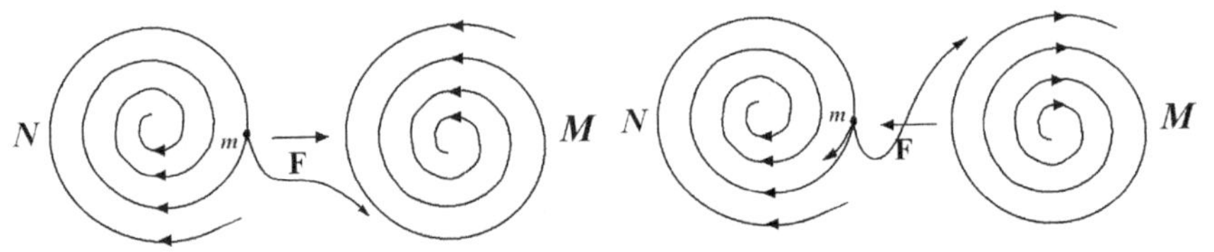

(c) Object m is located in a converging eddy and pulled by a converging eddy M

(d) Object m is located in a converging eddy and pushed or pulled by a diverging eddy M

Fig. C.13 Acting and reacting models with yoyo structures of inharmonic spinning patterns

The Fourth Law on State of Motion: *When the spin field M acts on an object m, rotating in the spin field N, the object m experiences equal, but opposite, action and reaction, if it is either thrown out of the spin field N and not accepted*

by that of M (Figure C.12(a), (d), Figure C.13(b) and (c)) or trapped in a sub-eddy motion created jointly by the spin fields of N and M (Figure C.12(b), (c), Figure C.13(a) and (d)). In all other possibilities, the object m does not experience equal and opposite action and reaction from N and M.

C.4 Uncertainties Everywhere

Based on what are presented in the previous three sections, we see in this section how in the ocean of spinning yoyo structures uncertainties appear, even in such an area of learning as mathematics that has been traditionally believed to be a realm of the ultimate universal truths.

C.4.1 Artificial and Physical Uncertainties

According to what has been discussed, the physical world is an ocean of yoyo fields fighting against each other. So, in the discontinuous zones that naturally exist between the spinning fields uncertainties appear especially when we are limited to using calculus-based or stochastics-based mathematical tools to describe what is happening in nature. Here, the uncertainties can be either mathematical or physical or both. In particular, when we limit ourselves to employ only certain theoretical tools to study nature, because these theoretical tools are developed in Euclidean spaces, which, as discussed earlier, are fundamentally linear in nature, and simply not rich and powerful enough to adequately describe nonlinear phenomena and movements in curvature spaces, we are inevitably trapped in uncertainty. Such uncertainties, caused by the inadequate tools we employ, are consequences of human ignorance. For instance, the boxed regions in Figure C.3 stand for occasions when theoretically linear tools generally lose their validity. In particular, in these regions, the mathematically linear differences of the acting and reacting forces are very small. However, large amounts of energies are accumulated so that as soon as one side of the action-reaction pair gains a little initial control, the situation will soon erupt into a landslide victory to the initial-controlling field. If there are only two such fields interacting with each other, there might not be any uncertainty involved. However, if we only employ quantities or variable mathematics to study the acting-reaction pair, our work will likely suffer from uncertainties, such as those considered in Lorenz's chaos (Lin, 1998).

However, in the physical world, between any two spinning fields, it is extremely likely that there are other rotational fields located in more distant areas. Their slight effects on the interaction of two spinning fields can lead to drastically different outcomes. That is where natural uncertainties appear. Such uncertainties naturally existing in the physical world have nothing to do with artificial uncertainties. In terms of the systemic yoyo model, the so-called uncertainties that naturally exist in the physical world come from the creation and disappearance of the sub- and sub-sub-eddies in the discontinuous zones between spinning yoyo fields. For instance, in Figure C.2, due to the interactions between the spinning fields N and M, two relatively stable subeddies are created. Now, between the subeddies and the mother fields N and M, infinitely many instable sub-subeddies are created. Although these sub-subeddies are instable and existing ones are constantly absorbed by the overriding flow patterns, the flow patterns continuously create new sub-subeddies. Once again, the life-spans and how they short-living sub-subeddies are created are very much dependent on how the specific regions are affected by other nearby spinning fields. For instance, an action-reaction pair as depicted in Figure C.2 is affected by some more drastic movements of spinning fields as described in Figures C.8 and C.9, then the way how these sub-subeddies are created and destroyed becomes truly unpredictable and uncertain.

C.4.2 Uncertainties That Exist in the System of Modern Mathematics

Mathematics has been treated as the last part of human knowledge that discovers the ultimate truths of nature in more or less definite ways (Kline, 1972). However, throughout history, such unproven treatment has not been adequately shown. Instead, in recent years, the opposite is proven. That is, mathematics contains some widespread inconsistencies and suffers from uncertainties.

C.4.2.1 Uncertainties of Mathematics

In terms of mathematics, a theory is a set of sentences expressed in a formal language. Some statements in the theory are included without any proof regarding their validity. These statements are known as axioms of the theory. Other statements are included because they can be derived from the axioms using the formal language. These provable statements are known as theorems. A formal theory is said to be *effectively generated* if its set of axioms is a recursively enumerable set. In particular, Peano arithmetic and Zermelo–Fraenkel set theory are two examples in each of which an infinite number of axioms exist and each is effectively generated.

When establishing a set of axioms for a theory, one goal is to be able to prove based on the set of axioms as many correct results as possible while excluding incorrect results. A set of axioms is said to be complete if any statement written in the language of the axioms or its negation is provable from the axioms. A set of axioms is said to be consistent if there is no such a statement that both itself and its negation are provable from the axioms. In the standard system of first-order logic, an inconsistent set of axioms will prove every statement in its language, which is known as the principle of explosion, so that it is a complete set. A set of axioms that is both complete and consistent proves a maximal set of non-contradictory theorems. To this end, Gödel's two incompleteness theorems show that in certain cases it is not possible to obtain an effectively generated, complete, consistent theory.

Gödel's First Incompleteness Theorem (Kleene 1967, p. 250): Any effectively generated theory capable of expressing elementary arithmetic cannot be both consistent and complete. In particular, for any consistent, effectively generated formal theory that proves certain basic arithmetic truths, there is an arithmetical statement A that is true, but not provable in the theory.

Gödel's second incompleteness theorem: For any formal effectively generated theory T including basic arithmetical truths and also certain truths about formal provability, T includes a statement of its own consistency if and only if T is inconsistent.

Ever since these incompleteness theorems were published in 1931, they have created major shocks in the communities of mathematics, logic, and philosophy. Because neither A nor (not A) can be proved, does it mean that the standards of truths are lost? Even if these results and their implications are limited to the realm of mathematics, it still means that the certainties of mathematics are forever lost.

The proofs of Gödel's theorems are extremely complicated and long. When they were initially publicized, John von Neumann twice questioned their validity. After some struggle, he eventually accepted the validity of the results and Kurt Gödel became known as a great logician who developed and established the second milestone in the history of logic.

Along this line of study of the potential loss of mathematical certainty and standards of truths, logician Bangyan Wen (2005; 2006; 2008a; 2008b; 2009a; 2009b) recently published a series of important works. He pinpoints out how a double standard for contradictions was employed in the proofs of the Gödel's theorems, how the self-referencing propositions Gödel constructed and their negations were not theorems, how various explanations of the Gödel's theorems become problematic by correcting the way of comprehending the concept of "undecidability", and why these Gödel's theorems would not lead to the loss of mathematical standards of truths and certainty by introducing the need to discuss decidability in different levels. For details, please consult with the relevant publications.

C.4.2.2 Inconsistencies in the System of Mathematics

Throughout the recorded history, there have appeared three crises in the foundations of mathematics. In particular, in the fifth century BC, the unexpected discovery of irrational numbers, called the Hippasus paradox, together with other paradoxes, such as those constructed by Zeno, gave rise to the first crisis to the foundations of mathematics. This discovery is one of the greatest achievements of the Pythagoreans. The rigorous proof for the existence of irrational numbers was surprising and disturbing to the Pythagoreans, because it was a mortal blow to the Pythagorean philosophy that all in the world depend on whole numbers. The final resolution of this crisis in the foundations of mathematics was achieved in about 370 BC by Eudoxus with a treatment of incommensurables that coincides essentially with the modern exposition of irrational numbers that was first given by Richard Dedekind in 1872. This crisis in the foundations of mathematics is largely responsible for the subsequent formulation and adoption of the axiomatic method in mathematics.

Toward the end of the 17th century, Isaac Newton and Gottfried Wilhelm Leibniz created calculus. As it turned out, calculus was remarkably powerful and capable of attacking problems quite unassailable in earlier days. The processes employed in the research papers of the day were justified largely on the ground that they worked. Because the early calculus was established on the ambiguous and vague concept of infinitesimals without a solid foundation, many operations and applications of the theory were reproached and attacked from various angles. Among all the reproaches, the most central and able about the faulty foundation of the early calculus was brought forward by Bishop George Berkeley (1685–1753). In 1734, he published a book entitled *The Analyst: A Discourse Addressed to an Infidel Mathematician*. It directly attacked the foundation and principles of calculus. With the passage of time, conflicts between achievements of calculus in applications and the ambiguity in the foundation became incisive so that a serious crisis in the foundations of mathematics became evident. After many unsuccessful attempts, a great stride was made in 1821 when Augustin-Louis Cauchy (1789–1857) developed an acceptable theory of limits and then defining continuity, differentiability, and definite integral in terms of the concept of limits. Afterward, on the basis of Cauchy's theory of limits, Dedekind proved the fundamental theorems in the theory of limits using the rigorized real number theory. With the combined efforts of many mathematicians, the methods of $\varepsilon - N$ and $\varepsilon - \delta$ became widely accepted, so that infinitesimals and infinities could be entirely avoided. Therefore, the second crisis in the foundations of mathematics has been considered successfully resolved.

In 1874, Georg Cantor commenced his set theory. Because so much of mathematics is permeated with set concepts, the superstructure of mathematics can actually be made to rest upon set theory as its foundation. However, starting in 1897 various paradoxes of set theory began to emerge. The existence of these paradoxes suggests that there must be something wrong with set theory. However, what is more devastating is that now the foundation of mathematics faces another major crisis, called the third crisis of mathematics, because these paradoxes naturally cast doubt on the validity of the entire fundamental structure of mathematics and severely challenge the common belief that mathematics and logic are two most rigorous and exact scientific disciplines. What needs to be noted is that to a certain degree, this third crisis is in fact a deepening evolution of the previous two crises, because the problems touched on by these paradoxes are more profound and involved a much wider area of human thought. To resolve this crisis, mathematicians thought of giving up the naive set theory and identifying another theory as the foundation of mathematics. However, after careful analysis, it was realized how difficult it would be to adopt this approach. So instead, these mathematicians focused on modifying the naive set theory in an attempt to make it plausible. One successful attempt was Zermelo's axiomatic set theory. In 1908, Zermelo established his system of axioms for his new set theory. After several rounds of modification, Fraenkel and Skolem (1912–1923) provided a rigorous interpretation and formed the present day Zermelo-Fraenkel (ZFC) system, where C stands for the axiom of choice. The ultimate goal of this new set theory is still about establishing a rigorous foundation for mathematical analysis. The specific thinking and technical details are:

1) By introducing the axioms of the empty set and of infinity, the legality of the set of all natural numbers is warranted.
2) The legality of the set of all real numbers is derived by using the axiom of power sets. Then,
3) The legality of each subset of those elements, satisfying a given property P, of the real numbers is based on the axiom of subsets.

Therefore, as long as the ZFC system is consistent or contains no contradiction, the theory of limits, as a rigorous theoretical foundation of calculus, can be satisfactorily constructed on the ZFC axiomatic system. However, as of this writing, we still cannot show in theory that no paradox whatever kind will ever be constructed in the ZFC system. That is, the consistency of the ZFC system still cannot be shown. On the other hand, what is worth celebrating is that for a century, since the time when the ZFC system was initially suggested, there had been no new paradoxes found in the ZFC system until 2008 when Henri Poincare's worries over a century ago (Kline, 1972), Felix Hausdorff's (1935) opinion, and Abraham Robinson's (1964) belief about infinity were shown to be correct (Lin, 2008b).

In particular, by actual infinity, it means a nonfinite process that definitely reaches the very end. For example, when a variable x approaches the endpoint b from within the interval $[a, b]$, x actually reaches its limit b. In history, Plato was the first scholar to clearly recognize the existence of actual infinities (IFHP, 1962).

By potential infinity, it stands for a nonfinite, nonterminating process that will never reach the very end. For example, let us consider a variable x defined on an open interval (a, b), and let x approach the endpoint b from within this open interval. Then, the process for x to get indefinitely close to its limit b is a potential infinity, because even though x can get as close to b as any predetermined accuracy, the variable x will never reach its limit b. In history, Aristotle was the first scholar to acknowledge the concept of potential infinities – he never accepted the existence of actual infinities. Similarly, Plato did not believe in the existence of potential infinities (IFHP, 1962).

Although these two kinds of infinites are different, where each actual infinity stands for a perfect tense and a potential infinity a present progressive tense, in the modern mathematics system, they have been seen as the same or applied hand by hand without discrimination. For example, in the theory of limits, to avoid the Berkeley paradox, the definitions of limits, infinite or finite, using $\varepsilon - N$ and $\varepsilon - \delta$ methods, are completely based on the thinking logic of potential infinities. In the naive set theory and the modern (ZFC) axiomatic set theories, from Cantor to Zermelo, the existence and construction of infinite sets have been established on the concept of actual infinities. And, the vase puzzle (Lin, 1999) shows that these two kinds of infinities can in fact lead to different consequences. That is, the concepts of actual and potential infinities are indeed different. However, it is shown (Lin, 2008b) that both of the assumptions that actual infinities are the same as potential infinities and that actual infinites are different of potential infinities have been long held in the system of modern mathematics.

Considering how much a portion of mathematics is affected by this new discovery, a major crisis in the foundations of mathematics has been unearthed. In fact, if the system of mathematics is seen as an abstract yoyo field, then Fultz's (1959) dishpan experiment indicates that inconsistencies in this system of knowledge should appear periodically in the past, at the present, and in the future, for details, see (Lin, 2008, pp. 275 – 294).

What is worth mentioning is that recently Bangyan Wen (2008c; 2009c; 2009d) walks along a different approach to address some of the problems existing with the ZFC set theory. In his new set theory, he introduces five axioms to describe the structures of elements that make up different sets, five axioms to construct sets, where he clearly addresses the concept of potential infinity while avoiding the situation that the whole is equal to its part, and another five axioms to establish natural numbers. In his new set theory, Wen shows that Cantor's uncountable infinite sets do not exist so that the inconsistency between actual infinite = potential infinity and actual infinite ≠ potential infinity is avoided. At the same time, the Cantor paradox that the largest cardinal (ordinal) number must be greater than itself is avoided. By comparing with Peano's axioms of natural numbers, Cantor's naïve set theory, the axiomized ZFC set theory, Wen's theory does seem to have some advantages. For details, please consult with the relevant references.

C.5 A Few Final Words

What is presented in this appendix shows the fact that due to the characteristics of the underlying systemic structure of the physical world uncertainty exists constantly in the discontinuous zones between systemic yoyo fields and periodically within each and every yoyo field. And, because systems of knowledge also possess the spinning yoyo field structure, each scientific theory, including mathematics, has to suffer from uncertainties and internal turmoil. As a matter of fact, historically speaking, meeting challenges of uncertainties or unsolvable problems have been one of the major stimuli for the continued progress of science and technology from the antiquity to modern times. So, to a degree, the conclusions presented in this paper have been intuitively felt except that here in this paper we presented a cohesive theory to prove the validity of this intuition.

References

Brandelik, R., et al.: Evidence for planar events in e^+e^- annihilation at high energies. Phys. Lett. B 86, 243–249 (1979)

Chen, G.R.: The Original State of the World: The Theory of Ether Whirltrons. Tianma Books Limited, Hong Kong (2007)

Comins, N.F.: Discovering the Essential Universe. W.H. Freeman, New York (2006)

Einstein, A.: Complete Collection of Albert Einstein (translated by L. Y. Xu). Commercial Press, Beijing (1983)

Fultz, D., Long, R.R., Owens, G.V., Bohan, W., Kaylor, R., Weil, J.: Studies of Thermal Convection in a Rotating Cylinder with Some Implications for Large-Scale Atmospheric Motion. Meteorol. Monographs (American Meteorological Society) 21(4) (1959)

Hausdorff, F.: Mengenlehve, Watter ac. Hruyle (1935)

Hendrix, H.: Getting the love you want: A Guide for Couples. Owl Books, New York (2001)

IFHP (Institute of Foreign History of Philosophy). The philosophy of ancient Greeks. Commercial Press, Beijing (1962) (in Chinese)

Kleene, S.C.: Mathematical Logic (1967); Reprinted by Dover (2002)

Kline, M.: Mathematical Thought from Ancient to Modern Times. Oxford University Press, Oxford (1972)

Klir, G.: Facets of Systems Science. Springer, New York (2001)

Lin, Y.: Can the world be studied in the viewpoint of systems?? Mathl. Comput. Modeling 11, 738–742 (1988)

Lin, Y. (guest editor): Mystery of nonlinearity and Lorenz's chaos. A special double issue. Kybernetes: The International Journal of Cybernetics, Systems and Management Science 27(6-7), 605–854 (1998)

Lin, Y.: General systems theory: A mathematical approach. Plenum and Kluwer Academic Publishers, New York (1999)

Lin, Y.: Systemic yoyo model and applications in Newton's, Kepler's laws, etc. Kybernetes: The International Journal of Cybernetics, Systems and Management Science 36(3-4), 484–516 (2007)

Lin, Y.: Systemic Yoyos: Some Impacts of the Second Dimension. CRC Press, an imprint of Taylor and Francis, New York (2008)

Lin, Y. (guest editor): Systematic studies: The infinity problem in modern mathematics. Kybernetes; The International Journal of Systems, Cybernetics and Management Science 37(3-4), 387–578 (2008b)

Lin, Y., Forrest, B.: Systemic Structure behind Human Organizations: From Civilizations to Individuals. Springer, Dordrecht (2010)

Lin, Y., Ma, Y., Port, R.: Several epistemological problems related to the concept of systems. Mathl. Comput. Modeling 14, 52–57 (1990)

Lin, Y., Wu, Y.: Blown-ups and the concept of whole evolution in systems science. Problems of Nonlinear Analysis in Engineering systems 7(4), 16–31 (1998)

Liu, S.F., Lin, Y.: Grey Information: Theory and Practical Applications. Springer, London (2006)

Lorenz, E.N.: The Essence of Chaos. Washington University Press, Seattle (1993)

OuYang, S.C., Chen, Y.G., Lin, Y.: Digitization of Information and Prediction. China Meteorological press, Beijing (2009)

OuYang, S.C., Miao, J.H., Wu, Y., Lin, Y., Peng, T.Y., Xiao, T.G.: Second stir and incompleteness of quantitative analysis. Kybernetes: The International Journal of Cybernetics, Systems and Management Science 29, 53–70 (2000)

Ren, Z.Q., Lin, Y., OuYang, S.C.: Conjecture on law of conservation of informational infrastructures. Kybernetes: The International Journal of Systems and Cybernetics 27, 543–552 (1998)

Robinson, A.: Formalism 64, logic, methodology, and philosophy of science. In: Bar-Hillel, Y. (ed.) Proceedings of the 1964 International Congress, pp. 228–246. North-Holland and Pub. Co., Amsterdam (1964)

Soros, G.: The Alchemy of Finance. Wiley, New York (2003)

Walt, M.: Introduction to Geomagnetically Trapped Radiation. Cambridge University Press, Cambridge (2005)

Wen, B.Y.: A brief discussion on innovation and logic. Journal of Chinese People's University (1), 95–101 (2005)

Wen, B.Y.: Use of self-referencing propositions is prohibited: Elimination of liar's paradox and discussion of Gödel's theorems. Journal of Anhui University 30(5), 13–20 (2006)

Wen, B.Y.: Elimination of liar's paradox and questions about Gödel's theorems. Journal of Chongqing Institute of Technology (Philosophy and Social Science Edition) 22(3), 13–17 (2008a)

Wen, B.Y.: New choice of the primitive concepts of mathematics. Journal of Chongqing Institute of Technology (Natural Science Edition) 22(5), 135–144 (2008b)

Wen, B.Y.: Is the number of elements in natural number set equal to that in even number set? New Scheme for infinite theory (I). Journal of Chongqing Institute of Technology (Natural Science Edition) 22(11), 70–77 (2008c)

Wen, B.Y.: Also about correctly understanding Gödel's incompleteness theorems: A discussion with Mr. Chen Muzhe. Journal of Chongqing Institute of Technology (Social Science Edition) 23(4), 8–13 (2009a)

Wen, B.Y.: The elimination of the liar paradox and discussion of Gödel's incompleteness theorems. In: Tarasti, E. (ed.) Communication: Understanding / Misunderstanding, Proceedings of the 9th congress of the IASS/AIS, Helsinki-Imatra, June 11-17, pp. 1904–1914 (2009b) (printed by Greif Tartu 2009)

Wen, B.Y.: Does the larget natural number exist? A new method in infinity theory to eliminate the paradox of transfinite numbers (2). Journal of Chongqing Institute of Technology (Natural Science Edition) 23(7), 146–152 (2009c)

Wen, B.Y.: What is uncountable set by Georg Cantor? A new solution for the infinite theory (3). Journal of Chongqing Institute of Technology (Natural Science Edition) 23(11), 145–153 (2009d)

Wilhalm, R., Baynes, C.: The I Ching or Book of Changes, 3rd edn. Princeton University Press, Princeton (1967)

Wu, X.M.: The Pansystems View of the World. People's University of China Press, Beijing (1990)

Wu, Y., Lin, Y.: Beyond Nonstructural Quantitative Analysis: Blown-Ups, Spinning Currents and the Modern Science. World Scientific, River Edge (2002)

Zhu, Y.Z.: Albert Einstein: The Great Explorer. Beijing People's Press, Beijing (1985)

References

Bertranffy, L.V.: General Systems Theory: Foundation, Development, Applications. George Braziller, New York (1968)

Chan, J.W.K., Tong, T.K.L.: Multi-criteria material selections and end-of-life product strategy: Grey relational analysis approach. Materials & Design 28(5), 1539–1546 (2007)

Chang, B.R., Tsai, H.F.: Forecast approach using neural network adaptation to support vector regression grey model and generalized auto-regressive conditional heteroscedasticity. Expert Systems with Applications 34(2), 925–934 (2008)

Chang, S.C., Lai, H.C., Yu, H.C.: A variable P value rolling grey forecasting model for Taiwan semiconductor industry production. Technological Forecasting and Social Change 72(5), 623–640 (2005)

Chen, C.I., Chen, H.L., Chen, S.P.: Forecasting of foreign exchange rates of Taiwan's major trading partners by novel nonlinear Grey Bernoulli model NGBM (1,1). Communications in Nonlinear Science and Numerical Simulation 13(6), 1194–1204 (2008)

Chen, C.L., Dong, D.Y., Chen, Z.H., Wang, H.B.: Grey systems for intelligent sensors and information processing. Journal of Systems Engineering and Electronics 19(4), 659–665 (2008)

Chen, D.J., Zhang, Y.L., Chen, M.Y.: Grey macroscopic control prediction model of systems clouds and applications. Control and Decision Making 20(5), 553–557 (2005)

Chen, H.S., Chang, W.C.: Optimization of GM(1,N) Modeling. Journal of Grey System 12(3), 299–302 (2000)

Chen, K.X.: The epistemology of grey medical analysis. PhD Dissertation of Huazhong University of Science and Technology, Wuhan (1999)

Chen, M.Y.: Grey dynamic control of boring machine. Journal of Huazhong University of Science and Technology 10(2), 7–11 (1982)

Chen, M.Y.: Stability of grey systems and problem of calmness. Fuzzy Mathematics 5(2), 54–58 (1985)

Chen, X.X., Liu, S.F.: On grey multi-attribute group decision-making with completely unknown weight information. Management science of China 16(5), 146–152 (2008)

Chen, Y.J., Fu, Y.H., Sun, H.: Grey theory applied in the construction control of steel-pipe-concrete arch bridges. Journal of Harbing University of Technology 39(4), 546–548 (2007)

Chen, Y.K., Tan, X.R.: Grey Relational Analysis on Serum Markers of Liver Fibrosis. The Journal of Grey System 7(1), 63–68 (1995)

Chen, Z.B., Liu, J.G.: Floatation of the feedback extrema of grey systems' states. Journal of Huazhong University of Science and Technology 12(3), 8–14 (1984)

Chen, Z.B.: Recognition of recurrence errors of grey interval functions. Journal of Huazhong University of Science and Technology 12(1), 67–70 (1984)

Cheng, Q.Y., Qiu, W.H.: Grey System Forecast for Firing Accuracy of Gun. Journal of Systems Science and Systems Engineering 10(2), 205–211 (2001)

Chiang, J.Y., Cha'o-Kuang, C.: Application of grey prediction to inverse nonlinear heat conduction problem. International Journal of Heat and Mass Transfer 51(3), 576–585 (2008)

Chiang, Y.M., Hsieh, H.H.: The use of the Taguchi method with grey relational analysis to optimize the thin-film sputtering process with multiple quality characteristic in color filter manufacturing. Computers & Industrial Engineering 56(2), 648–661 (2009)

Chiao, C.H., Wang, W.Y.: Reliability improvement of fluorescent lamp using grey forecasting model. Microelectronics Reliability 42(1), 127–134 (2002)

Chiu, N.H.: An early software-quality classification based on improved grey relational classifier. Expert Systems with Applications 36(7), 10727–10734 (2009)

Chu, Y.F.: Predictive research on the development scale of Chinese logistics based on grey systems theory. Management Review 20(3), 58–64 (2008)

Churchman, G.W.: The Systems Approach and Its Enemies. Basic Books, New York (1979)

Cui, J., Dang, Y.G.: On the GM(1,1) prediction accuracy based on another class of strengthening buffer operators. Control and Decision-Making 24(1), 44–48 (2009)

Cui, J., Dang, Y.G., Liu, S.F.: An improvement method for solving criteria weights based on degrees of grey incidence. Management Science of China 16(5), 141–145 (2008)

Cui, J., Dang, Y.G., Liu, S.F.: Precision analysis of the GM(1,1) model based on a new weakening operator. Systems Engineering: Theory and Practice 29(7), 132–138 (2009)

Cui, L.Z., Liu, S.F., Wu, Z.P.: Some new weakening buffer operators and their applications. Control and Decision-Making 24(8), 1252–1256 (2009)

Dai, W.Z., Li, J.F.: Modeling with non-equal-distant GM(1,1). Systems Engineering: Theory and Practice 25(9), 89–93 (2005)

Dai, W.Z.: An nth order accumulation grey model with remnant corrections. Systems Engineering: Theory and Practice 17(12), 121–124 (1997)

Dang, Y.G.: Valuation of grey clustering with human experience. The Journal of Grey System 7(2), 179–184 (1995)

Dang, Y.G., Liu, B., Guang, Y.Q.: On strengthening buffer operators. Control and Decision-Making 20(12), 1332–1336 (2005)

Dang, Y.G., Liu, S.F.: The GM models that x(n) is taken as initial value. Kybernetes: The International Journal of Systems & Cybernetics 33(2), 247–254 (2004)

Dang, Y.G., Liu, S.F.: Multi-attribute grey incidence decision model for interval number. Kybernetes, The International Journal of Systems & Cybernetics 35(7), 1265–1272 (2006)

Dang, Y.G., Liu, S.F., Fang, Z.G.: The matrix method for computing the cross-section set of maximum flows in networks. Systems Engineering: Theory and Practice 23(9), 125–128 (2003)

Dang, Y.G., Liu, S.F., Liu, B.: The GM model with x(n) as the initial value. Management Science of China 13(1), 132–135 (2005)

Dang, Y.G., Liu, S.F., Li, B.J., Chen, Z.: Predictive analysis on the contribution of technology in the dominant industries. Systems Engineering: Theory and Practice 20(8), 95–99 (2000)

Dang, Y.G., Liu, S.F., Liu, B., Mi, C.M.: An improvement on the slope degree of grey incidence. Engineering Science of China 6(3), 41–44 (2004)

Dang, Y.G., Liu, S.F., Liu, B., Tang, X.W.: About weakening buffer operators. Management Science of China 12(2), 108–111 (2004)

Dang, Y.G., Liu, S.F., Liu, B., Yu, Y.W.: Study on multi-criteria decision-making based on interval number incidence. Journal of Nanjing University of Aeronautics and Astronautics 36(3), 403–406 (2004)

Dang, Y.G., Liu, S.F., Ping, X.L.: The Study of grey synthetic decision appraisal model. In: Proceedings of the 32nd International Conference on Computers & Industrial Engineering, pp. 542–547 (2003)

Dang, Y.G., Wang, L.H.: Grey clustering analysis of country agriculture. In: Liu, S.F., Xu, Z.X. (eds.) New Advances in Grey Systems Research, Wuhan: Press of Huazhong University of Science and Technology, pp. 221–223 (1996)

David, E.D.: Science Development: An Evaluation Study. NAS Publications Office, Washington (1975)

David, N.: The Practice of Policy Evaluation. St. Martin's Press, New York (1990)

Deng, J.L.: Grey dynamic models and their application in the long-term prediction of food production. Exploration of Nature 3(3), 7–43 (1984)

Deng, J.L.: The incidence space of grey systems. Fuzzy Mathematics 5(2), 1–10 (1985)

Deng, J.L.: Generation functions of grey systems. Fuzzy Mathematics 5(2), 11–22 (1985)

Deng, J.L.: The GM model of grey systems theory. Fuzzy Mathematics 5(2), 23–32 (1985)

Deng, J.L.: Five kinds of grey predictions. Fuzzy Mathematics 5(2), 33–42 (1985)

Deng, J.L.: Grey situational decision making. Fuzzy Mathematics 5(2), 43–50 (1985)

Deng, J.L.: Grey Control Systems. Press of Huazhong University of Science and Technology, Wuhan (1985)

Deng, J.L.: Grey Social and Economic Systems. Press of National Defense Industry, Beijing (1985)

Deng, J.L.: Grey Prediction and Decision-Making. Press of Huazhong College of Science and Technology, Wuhan (1986)

Deng, J.L.: Elementary methods of grey systems. Press of Huazhong University of Science and Technology, Wuhan (1987)

Deng, J.L.: Grey exponential law of accumulating generation: the optimization information problem of grey control systems. Journal of Huazhong College of Science and Technology 12(5), 11–16 (1987)

Deng, J.L.: Grey Multi-Dimensional Planning. Press of Huazhong University of Science and Technology, Wuhan (1988)

Deng, J.L.: Introduction to grey system theory. The Journal of Grey System 1(1), 1–7 (1989)

Deng, J.L.: Grey information space. The Journal of Grey System 1(2), 103–118 (1989)

Deng, J.L.: The point-set topological properties of grey series. The Journal of Grey System 2(4), 283–295 (1990)

Deng, J.L.: A Course in Grey Systems Theory. Press of Huazhong University of Science and Technology, Wuhan (1990)

Deng, J.L.: Representational forms of information in grey systems theory. Grey Systems Theory and Practice 1(1), 1–12 (1991)

Deng, J.L.: The unit of information representation in grey system theory. The Journal of Grey System 3(2), 87–106 (1991)

Deng, J.L.: Grey causality and information cover. The Journal of Grey System 4(2), 99–119 (1992)

Deng, J.L.: Grey hazy sets. The Journal of Grey System 4(2), 13–29 (1992)

Deng, J.L.: Grey inclusion in set. The Journal of Grey System 4(3), 195–220 (1992)

Deng, J.L.: Grey units. The Journal of Grey System 4(1), 3–11 (1992)

Deng, J.L.: Grey hazy Sets: Introduction to Grey Mathematics. Press of Huazhong University of Science and Technology, Wuhan (1992)

Deng, J.L.: Comparability of information representations. Grey Systems Theory and Practice 2(1), 1–10 (1992)

Deng, J.L.: Grey differential equation. The Journal of Grey System 5(1), 1–14 (1993)

Deng, J.L.: Extent information cover in grey system theory. The Journal of Grey System 7(2), 131–138 (1995)

Deng, J.L.: Several problems in grey systems theory and development. In: Liu, S.F., Xu, Z.X. (eds.) New Advances in Grey Systems Research, pp. 1–10. Press of Huazhong University of Science and Technology, Wuhan (1996)

Deng, J.L.: Negative power AGO in grey theory. Journal of Grey System 13(3), 31–38 (2001)

Deng, J.L.: On boundary of grey input b in GM(1,1). Journal of Grey System 13(1) (2001)

Deng, J.L.: On boundary of parameters in GM(1,1| tau,r). Journal of Grey System 13(2), 199–200 (2001)

Deng, J.L.: Foundations of Grey Theory. Press of Huazhong University of Science and Technology, Wuhan (2002)

Deng, J.L.: Grey Information Coverage in Grey Theory. Gaoli Books, Taiwan (2002)

Deng, J.L.: Grey Prediction and Grey Decision-Making. Press of Huazhong University of Science and Technology, Wuhan (2002)

Deng, J.L.: Proving GM(1,1) modeling via four data (at least). Journal of Grey System 16(1), 1–4 (2004)

Deng, J.L.: Figure on difference information space in grey relational analysis. Journal of Grey System 16(2), 96–100 (2004)

Deng, J.L.: Introduction to Grey Mathematical Physical Resource Science. Press of Huazhong University of Science and Technology, Wuhan (2007)

Deng, J.L., Ng, D.K.W.: Chaos in Grey Model GM(1,N). The Journal of Grey System 8(1), 1–10 (1996)

Deng, J.L., Zhou, C.S.: Sufficient conditions for the stability of a class of interconnected dynamic systems. Systems & Control Letters 5(2), 105–108 (1986)

Deng, Q., Feng, Q.Y., Deng, J.L.: A control system for underground water and optimal grey decision-making. Journal of Automation 12(2), 125–132 (1985)

Dong, F.Y.: Development prediction on the finance of Chinese firms based on newly improved GM(1,1) model. Management Science of China 15(4), 93–97 (2007)

Dunn, R.A.: Management Science: A Practical Approach to Decision-Making. Macmillan, New York (1981)

Fan, T.G., Zhang, Y.C., Quan, X.Z.: Grey model GM(1,1,t) and its chaos feature. Journal of Grey System 13(2), 117–120 (2001)

Fang, Z.G., Liu, S.F.: Grey matrix model based on pure strategy. In: Dessouky, M.I., Heavey, C. (eds.) Proceedings of the 32nd International Conference on Computers & Industrial Engineering, pp. 520–525. Gemini International Limited Dublin, Ireland (2003)

Fang, Z.G., Liu, S.F.: Grey matrix game models based on pure strategies (1): Construction of standard grey matrix game models. Journal of Southeast University (Natural Science Edition) 33(6), 796–800 (2003)

Fang, Z.G., Liu, S.F.: Two-player zero-sum game models based on the grey matrices of pure strategies. Journal of Nanjing Aeronautics and Astronautics 35(4), 441–445 (2003)

Fang, Z.G., Liu, S.F.: A study on the GM(1,1) (GMBIGN(1,1)) model based on interval grey sequences. Engineering Science of China 7(2), 57–61 (2005)

Fang, Z.G., Liu, S.F., Dang, Y.G.: Analysis of the concealed allocation model with the optimal grey information entropy of military traffic flow. Management Science of China 11(3), 56–61 (2003)

Fang, Z.G., Liu, S.F., Dang, Y.G.: Algorithm model research of the logical cutting tree on the network maximum flow. Kybernetes: The International Journal of Systems & Cybernetics 33(2), 255–262 (2004)

Fang, Z.G., Liu, S.F., Zhang, Y.J., Zhang, L.J.: Consistency analysis on international free trades and the development of Chinese national economic interests. Management Science of China 10, 221–225 (2002) (special issue)

Feng, Y.C., Li, D., Wang, Y.Z., Cao, Z.S.: A grey model for response time in database system DM3. Journal of Grey System 13(3) (2001)

Feng, Y.G.: Grey clustering and comprehensive evaluation of contaminated water. Systems Engineering: Theory and Practice 12(6), 46–48 (1992)

Feng, Z.Y.: Direct grey modeling. Journal of Applied Mathematics 15(3), 345–354 (1992)

Forrester, J.W.: Principles of Systems. Wright-Allen Press, Cambridge (1968)

Fu, L.: Grey systems theory and applications. Press of Scientific and Technological Literature, Beijing (1992)

Galison, P.L., Hevly, B.W.: Creative Management. Sage Publications, London (1991)

Guang, Y.Q., Liu, S.F.: Strengthening buffer operator sequences constructed on fixed points and applications. Control and Decision Making 22(10), 1189–1193 (2007)

Guo, H.: Distinguishing coefficients of grey incidences. Fuzzy Mathematics 5(2), 55–58 (1985)

Guo, H.: Grey exponential law attributed to mutual complement of series. The Journal of Grey System 3(2), 153–162 (1991)

Guo, H.F., Mo, L.G.: Theory of grey incidences applied to the study of farmers' incomes. Research on Finance and Trade 27(1), 31–37 (2007)

Guo, P.: MAC control on grey error corrections and application in workloads of machine groups. Journal of System Simulation 19(18), 4326–4330 (2007)

Guo, R.K.: Modeling imperfectly repaired system data via grey differential equations with unequal-gapped times. Reliability Engineering & System Safety 92(3), 378–391 (2007)

Hao, Q., Zhu, M.L., Li, B.: Methods of ordering fuzzy and grey incidences and applications. Systems Engineering: Theory and Practice 16(6), 44–49 (1996)

Hao, Y.H., Wang, X.M.: Period residual modification of GM(1,1) modeling. Journal of Grey System 12(2), 181–183 (2000)

He, R.S., Hwang, S.F.: Damage detection by a hybrid real-parameter genetic algorithm under the assistance of grey relation analysis. Engineering Applications of Artificial Intelligence 20(7), 980–992 (2007)

He, W.Z., Song, G.X., Wu, A.D.: A class of schemes for computing the parameters of the GM(1,1) model. Systems Engineering: Theory and Practice 25(1), 69–75 (2005)

He, W.Z., Wu, A.D.: Linear programming methods for estimating the parameters of Verhulst model and applications. Systems Engineering: Theory and Practice 26(8), 141–144 (2006)

He, X.J., Sun, G.Z.: A non-equigap grey model NGM(1,1). Journal of Grey System 13(2), 189–192 (2001)

He, Y.: Remarks on the solution of GM(1,1) model and test of its accuracy. Systems Engineering 12(3), 36–70 (1993)

He, Y.: Multi-layered grey evaluation models and application. Systems Engineering: Theory and Practice 14(4), 72–74 (1994)

He, Y., Bao, Y.D.: Grey Markov prediction model and applications. Systems Engineering: Theory and Practice 12(4), 59–63 (1992)

Hoover, E.M.: An Introduction to Regional Economics. MaGraw-Hill, New York (1975)

Hsieh, C.H.: Grey data fitting model and its application to image coding. Journal of Grey System 13(1), 245–254 (2001)

Hsieh, C.H., Jyh-Horng, C., Ying-Jeng, W.: Taguchi-genetic algorithms for optimizing GM(1,1). Journal of Grey System 12(3), 18–23 (2000)

Hsu, C.C., Chen, C.Y.: A modified grey forecasting model for long-term prediction. Journal of the Chinese Institute of Engineers 26(3), 301–308 (2003)

Hsu, C.C., Chen, C.Y.: Applications of improved grey prediction model for power demand forecasting. Energy conversion and management 44(14), 2241–2249 (2003)

Hsu, L.C., Wang, C.H.: Forecasting the output of integrated circuit industry using a grey model improved by the Bayesian analysis. Technological Forecasting and Social Change 74(6), 843–853 (2007)

Hu, D.H.: A multi-layered grey evaluation method using subjective criteria. Systems Engineering: Theory and Practice 16(1), 12–20 (1996)

Hu, F., Huang, J.G., Zhang, Q.F.: Research on effect evaluation of underwater navigation devices based on grey systems theory. Journal of Northwest University of Technology 25(3), 411–415 (2007)

Hu, Q.Z.: Decision-making model of grey incidences that optimizes urban traffic networks. Journal of Systems Engineering 22(6), 607–612 (2007)

Hu, Y.C.: Grey relational analysis and radial basis function network for determining costs in learning sequences. Applied Mathematics and Computation 184(2), 291–299 (2007)

Huang, F.Y.: Transformation method of grey systems modeling. Systems Engineering: Theory and Practice 14(6), 35–38 (1994)

Huang, J., Zhong, X.L.: A general accumulation grey predictive control model and its optimal computational scheme. Systems Engineering: Theory and Practice 29(6), 147–156 (2009)

Huang, K.Y., Jane, C.J.: A hybrid model for stock market forecasting and portfolio selection based on ARX, grey system and RS theories. Expert Systems with Applications 36(3), 5387–5392 (2009)

Huang, S.J., Chiu, N.H., Chen, L.W.: Integration of the grey relational analysis with genetic algorithm for software effort estimation. European Journal of Operational Research 188(3), 898–909 (2008)

Huang, Y.L., Fang, B.: Systems' quasi-quantitative incidence matrices based on grey models. Journal of System Simulation 19(3), 474–477 (2007)

Huang, Y.P., Chu, H.C.: Practical consideration for grey modeling and its application to image processing. The Journal of Grey System 8(3), 217–233 (1996)

Huang, Y.P., Chu, H.C.: Simplifying fuzzy modeling by both gray relational analysis and data transformation methods. Fuzzy Sets and Systems 104(2), 183–197 (1999)

Ji, P.R., Huang, W.S., Hu, X.Y.: On characteristics of grey prediction models. Systems Engineering: Theory and Practice 21(9), 105–108 (2001)

Jia, H.F., Zheng, Y.Q.: Combined prediction models of grey time series and its application in predicting annual precipitations. Systems engineering: Theory and Practice 18(8), 42–45 (1998)

Jiang, C.W., Hou, Z.J., Zhang, Y.C.: Chaos analysis of the grey forecasting model. Journal of Grey System 12(4), 319–322 (2000)

Jiang, Y.Q., Yao, Y., Deng, S.M., Ma, Z.L.: Applying grey forecasting to predicting the operating energy performance of air cooled water chillers. International Journal of Refrigeration 27(4), 385–392 (2004)

Jie, J.X., Song, B.F., Liu, D.X.: A grey incidence method designed for selecting the optimum design of the top layer of aircrafts. Journal of Systems Engineering 19(4), 350–354 (2004)

Jin, Y.F., Zhu, Q.S., Xing, Y.K.: The grey logical method for inserting the missing values in sequential data. Control and Decision Making 21(2), 236–240 (2006)

Jing, X.: A rigorous differential simulation method of GM(1,1) and application. Systems Engineering 11(6), 31–36 (1993)

John, G.B.: Information Systems: Theory and Practice. Wiley, New York (1979)

Karmakar, S., Mujumdar, P.P.: A two-phase grey fuzzy optimization approach for water quality management of a river system. Advances in Water Resources 30(5), 1218–1235 (2007)

Kendrick, J., David, A.: Stochastic Control for Economic Models. McGraw-Hill, New York (1981)

Klir, G.J.: Facets of Systems Science. Plenum Press, New York (1991)

Ko, A.S., Chang, N.B.: Optimal planning of co-firing alternative fuels with coal in a power plant by grey nonlinear mixed integer programming model. Journal of Environmental Management 88(1), 11–27 (2008)

Ko, J.Z.: Data sequence transformation and GM(1,1) modeling precision. In: Liu, S.F., Xu, Z.X. (eds.) New Advances in Grey systems Research, pp. 233–235. Press of Huazhong University of Science and Technology, Wuhan (1996)

Ku, L.L., Huang, T.C.: Sequential monitoring of manufacturing processes: an application of grey forecasting models. International Journal of Advanced Manufacturing Technology 27(5), 543–546 (2006)

Kung, C.Y., Wen, K.L.: Applying grey relational analysis and grey decision-making to evaluate the relationship between company attributes and its financial performance: A case study of venture capital enterprises in Taiwan. Decision Support Systems 43(3), 842–852 (2007)

Kuo, Y.Y., Yang, T.H., Huang, G.W.: The use of grey relational analysis in solving multiple attribute decision-making problems. Computers & Industrial Engineering 55(1), 80–93 (2008)

Lai, H.H., Lin, Y.C., Yeh, C.H.: Form design of product image using grey relational analysis and neural network models. Computers & Operations Research 32(10), 2689–2711 (2005)

Lee, W.M., Liao, Y.S., Chang, S.H.: Neural-based WEDM optimization via grey analysis. Journal of Grey System 13(2), 135–146 (2001)

Li, B., Wei, Y.: A new GM(1,1) model established with optimized grey derivatives. Systems Engineering: Theory and Practice 29(2), 100–105 (2009)

Li, B.Z., Zhu, X.X.: On the RIS evolutionary mechanism based on dissipation and grey incidence entropy. Studies of Science of Sciences 25(6), 1239–1244 (2007)

Li, B.L., Deng, J.L.: A grey model for the biological prevention and treatment system of cotton aphids. Exploration of Nature 3(3), 44–49 (1984)

Li, B.J., Liu, S.F., Zhu, Y.D.: A method to determine the reliability of grey intervals. Systems Engineering: Theory and Practice 20(4), 104–106 (2000)

Li, C.H., Tsai, M.J.: Multi-objective optimization of laser cutting for flash memory modules with special shapes using grey relational analysis. Optics & Laser Technology 41(5), 634–642 (2009)

Li, F.: Theory of grey prediction and grey evaluation and its applications. Ph.D Dissertation of Huazhong University of Science and Technology (2002)

Li, G.D., Yamaguchi, D., Nagai, M.: A grey-based decision-making approach to the supplier selection problem. Mathematical and Computer Modelling 46(3), 573–581 (2007)

Li, G.D., Yamaguchi, D., Nagai, M.: A GM(1,1)–Markov chain combined model with an application to predict the number of Chinese international airlines. Technological Forecasting and Social Change 74(8), 1465–1481 (2007)

Li, G.D., Yamaguchi, D., Nagai, M.: A grey-based decision-making approach to the supplier selection problem. Mathematical and Computer Modelling 46(3), 573–581 (2007)

Li, G.Y., Li, Q.F., Zhao, G.F.: An improvement of the grey GM(1,1) model. Systems Engineering 10(6), 27–31 (1992)

Li, J.B., Zhao, J.Y., Zheng, X.X., Guang, H.J.: A new evaluation method of the state of electric transformers based on grey target theory. Journal of Jilin University (Industry Edition) 38(1), 201–205 (2008)

Li, J.W., Chen, M.Y., Dong, P.Y., Zhang, S.H.: Simplified SCGM(1,1) model and applications. Journal of Wuhan University of transportation Science and Technology 24(6), 615–618 (2000)

Li, J.F., Dai, W.Z.: Multi-period grey modeling method based on stepwise ratios and its application in Chinese GDP modeling. Systems engineering: Theory and Practice 24(9), 98–102 (2004)

Li, Q.F., Zhang, Q.K., Zhang, Y.L., Liu, X.H.: Research on the irrigation planning of irrigation areas. Systems engineering: Theory and Practice 18(5), 131–136 (1998)

Li, Q.X.: Grey dynamic input–output analysis. Journal of Mathematical Analysis and Applications 359(2), 514–526 (2009)

Li, Q.X., Liu, S.F.: The foundation of the grey matrix and the grey input–output analysis. Applied Mathematical Modelling 32(3), 267–291 (2008)

Li, S.B.: Grey incidence projection model and application for evaluating the friendliness of regional environment. Science and Sciences and Management of Science and Technology 28(8), 103–107 (2007)

Li, T.J., Xiong, W., Wang, P., Zheng, Y.S.: Grey prediction methods of experiments without structural damages. Journal of Huazhong University of Science and Technology 23(5), 75–78 (1990)

Li, W.X.: Grey algebraic curve models. Systems Engineering 8(1), 32–36 (1990)

Li, W.X.: Clustering analysis methods and applications based on degrees of grey incidence. Systems Engineering 8(3), 37–48 (1990)

Li, W.X.: Degrees of incidence of continuous processes. Journal of Huazhong University of Science and Technology 19(3), 113–116 (1991)

Li, X.C.: Expansion of the validity range of the GM(1,1) model of the grey systems. Systems engineering: Theory and Practice 19(1), 97–101 (1999)

Li, X.F., Wang, J.J., Ma, J., Zhou, Y.: Grey incidence analysis for the quality of cotton fiber materials and the resultant yarns and single yarn strength. Journal of Applied Science 25(1), 100–102 (2007)

Li, X.Q.: A further investigation of quantified grey incidences. Systems Engineering 13(6), 58–61 (1995)

Li, X.Q., Li, S.R., Han, X.L.: A generalized application of grey systems GM(n,h) model. Systems Engineering: Theory and Practice 17(8), 82–85 (1997)

Liang, Q.W., Song, B.W., Jia, Y.: The grey Verhulst model of the costs on the research of torpedoes. Journal of System Simulation 17(2), 257–258

Lin, C.T., Yang, S.Y.: Forecast of the output value of Taiwan's opto-electronics industry using the Grey forecasting model. Technological Forecasting and Social Change 70(2), 177–186 (2003)

Lin, J.L., Lin, C.L.: The use of the orthogonal array with grey relational analysis to optimize the electrical discharge machining process with multiple performance characteristics. International Journal of Machine Tools and Manufacture 42(2), 237–244 (2002)

Lin, J.L., Wang, K.S., Yan, B.H., Tarng, Y.S.: Optimization of parameters in GM(1,1) model by Taguchi method. Journal of Grey System 11(4), 257–277 (1999)

Lin, W., Ho, Y.X.: Research and application of grey multi-target programming. In: Liu, S.F., Xu, Z.X. (eds.) New Advances in Grey Systems Research, pp. 373–375. Press of Huazhong University of Science and Technology, Wuhan (1996)

Lin, Y., Liu, S.F.: Law of exponentiality and exponential curve fitting. Systems Analysis Modelling Simulation 38(1), 621–636 (2000)

Lin, Y., Liu, S.F.: Regional economic planning based on systemic analysis of small samples (I). Problems of Nonlinear Analysis in Engineering Systems 2(10), 24–33 (1999)

Lin, Y., Liu, S.F.: Regional economic planning based on systemic analysis of small samples (II). Problems of Nonlinear Analysis in Engineering Systems 6(11), 33–49 (2000)

Lin, Y., Liu, S.F.: Several programming models with unascertained parameters and their application. Journal of Multi-Criteria Decision Analysis 8, 206–220 (1999)

Lin, Y., Liu, S.F., McNeil, D.H.: The past, present, and future of systems science research. In: Systems Science and Sustainable Development, pp. 245–252. Press of Scientific and Technological Literature, Beijing (1998)

Lin, Y.H., Lee, P.C.: Novel high-precision grey forecasting model. Automation in Construction 16(6), 771–777 (2007)

Lin, Y.H., Lee, P.C., Chang, T.P.: Adaptive and high-precision grey forecasting model. Expert Systems with Applications 36(6), 9658–9662 (2009)

Liu, B., Liu, S.F., Dang, Y.G.: The GST-based data mining techniques of time sequences. In: Proceedings of the 32nd International Conference on Computers & Industrial Engineering, pp. 548–553 (2003)

Liu, B., Liu, S.F., Dang, Y.G.: Time series data mining techniques based on grey systems theory. Engineering Science of China 5(9), 32–35 (2003)

Liu, B., Liu, S.F., Cui, Z.J., Dang, Y.G., et al.: Optimization of time response functions of the GM(1,1) model. Management Science of China 11(4), 54–57 (2003)

Liu, J.P., Ji, C.S., Li, H.: Application of fixed weight grey clustering analysis in evaluating coal mining methods. Journal of Coal Industry 26(5), 493–495 (2001)

Liu, S.A., Du, H.T., Wang, X.L.: Rough set theory and applications. Systems Engineering: Theory and Practice 21(10), 77–82 (2001)

Liu, S.F., Chen, K.J.: Scientific and technological funds allocation in Jiangsu Province and their efficiency. Journal of Nanjing University of Aeronautics and Astronautics 33(3), 24–27 (2001)

Liu, K.L., Lin, Y.X., Dal, G.: Grey predicting the structural failure load. Journal of Grey System 13(2) (2001)

Liu, P.L.: Stability of continuous and discrete time-delay grey systems. International Journal of Systems Science 32(7), 29–39 (2001)

Liu, S.F.: The P-F theorem of grey non-negative matrices. Journal of Henan Agricultural University 21(4), 376–380 (1987)

Liu, S.F.: On Perron-Frobenius theorem of grey nonnegative matrix. Journal of Grey System 1(2), 157–166 (1989)

Liu, S.F.: The three axioms of buffer operator and their application. Journal of Grey System 3(1), 39–48 (1991)

Liu, S.F.: Generalized degree of grey incidence. In: Shengkai, Z. (ed.) Information and Systems, pp. 113–116. DMU Publishing House, Dalian (1992)

Liu, S.F.: Buffer operators and applications. Grey Systems Theory and Application 2(1), 45–50 (1992)

Liu, S.F.: Fixed weight grey clustering evaluation analysis. In: New Methods of Grey systems, pp. 178–184. Agriculture Press, Beijing (1993)

Liu, S.F.: Fixed position solutions of linear programming with grey parameters. Grey Systems Theory and Application 3(2), 23–28 (1993)

Liu, S.F.: Von Neumann's economic growing turnpike of Henan Province. In: Weimin, Z. (ed.) Systems Science and Systems Engineering, pp. 316–319. International Academic Publishers, Beijing (1993)

Liu, S.F.: Grey forecast of drought and inundation in Henan Province. The Journal of Grey System 6(4), 279–288 (1994)

Liu, S.F.: On measure of grey information. The Journal of Grey System 7(2), 97–101 (1995)

Liu, S.F.: Axiom on Grey Degree. The Journal of Grey System 8(4), 397–400 (1996)

Liu, S.F.: The rule and position of grey system theory in the development of science. In: Liu, S.F., Xu, Z.X. (eds.) New Advances in Grey Systems Research, pp. 230–232. Press of Huazhong University of Science and Technology, Wuhan (1996)

Liu, S.F.: Prediction trap of shock-disturbed systems and buffer operators. Journal of Huazhong University of Science and Technology 25(1), 28–31 (1997)

Liu, S.F.: Methods of grey mathematics and systems analysis for science and technology management. PhD Dissertation of Huazhong University of Science and Technology, Wuhan (1998)

Liu, S.F.: Evaluation criteria for comprehensive scientific and technological strength and mathematical models. Journal of Nanjing University of Aeronautics and Astronautics 32(5), 409–412 (2002)

Liu, S.F.: The appearance, development and current state of grey systems theory. Management Science of China 16(4), 14–17 (2003)

Liu, S.F.: Emergence and development and forward trends of grey systems theory. In: Proceedings of the 32nd International Conference on Computers& Industrial Engineering, pp. 526–531 (2003)

Liu, S.F.: Appearance and development of grey systems theory. Journal of Nanjing University of Aeronautics and Astronautics 36(2), 267–272 (2004)

Liu, S.F.: On index system and mathematical model for evaluation of scientific and technical strength. Kybernetes: The international journal of systems & cybernetics 35(7-8), 1256–1264 (2006)

Liu, S.F., Chen, K.J., Liu, J.: Quantitative assessment of SCM in the air conditioning industry in China. In: Proceedings of 4th International Conference on Industrial Engineering and Management Systems, pp. 365–369. RMIT Press (2002)

Liu, S.F., Dang, Y.G.: The positioned solution of the linear programming with grey parameters. Advances in Systems Science and Applications 3(1), 371–377 (1997)

Liu, S.F., Dang, Y.G.: Technical change and the funds for science and technology. Kybernetes: The International Journal of Systems & Cybernetics 33(2), 295–302 (2004)

Liu, S.F., Dang, Y.G., Li, B.J.: The G-C-D model and contribution of technological advances. Management Science of China 7(2), 76–80 (1999)

Liu, S.F., Dang, Y.G., Lin, Y.: Synthetic utility index method and venturous capital decision-making. Kybernetes: The International Journal of Systems & Cybernetics 33(2), 288–294 (2004)

Liu, S.F., Dang, Y.G.: The degree of satisfaction of the floating and fixed positional solutions of LPGP. Journal of Huazhong University of Science and Technology 25(1), 24–27 (1997)

Liu, S.F., Deng, J.L.: Range of applicability of the GM(1,1) model. Systems Engineering: Theory and Practice 20(5), 121–124 (2000)

Liu, S.F., Dang, Y.G., Li, B.J., et al.: Computational analysis on the periodic contributions of technological advances in Henan Province. Journal of Henan Agricultural University 32(3), 203–207 (1998)

Liu, S.F., Deng, J.L.: GM(1,1) coding for exponential series. The Journal of Grey System 11(2), 147–152 (1999)

Liu, S.F., Deng, J.L.: The range suitable for GM(1,1). The Journal of Grey System 11(1), 131–138 (1999)

Liu, S.F., Fang, Z.G., Lin, Y.: Study on a new definition of degree of grey incidence. Journal of Grey System 9(2), 115–122 (2006)

Liu, S.F., Forrest, J.: The role and position of grey system theory in science development. The Journal of Grey System 9(4), 351–356 (1997)

Liu, S.F., Guo, T.B.: Grey Systems Theory and Applications. Press of Henan University, Kaifeng (1992)

Liu, S.F., Guo, T.B., Dang, Y.G.: Grey Systems Theory and Applications (2nd edition). Science Press, Beijing (1999)

Liu, S.F., Li, B.J., Dang, Y.G.: The G-C-D model and technical advance. Kybernetes: The International Journal of Systems & Cybernetics 33(2), 303–309 (2004)

Liu, S.F., Li, B.J., et al.: Evaluation criteria of regional leading industries and mathematical models. Management Science of China 6(2), 8–13 (1998)

Liu, S.F., Lin, Y.: An Introduction to Grey Systems: Foundations, Methodology and Applications. IIGSS Academic publisher, Grove City (1998)

Liu, S.F., Lin, Y.: An axiomatic definition for the degree of greyness of grey numbers. IEEE System Man and Cybernetics, 2420–2424 (2004)

Liu, S.F., Lin, Y.: Grey Information: Theory and Practical Applications. Springer, London (2006)

Liu, S.F., Lin, Y., et al.: On measures of information content of grey numbers. Kybernetes: The International Journal of Systems & Cybernetics 35(5), 899–904 (2006)

Liu, S.F., Shi, K.Q. (eds.): Collected Papers of Grey Systems. Press of Henan University, Kaifeng (1993)

Liu, S.F., Tang, X.W., Yuan, Z.Q., Dang, Y.G.: On the order of Chinese industrial structures. Economic Advances (5), 53–56 (2004)

Liu, S.F., Wan, S.Q., et al.: Models for evaluating the importance of the rescuers' locations in terms of accidents in complex traffic networks. Management Science of China 17(1), 119–124 (2009)

Liu, S.F., Wang, Z.Y.: Entropy of grey evaluation coefficient vector. The Journal of Grey System 12(3) (2000)

Liu, S.F., Wang, R.L.: Grey incidence analysis of the third industries during the ninth fifth period in Nanjing City. Journal of Nanjing University of Science and Technology 16(5), 55–58 (2003)

Liu, S.F., Xie, N.M.: A new grey evaluation method based on improved triangular whitenization weight functions. Journal of Systems Engineering (2009) (in press)

Liu, S.F., Xie, N.M., Forrest, J.: A class of new models of grey incidence analysis. Systems Engineering: Theory and Practice (2009) (in press)

Liu, S.F., Xu, Z.X. (eds.): New Advances in Grey Systems Research. Press of Huazhong University of Science and Technology, Wuhan (1996)

Liu, S.F., Yang, L.: Evaluation, Advanced Warming and Control of Regional Economics. Henan People's Press, Zhengzhou (1994)

Liu, S.F., Yang, L., Zheng, G.Q.: Evaluation and advanced warming of regional economics. Henan Economy 10(1), 41–44 (1994)

Liu, S.F., Yuan, W.F.: A new multi-target intelligent weighted decision-making model. Control and Decision Making (2009) (in press)

Liu, S.F., Zhao, L., Wang, Z.Y.: A new method for evaluating risky investment. Management Science of China 9(2), 22–26 (2001)

Liu, S.F., Zhao, L., Zhu, Y.D.: The G-E prediction model for Henan Province's food production. Journal of Henan Agricultural University 27(1), 90–94 (1993)

Liu, S.F., Zhao, L., Zhu, Y.D.: Grey econometrics combined models. In: Liu, S.F., Shi, K.Q. (eds.) Collected Papers of Grey Systems, pp. 133–138. Press of Henan University, Kaifeng (1993)

Liu, S.F., Zhu, Y.D.: Models based on criteria's triangular membership functions for the evaluation of regional economics. Journal of Agricultural Engineering 9(2), 8–13 (1993)

Liu, S.F., Zhu, Y.D.: Study on triangular model and indexes in synthetic evaluation of regional economy. In: Mianyun Chen, Systems Control Information Methodologies & Applications, pp. 1274–1279. HUST Publishing House (1994)

Liu, S.F., Zhu, Y.D.: Grey-econometrics combined model. The Journal of Grey System 8(1), 103–110 (1996)

Liu, X.Q., Wang, Z.M.: Grey Econometrics Models and Applications. Yellow River Press, Jinan (1996)

Liu, X.Q., Wang, S.Z., Song, Z.M.: Introduction to Grey Incidence Spaces. Guizhou People's Press, Guizhou (1993)

Liu, X.W.: Evaluations of agricultural environmental quality of the typical regions located in Yangtze River Delta. Systems Engineering: Theory and Practice 25(6), 133–138 (2005)

Liu, Z.Y.: Grey production functions with n kinds of input factors. Systems Engineering: Theory and Practice 15(9), 6–8 (1995)

Liu, Z.Y.: Two fundamental problems of grey systems analysis. Systems Engineering: Theory and Practice 20(9), 123–124 (2000)

Luo, B.X.: Input-Output Analysis and Its Mathematical Foundations. Press of Huazhong University of Science and Technology, Wuhan (1986)

Luo, D.: A characteristic vector method for grey decision making. Systems Engineering: Theory and Practice 24(4), 67–71 (2006)

Luo, D., Liu, S.F.: combined grey and rough decision making models. Journal of Xiamen University 43(1), 26–30 (2004)

Luo, D., Liu, S.F.: On grey dynamic programming. Systems Engineering: Theory and Practice 24(4), 56–62 (2004)

Luo, D., Liu, S.F.: Grey incidence decision-making methods for systems with incomplete information. Journal of Applied Science 23(4), 408–412 (2005)

Luo, D., Liu, S.F.: Study on grey incidence decision-making methods. Management Science of China 13(1), 101–106 (2005)

Luo, D., Liu, S.F., Dang, Y.G.: Optimization of grey GM(1,1) model. Engineering Science of China 5(8), 50–53 (2003)

Luo, M.F.: Fault detection, diagnosis and prognosis using grey system theory. Monash University, A Dissertation Submitted for the Degree of Ph.D in Engineering (1994)

Luo, Q.C.: Design and application of grey optimized input-output models. Systems Engineering: Theory and Practice 9(5), 55–59 (1989)

Luo, Q.C., Xu, G.X.: Grey Incidence Analysis and Applications. Press of Jiangsu Science and Technology, Nanjing (1989)

Luo, X.M., Yang, H.H.: Grey comprehensive evaluation models. Systems Engineering and Electronic Technology 16(9), 18–25 (1994)

Lv, A.L., Shao, Y.: Diagnosis models for difficult diseases. In: Liu, S.F., Xu, Z.X. (eds.) New Advances in Grey Systems Research, Press of Huazhong University of Science and Technology, Wuhan (1996)

Lv, A.L.: A grey plan for the physiological study of digest systems. In: Liu, S.F., Xu, Z.X. (eds.) New Advances in Grey Systems Research, pp. 149–150. Press of Huazhong University of Science and Technology, Wuhan (1996)

Lv, A.L.: Theory and application of incidence spaces of medical grey information. PhD. Dissertation of Huazhong University of Science and Technology, Wuhan (2000)

Lv, D.G., Wang, L., Zhang, P.: A grey incidence method for fuzzy multi-attribute decision-making regarding structural design. Journal of Harbin University of Technology 39(6), 841–844 (2007)

Lv, F.: Study on the distinguishing coefficients of grey systems incidence analysis. Systems Engineering: Theory and Practice 17(6), 49–54 (1997)

Lv, L.Z., Wu, W.J.: Optimization of the grey GM(1,1) model. Systems Engineering: Theory and Practice 21(8), 92–96 (2001)

Lv, Z.L., Fan, Z.P.: Grey systems modeling for development tendencies with jumps. Systems Engineering 8(5), 55–58 (1990)

Ma, F.Y., Wei, H.W.: Prediction of pitting corrosion behavior for stainless SUS 630 based on grey system theory. Materials Letters 61(4), 998–1001 (2007)

Mi, C.M., Liu, S.F., Yang, J.: Grey incidence analysis of Jiangsu Province's investment in science and technology and economic growth. Science of Sciences and Management of Science and Technology 25(1), 34–36 (2004)

Mi, C.M., Liu, S.F., Wu, Z.P., Wang, J.L.: Study on the sequence of strengthening buffer operators of the opposite directional accumulation. Control and Decision Making 24(3), 352–355 (2009)

Miller, J.G.: Living Systems. McGraw-Hill, New York (1978)

Mills, E.D.: Studies in the Structure of the Urban Economy. Johns Hopkins University Press, Baltimore (1972)

Morán, J., Granada, E., Míguez, J.L., Porteiro, J.: Use of grey relational analysis to assess and optimize small biomass boilers. Fuel Processing Technology 87(2), 123–127 (2006)

Mu, H.S., Yang, J.B.: Grey systems theory and economic cyclic analysis. Systems Engineering 10(2), 25–28 (1992)

Mu, R., Zhang, J.Q.: Layered comprehensive evaluation based on grey incidence analysis. Systems Engineering: Theory and Practice 28(10), 125–130 (2008)

Nagel, S.S.: Policy Evaluation: Making Optimum Decision. Pyaeger, New York (1982)

Nelson, S., Arthur, C.: Growth Management Principles and Practices. Planners Press, Washington (1995)

Nikaido, H.: Convex Structures and Economic Theory. Academic Press, New York (1968)

Ning, X.X., Liu, S.F.: Methods for Management Prediction and Decision-Making. Science Press, Beijing (2003)

OECD, The Measurement of Scientific Activities. Frascati Manual, Paris (1981)

Olson, D.L., Wu, D.S.: Simulation of fuzzy multi-attribute models for grey relationships. European Journal of Operational Research 175(1), 111–120 (2006)

Pan, H.P., Huang, Z.H.: Recognizing lithofacies via grey clustering. The Journal of Grey System 8(1), 93–102 (1996)

Patton, M.Q.: Qualitative Evaluation and Research Methods. Sage Publications, Calif (1990)

Pawlak, Z.: Rough Sets: Theoretical Aspects of Reasoning about Data. Kluwer Academic Publishers, Dordrecht (1991)

Qi, M., Deng, L., Ge, D., Xuan, H., Hao, Z.Y.: Decomposition method for colored medical figures designed by using the maximum threshold entry value of the grey incidence space. Journal of Chinese University of Science and Technology 37(12), 1543–1545 (2007)

Qian, W.Y., Dang, Y.G.: The GM(1,1) model based on fluctuating sequences. Systems Engineering: Theory and Practice 29(3), 149–154 (2009)

Qiu, W., Li, S., Zhao, Q.L., Zhang, J.Q.: Grey evaluation and model prediction of the forest coverage in Heilongjiang Province. Journal of Harbin University of Technology 39(10), 1650–1652 (2007)

Qiu, X.J.: Grey clustering incidence analysis and applications. Systems Engineering: Theory and Practice 15(1), 15–21 (1995)

Rao, C.J., Xiao, X.P., et al.: Generalized accumulated generating operation and its generating space. Dynamics of Continuous Discrete and Impulsive Systems-series B: Applications & Algorithms 13, 517–521 (2006)

Rao, R., Yadava, V.: Multi-objective optimization of Nd:YAG laser cutting of thin superalloy sheet using grey relational analysis with entropy measurement. Optics & Laser Technology 41(8), 922–993 (2009)

Ruan, A.Q., Liu, S.F.: Grey GERT networks and precision of grey estimates based on customers' demand. Systems Engineering 25(12), 100–104 (2007)

Shen, D.X.: Stochastic processes. Press of Huazhong University of Science and Technology, Wuhan (1993)

Shen, V., Chung, Y.F., Chen, T.S.: A novel application of grey system theory to information security. Computer Standards & Interfaces 31(2), 277–281 (2009)

Shi, K.Q., Liu, S.F.: Grey relation G(2) on (X,Γ). The Journal of Grey System 6(2), 109–118 (1994)

Shui, N.X., Dong, T.H., Sha, Z.: About grey systems' GM(2,1) model. Systems Engineering 8(6), 31–35 (1990)

Shui, N.X., Dong, T.H., Sha, Z.: Some theoretical problems about grey incidences. Systems Engineering 10(6), 23–26 (1992)

Shui, N.X., Qing, T.C.: Some theoretical problems about grey systems' GM(1,1) model. Systems Engineering: Theory and Practice 18(4), 59–63 (1998)

Sjoberg, J.A.: Nonlinear grey-box example using a stepwise system identification approach. In: Presented at the 12th Symposium on System Identification, Santa Barbara, CA, USA (2000)

Smith, N.L.: Varieties of Investigative Evaluation. Jossey-Bass, San Francisco (1992)

Song, Z.M.: Order structures of grey systems. PhD Dissertation of Huazhong University of Science and Technology, Wuhan (2003)

Song, Z.M., Liu, X.Q., Wang, S.Z.: Curvature simulation method of grey exponential curves. Systems Engineering: Theory and Practice 17(6), 55–57 (1997)

Song, Z.M., Tong, X.J., Xiao, X.P.: Grey GM(1,1) model of the type of central approximation. Systems Engineering: Theory and Practice 21(5), 110–113 (2001)

Song, Z.Q., Tan, C.Q.: Analysis criteria, treatments and applications of refined evaluation of petroleum deposits using grey systems theory. Systems Engineering: Theory and Practice 17(3), 74–82 (1997)

Song, Z.M., Ning, X.X.: Construction and operation of a grey evaluation system of virtual firms. Systems Engineering 26(1), 20–24 (2008)

Sun, H.Y., Cao, Q.J.: Modeling and prediction of chaotic time series. Systems Engineering: Theory and Practice 21(5), 106–109 (2001)

Sun, J.J.: Study on harmonic development between science, technology, economy, and society. Management Science of China 4(2), 13–18 (1996)

Sun, X.D., Jiao, Y., Hu, J.S.: A decision-making method based on grey incidences and ideal solutions. Management Science of China 13(4), 63–68 (2005)

Sun, Y.G., Dang, Y.G.: An improvement of the grey T-type incidence. Systems Engineering: Theory and Practice 28(4), 135–139 (2008)

Tamura, Y., Zhang, D.P., Umeda, N., Sakeshita, K.: Load Forecasting Using Grey Dynamic Model. The Journal of Grey System 4(4), 49–58 (1992)

Tan, J.B., Lv, Y.J.: Weak uniformity and uniformity of grey judgment matrices and their properties. Systems engineering: Theory and Practice 28(3), 159–165 (2008)

Tan, X.R.: Representative summary of grey incidence spaces and applications in medicine. In: Liu, S.F., Xu, Z.X. (eds.) New Advances in Grey Systems Research, pp. 130–132. Press of Huazhong University of Science and Technology, Wuhan (1996)

Tan, X.R.: Grey medical incidence theory and applications. PhD Dissertation of Huazhong University of Science and Technology, Wuhan (1997)

Tang, W.M.: A new prediction model based on grey support vector machine. Journal of Systems Engineering 21(4), 410–413 (2006)

Tang, W.X.: A new method for estimating the parameters of the GM(1,1) model and hypothesis testing. Systems Engineering: Theory and Practice 15(3), 20–49 (1995)

Tang, W.X.: Type T degree of incidence and its computational scheme. Mathematical Statistics and Management 14(1), 34–37 (1995)

Tang, X.W., Liu, S.F., Liu, B.: Predicting the total investment in Jiangsu with grey system model. In: Proceedings of the 32nd International Conference on Computers& Industrial Engineering, pp. 538–541 (2003)

Thananchai, L.: Grey prediction on indoor comfort temperature for HVAC systems. Expert Systems with Applications 34(4), 2284–2289 (2008)

Tinbergen, J.: Econometrics. Prentice-Hall, Englewood Cliffs (1964)

Tinbergen, J.: Development Planning. Weidenfeld and Nicolson, London (1967)

Utterback, J.M.: Innovation and corporate strategy. International Journal of Technology Management 1(1-2), 119–132 (1986)

Wang, H.Y., Liu, L., Liu, L.: Applied study of BP neural networks based on GRA and PCA. Management Review 19(10), 50–55 (2007)

Wang, J.J., Jing, Y.Y.: Integrated evaluation of distributed triple-generation systems using improved grey incidence approach. Energy 33(9), 1427–1437 (2008)

Wang, J.Q., Ren, S.X., Chen, X.H.: Preference ordering of grey stochastic multi-criteria decision makings. Control and Decision Making 24(5), 701–705 (2009)

Wang, Q.Y., Cui, Y.M.: Fundamental problems regarding uncertain systems predictions. Systems Engineering: Theory and Practice 20(3), 93–98 (2000)

Wang, Q.Y., Liu, K.D., et al.: Methods and Applications of Grey Mathematics. Press of Southwest Jiaotong University, Chengdu (1990)

Wang, Q.Y., Zhao, R.H.: Pan-grey determinants and pan-grey matrices. Journal of Hebei College of Coal Industry 9(3), 27–33 (1992)

Wang, Q.Y., Lv, R.H.: Inherent connection between UG sets and vague sets. Journal of Huazhong University of Science and Technology 33(7), 115–118 (2005)

Wang, Q.Y., Zhao, X.H., Wang, F.L.: Concept of general sets and analysis of its intensions. Fuzzy Systems and Mathematics 20(4), 61–67 (2006)

Wang, W.P.: Analysis of optimal values in GLP. The Journal of Grey System 5(4), 315–317 (1993)

Wang, W.P.: Study on grey linear programming. The Journal of Grey System 9(1), 41–46 (1997)

Wang, W.P., Deng, J.L.: Intension measurement of information contained in grey systems propositions. Grey Systems Theory and Practice 2(1), 41–43 (1992)

Wang, W.P.: Theory and methods on dealing with grey information. PhD. Dissertation of Huazhong University of Science and Technology, Wuhan (1994)

Wang, X.C.: On relationship between rough sets and grey systems. Fuzzy Systems and Mathematics 20(6), 129–135 (2007)

Wang, X.Y., Yang, X.: Grey systems methods for quality predictive control and diagnosis. Systems Engineering: Theory and Practice 17(5), 105–108 (1997)

Wang, X.M.: Grey dynamic models for analyzing economic growths and cycles. Systems Engineering: Theory and Practice 13(1), 42–47 (1993)

Wang, X.M., et al.: Grey Systems Models Applied in Countryside Economies. Press of Huazhong University of Science and Technology, Wuhan (1992)

Wang, X.M., Luo, J.J.: Collected Programs for Grey Systems Prediction, Decision-Making, and Modeling. Press of Science Dissemination, Beijing (1986)

Wang, X.M., Zhang, J.Z., Wang, R.Z.: Grey Systems Analysis and Practical Computer Procedures. Press of Huazhong University of Science and Technology, Wuhan (2001)

Wang, Y.M., Dang, Y.G., Wang, Z.X.: Optimization of the background values of non-equal-distant GM(1,1) model. Management Science of China 16(4), 159–162 (2008)

Wang, Y.N., Liu, G.Z., Liu, K.D.: A step-by-step directly optimizing method of GM(1,1) modeling. Systems Engineering: Theory and Practice 20(9), 99–104 (2000)

Wang, Y.N., Liu, K.D., Li, Y.C.: The GM(1,1) modeling method that optimizes the whitenization values of grey derivatives. Systems Engineering: Theory and Practice 21(5), 124–128

Wang, Y.Z., Zhou, W.A.: An improved grey statistical method for evaluating the safety of mining worksites. Systems Engineering: Theory and Practice 17(9), 118–123

Wang, Y.H., Dang, Y.G.: A comprehensive posterior evaluation method based on the grey fixed weight clustering of the D-S evidence theory. Systems Engineering: Theory and Practice 29(5), 123–128 (2009)

Wang, Z.L., Liu, Y.L., Shi, K.Q.: On Kenning grey degree. The Journal of Grey System 7(2), 103–110 (1995)

Wang, Z.X., Dang, Y.G., Liu, S.F.: An optimized GM(1,1) model based on discrete exponential functions. Systems Engineering: Theory and Practice 28(2), 61–67 (2008)

Wang, Z.X., Dang, Y.G., Liu, S.F.: Variable buffer operators and their acting strengths. Control and Decision Making 29(8), 1218–1222 (2009)

Wang, Z.X., Dang, Y.G., Liu, S.F.: Two-piece grey models and applications. Systems Engineering: Theory and Practice 28(11), 109–114 (2008)

Wang, Z.X., Dang, Y.G., Song, C.P.: Multi-objective grey situational decision-making models based on interval numbers. Control and Decision Making 24(3), 388–392 (2009)

Wang, Z.L.: Application of Grey Systems Theory in Analysis of Cultivars. In: Collected Papers of Grey Systems Research, pp. 305–307. Press of Henan University, Kaifeng (1993)

Wang, Z.L.: A theory for grey modeling techniques. Ph.D. Dissertation of Huazhong University of Science and Technology, Wuhan (1998)

Wang, Z.L., Li, X.Z.: Ordering and inequalities of grey numbers. In: Liu, S.F., Xu, Z.X. (eds.) New Advances in Grey Systems Research, pp. 364–366. Press of Huazhong University of Science and Technology, Wuhan (1996)

Weinberg, G.M.: An Introduction to General Systems Thinking. John Wiley, New York (1975)

Wen, J.C., Chen, J.L., Yang, J.S.: Study of the non-equigap GM(1,1) modeling. Journal of Grey System 13(3), 205–214 (2001)

Wen, K.L., Chang, T.C., Chang, W.C., You, M.L.: The Study of Missing Point in GM(1,1) Model. In: Proceedings of IEEE International Conference on Systems. Man, and Cybernetics (2000)

Wong, H., Hu, B.Q., Xia, J.: Change-point analysis of hydrological time series using grey relational method. Journal of Hydrology 324(1), 323–338 (2006)

Wu, A.X., Xi, Y., Yang, B.H., Chen, X.S., Jiang, H.C.: Study on grey forecasting model of copper extraction rate with bioleaching of primary sulfide ore. Acta Metallurgica Sinica 20(2), 117–128 (2007)

Wu, C.C., Chang, N.B.: Grey input–output analysis and its application for environmental cost allocation. European Journal of Operational Research 145(1), 175–201 (2003)

Wu, J.H., Wen, K.L.: Rolling error in GM(1,1) modeling. Journal of Grey System 13(1), 77–80 (2001)

Wu, H.Q., Qiu, C.A., et al.: An Introduction to Grey Mathematics. Hebei People's Press, Shijiazhuang (1990)

Wu, Q., Liu, Z.T.: Real formal concept analysis based on grey-rough set theory. Knowledge-Based Systems 22(1), 38–45 (2009)

Wu, W.Y., Chen, S.P.: A prediction method using the grey model GMC(1, n) combined with the grey relational analysis: a case study on Internet access population forecast. Applied Mathematics and Computation 169(1), 198–217 (2005)

Wu, Y., Yang, S.Z., Tao, J.H.: Discussions on grey prediction and time series prediction. Journal of Huazhong University of Science and Technology 16(3), 27–33 (1988)

Wu, Z.P., Liu, S.F., Cui, L.Z., Mi, C.M., Wang, J.L.: Some new weakening buffer operators designed on monotonic functions. Control and Decision Making 24(7), 1055–1058 (2009)

Wu, Z.J., Wang, A.M.: Grey-fuzzy comprehensive evaluations for the planning of reconstructible manufacturing systems. Mechanic Engineering of China 18(19), 2313–2318 (2007)

Xia, J., Zhao, H.Y.: Grey artificial neural network models and applications in short-term prediction of runoffs. Systems Engineering: Theory and Practice 16(11), 82–90 (1996)

Xiao, X.P.: On parameters in grey models. Journal of Grey System 11(4), 315–324 (2000)

Xiao, X.P.: Grey relational analysis and application of hybrid index sequences. Dynamics of Continuous Discrete and Impulsive Systems-series B-Applications & Algorithms 13, 915–919 (2006)

Xia, X.T., Chen, X.H., Zhang, Y.Z., Liu, C.H.: Grey hypothesis testing for attributes of mechanic systems. Journal of Mechanical Engineering 26(12), 38–43 (2007)

Xiao, X.P.: Theoretical research and remarks on quantified models of grey degrees of incidence. Systems Engineering: Theory and Practice 17(8), 76–81 (1997)

Xiao, X.P.: On methods of grey systems modeling. PhD. Dissertation of Huazhong University of Science and Technology, Wuhan (2002)

Xiao, X.P., Xie, L.C., Huang, D.R.: A computational improvement of degrees of grey incidence and applications. Mathematical Statistics and Management 14(5), 27–30 (1995)

Xiao, X.P., Song, Z.M., Li, F.: Foundations of Grey Technology and Applications. Science Press, Beijing (2004)

Xiao, Y., Shao, D.G., Deng, R., Wu, Y.Q.: Comprehensive evaluation supporting systems for the effects of hydraulic engineering projects. Journal of Wuhan University 40(4), 49–52 (2007)

Xie, N.M., Liu, S.F.: Grey layered analysis method and its positional solutions. Journal of Jiangnan University 3(1), 87–89 (2004)

Xie, N.M., Liu, S.F.: Discrete GM(1,1) model and modeling mechanism of grey prediction models. Systems Engineering: Theory and Practice 28(4), 93–99 (2005)

Xie, N.M., Liu, S.F.: Research on discrete grey model and its mechanism. In: IEEE International Conference on Systems, Man and Cybernetics, v 1, IEEE Systems, Man and Cybernetics Society, Proceedings - 2005 International Conference on Systems, Man and Cybernetics, pp. 606–610 (2005)

Xie, N.M., Liu, S.F.: Generalized discrete grey models and their optimal solutions. Systems Engineering: Theory and Practice 26(6), 108–112 (2006)

Xie, N.M., Liu, S.F.: A class of discrete grey models and the effects of their predictions. Journal of Systems Engineering 21(5), 520–523 (2006)

Xie, N.M., Liu, S.F.: Similarity and consistency of several classes of incidence models. Systems Engineering 25(8), 98–103 (2007)

Xie, N.M., Liu, S.F.: Research on the multiple and parallel properties of several grey relational models. In: International Conference on Grey Systems and Intelligence Service, pp. 183–188 (2007)

Xie, N.M., Liu, S.F.: Discrete grey models of multi-variables and applications. Systems Engineering: Theory and Practice 28(4), 100–107 (2008)

Xie, N.M., Liu, S.F.: Characteristics of discrete grey models of approximately exponential sequences. Systems Engineering and Electronic Technology 30(5), 863–867 (2008)

Xie, N.M., Liu, S.F.: Affine characteristics of discrete grey models. Control and Decision Making 23(2), 200–203 (2008)

Xie, N.M., Liu, S.F.: Research on the Order-keeping Property of Several Grey Relational Models. Journal of Grey System 11(3), 157–164 (2008)

Xie, N.M., Liu, S.F.: Discrete grey forecasting model and its optimization. Applied Mathematical Modelling 33(1), 1173–1186 (2009)

Xie, N.M., Liu, S.F.: Research on evaluations of several grey relational models adapt to grey relational axioms. Journal of Systems Engineering and Electronics 20(2), 304–309 (2009)

Xie, N.M., Liu, S.F.: Ordering of grey numbers with the consideration of probability distributions. Systems Engineering: Theory and Practice 29(6), 169–175 (2009)

Xie, Y.M., Yu, H.P., Chen, J., Ruan, X.Y.: Design for square-cup deep-drawing based on grey systems theory. Journal of Mechanical Engineering 27(3), 10–15 (2007)

Xing, M.: A combined grey neural network model for seasonal predictions. Systems Engineering: Theory and Practice 21(1), 31–35 (2001)

Xiong, H.J., Chen, M.Y., Ju, T.: Two classes of grey models of control systems. Journal of Wuhan Jiaotong Science and Technology 23(5), 465–468 (1999)

Xiong, H.J., Xiong, P.F., Chen, M.Y., Ju, T.: A generalization of the grey systems SCGM(1,h)_c model. Journal of Huazhong University of Science and Technology 27(1), 1–3 (1999)

Xu, J.P.: Linear programming for grey predictions. Management Science of China 1(4), 35–42 (1993)

Xu, J.P.: Optimizing theory on working procedure with grey numbers. The Journal of Grey System 5(3), 189–196 (1993)

Xu, J.P.: On a kind of information grey number. The Journal of Grey System 7(2), 111–130 (1995)

Xu, J.P.: Grey drift linear programming. The Journal of Grey System 9(3), 243–246 (1997)

Xu, L.F., Hu, H.F., Son, Z.X., et al.: A credit report mechanism based on grey systems theory. Journal of Software 18(7), 1730–1737 (2007)

Xu, W.X., Zhang, Q.S.: An integrated computational scheme basedon on grey theory and fuzzy mathematics. Systems Engineering: Theory and Practice 21(4), 114–119 (2001)

Xu, Y.D., Wu, Z.Y.: Grey pricing of stocks based on limited rationality and inefficiency of stock markets. Journal of Management Engineering 17(2), 115–117 (2003)

Xu, Z.X., Wu, G.P.: Grey Systems Theory and Grey Prediction on Mineral Deposits. Press of Chinese University of Geology, Wuhan (1993)

Xue, X.W., Wang, D.: Grey input-output theory, models and methods. Systems Engineering: Theory and Practice 17(1), 56–61 (1997)

Yamaguchi, D., Li, G.D., Nagai, M.: A grey-based rough approximation model for interval data processing. Information Sciences 177(21), 4727–4744 (2007)

Yan, S.Y.: Systems' virtual evaluation of man-machine interfaces using multi-meters. Journal of Systems Simulation 19(12), 2720–2723 (2007)

Yan, L.Y., Mao, L.X.: An application of the GM(1,1) model. Systems Engineering: Theory and Practice 18(10), 104–106 (1998)

Yang, J., Zheng, L.G., et al.: Study on EDM grey control systems. Electrical Tools 5, 22–23 (1996)

Yang, J.H., Liu, J.L.: Grey information and applications. In: Liu, S.F., Xu, Z.X. (eds.) New Advances in Grey Systems Research, pp. 362–363. Press of Huazhong University of Science and Technology, Wuhan (1996)

Tang, J.H., Wang, Z.L.: Degree of grey information. In: Liu, S.F., Xu, Z.X. (eds.) New Advances in Grey Systems Research, pp. 360–361. Press of Huazhong University of Science and Technology, Wuhan

Yang, S., Ren, P., Dang, Y.G.: Opposite directional accumulation generation and optimization of grey GOM(1,1) model. Systems Engineering: Theory and Practice 29(8), 160–164 (2009)

Yang, Y.A., Wei, J., Feng, Z.R., Sun, L.Y.: Application of grey incidence analysis in the selection of the optimal initial orbit of satellites. Systems Engineering and Electronic Technology 30(2), 308–311 (2008)

Yang, Y., Wang, S.W., Hao, N.L., Shen, X.B., Qi, X.H.: On-line noise source identification based on the power spectrum estimation and grey relational analysis. Applied Acoustics 70(3), 493–497 (2009)

Yao, T.X., Liu, S.F.: Characteristics and optimization of discrete GM(1,1) model. Systems Engineering: Theory and Practice 29(3), 142–148 (2009)

Yeh, Y.L., Chen, T.C., Lin, C.N.: Effects of data characteristics for GM(1,1) modeling. Journal of Grey System 13(3), 121–130 (2001)

Yi, D.S., Guo, P.: Grey Theory and Methodology – Highlights, Solutions, Programs, and Applications. Press of Petroleum Industry, Beijing (1992)

Yi, D.S.: Grey models and prediction of human talents. Systems Engineering 5(1), 36–43 (1987)

Yi, C.H., Gu, P.L.: Predictions on energy using buffer operators of grey sequences. Journal of Systems Engineering 18(2), 189–192 (2003)

Ying, K.C., Liao, C.J., Hsu, Y.T.: Generalized admissible region of cass ratio for GM(1,1). Journal of Grey System 12(2), 153–156 (2000)

Yu, D.J.: About the methods of grey systems modeling. Systems engineering 9(5), 9–12 (1991)

Yuan, J.Z.: Grey Systems Theory and Applications. Science Press, Beijing (1991)

Zeng, G.M., Jiang, R., Huang, G.H., et al.: Optimization of wastewater treatment alternative selection by hierarchy grey relational analysis. Journal of Environmental Management 82(2), 250–259 (2007)

Zeng, X.Y., Xiao, X.P.: Ways to generalize the GM(1,1) model and applications. Control and Decision Making 24(7), 1092–1096 (2009)

Zhai, J., Sheng, J.M.: Grey MGM(1,N) model and applications. Systems Engineering: Theory and Practice 17(5), 109–113 (1997)

Zhang, C.K., Nie, M.L., Wu, J.B.: Partner selections for virtual firms based on improved grey evaluations. Systems Engineering: Theory and Practice 27(11), 54–61 (2007)

Zhang, D.H., Shi, K.Q., Jiang, S.F.: Grey predicting power load via GM(1,1) modified. Journal of Grey System 13(1), 65–67 (2001)

Zhang, J.J., Wu, D.S., Olson, D.L.: The method of grey related analysis to multiple attribute decision making problems with interval numbers. Mathematical and Computer Modelling 42(9), 991–998 (2005)

Zhang, M.A., Yuan, Y.B., Zhou, J., Feng, S.S.: Coordinated urban economic responsibility for the aftermath of potential disasters as analyzed using grey systems models. Systems Engineering: Theory and Practice 28(3), 171–176 (2008)

Zhang, N.L., Meng, X.Y., Li, Z.H., Teng, C.X.: Grey two-layered programming problem and its solution. Systems Engineering: Theory and Practice 27(11), 132–138 (2009)

Zhang, P., Sun, Q.: Prediction of the economic development of Baotou city using comparisons of grey theory and statistics. Mathematical Statistics and Management 26(4), 595–601 (2007)

Zhang, Q.S.: Deviate information theory of grey hazy sets. PhD Dissertation of Huazhong University of Science and Technology, Wuhan (1996)

Zhang, Q.S.: Difference information entropy in grey theory. Journal of Grey System 13(2), 111–116 (2001)

Zhang, Q.S., Deng, J.L., Fu, G.: On grey clustering in grey hazy set. The Journal of Grey System 7(4), 377–390 (1995)

Zhang, Q.S., Guo, X.J., Deng, J.L.: The entropy analysis method of grey incidences. Systems Engineering: Theory and Practice 16(8), 7–11 (1996)

Zhang, Q.S., Han, W.Y., Deng, J.L.: Information Eentropy of discrete grey number. The Journal of Grey System 6(4), 303–314 (1994)

Zhang, S.L., Zhang, G.L.: Comparison of the computational schemes of grey incidences and existing problems. Systems Engineering 14(3), 45–49 (1996)

Zhang, Y.G., Xu, Y., Wang, Z.P.: GM(1,1) grey prediction of Lorenz chaotic system. Chaos, Solitons & Fractals 42(2), 1003–1009 (2009)

Zhao, X.H.: Pattern mining in sequences of runoff data using grey Markov chains. Journal of Wuhan University (industry edition) 41(1), 1–4 (2008)

Zheng, Z.N., Wu, Y.Y., Bao, H.L.: Pathological problems of the GM(1,1) model. Management Science of China 9(5), 38–44 (2001)

Zhou, C.S., Deng, J.L.: Stability analysis of grey discrete-time systems. IEEE Trans. AC 34(2), 173–175 (1989)

Zhou, C.S., Deng, J.L.: The stability of grey linear system. Int. J. Control 43(1), 313–320 (1986)

Zhou, L.F., Xu, S.G., Sun, W.G.: A comprehensive evaluation of the underground water quality at Zalong wetland based on grey cluster method. Journal of Daliang University of Science and Technology 47(2), 240–245 (2007)

Zhou, P.: A trigonometric grey prediction approach to forecasting electricity demand. Energy 31(14), 2839–2847 (2006)

Zhou, T., Peng, C.H.: Application of grey model on analyzing the passive natural circulation residual heat removal system of HTR-10. Nuclear Science and Techniques 19(5), 308–313 (2008)

Zhou, Z.J., Hu, C.H.: An effective hybrid approach based on grey and ARMA for forecasting gyro drift. Chaos, Solitons & Fractals 35(3), 525–529 (2008)

Zhu, C.H., Li, N.P., Re, D., Guan, J.: Uncertainty in indoor air quality and grey system method. Building and Environment 42(4), 1711–1717 (2007)

Zhu, C.Q.: Grey prediction modeling via grey relational weighting. Journal of Grey System 13(3), 255–258 (2001)

Zhu, X.P., Zhi, X.Z., et al.: Grey predictive control of the vibration of rotor systems. Mechanical Science and Technology 21(1), 97–101 (2002)

Zhu, Z.Q., Wei, Y.X.: A grey multi-layered model for evaluating the implementation risk before establishing an ERP project. Industrial Engineering 11(3), 138–144 (2008)

Index

CPSIA information can be obtained at www.ICGtesting.com
Printed in the USA
LVOW01*1017150414
381775LV00010B/155/P

9 783642 161575